C O A E S

...mor...logy, c... society at the coast

'9

0

If the great geological labors of the ocean, such as the erosion of cliffs, the demolition of promontories, and the construction of new shores, astonish the mind of man by their grandeur, on the other hand, the thousand details of the strands and beaches charm by their infinite grace and marvelous variety. All those innumerable phenomena of the grain-of-sand and drop-of-water are produced by the same causes which determine the great changes of the shore.'

RECLUS, 1878, p. 176.

To David Stoddart for his continuing inspiration, and for introducing us to the wonders of The Tropics

COASTAL PROBLEMS

Geomorphology, ecology and society at the coast

by
Heather Viles,
St Catherine's College, University of Oxford

and
Tom Spencer,
Department of Geography, University of Cambridge and Magdalene College, Cambridge

A member of the Hodder Headline Group
LONDON • NEW YORK • SYDNEY • AUCKLAND

First published in Great Britain in 1995 by Edward Arnold,
Second impression 1996 by Arnold, a member of the Hodder Headline Group
338 Euston Road, London NW1 3BH

Copublished in the US, Central and South America by
John Wiley & Sons, Inc., 605 Third Avenue,
New York NY 10158-0012

British Library Cataloguing in Publication Data
A catalogue record for this book is available from the British Library

ISBN 0 340 53197 5
ISBN 0 470 23519 5 (Wiley)

Typeset in 10/13pt Photina by Phoenix Photosetting, Chatham, Kent
Printed and bound in Great Britain by St Edmundsbury Press, Bury St Edmunds, Suffolk and J W Arrowsmith Ltd, Bristol

Cover photograph: "Great Barrier Reef – Australia," by L. Isy-Schwart; reproduced by courtesy of The Image Bank, London.

CONTENTS

Acknowledgements ix

Chapter 1
The coastal context 1

What are coastal problems? 1
The coastal zone: some fundamental characteristics and principles 3
Coastal biogeomorphology 9
Coastal use and management 11
Approaches to the study and solution of coastal problems 15
Types of coastal problem 17
How this book is structured 18

Chapter 2
How coasts work 19

Introduction 19
Plate tectonics and coastal type 20
The coastal setting: Late Quaternary sea level changes 21
Wave processes in the shore zone 24
The influence of tides 41
Coastal ecology 44
Water on land at the coast: seaspray, fluvial inputs and groundwater 48
Physical disturbances to coastal biogeomorphology: storm surges and large scale pressure changes 51
Global warming, sea level rise and the coastal future 54

Chapter 3
Sandy coastlines: beaches and dunes 59

Introduction 59
Beaches 63
Dunes 67
Beach–dune interactions 73
What problems affect sandy coastlines? 75
Dune management 87
Barrier islands 93
Case study 3.1: Erosion problems along the barrier beach coasts of Nigeria 103
Case study 3.2: Dunes of the Dutch coast 106

Chapter 4
Rocky coasts: cliffs and platforms 110

Introduction 110
Distribution and nature of rocky coasts 110
Hard rock cliffs 117
Cliffs prone to failure 122
Shore platforms 133
Human influences on rocky coasts 144
Meeting the challenges on rocky coasts 145
Case study 4.1: Llantwit Major — blasted cliffs! 151
Case study 4.2: Jump-Off Joe: building on a disaster 153

Chapter 5
Coastal wetlands 156

Introduction 156
The magic of mud 157
Environmental settings of coastal wetlands 159
Types of coastal wetland 165
Sea level rise and wetland growth and destruction 176
Problems affecting coastal wetlands — upsetting the balance 181
Future sea level rise — towards a new balance? 193
Balancing the possibilities — effective coastal wetland management 197
Case study 5.1: The disappearing wetlands of south Louisiana, USA 198
Case study 5.2: War and the mangroves of Vietnam 203

Chapter 6
Coral reefs 206

Introduction 206
The biological scale: calcification, coral growth and reef accretion 214
The geological scale: reef construction, destruction and net accretion 221
The coral reef community 225
Corals and temperature 236
'Phase shifts' between hard and soft shallow water communities 240
Coral reefs and environmental degradation 241
Reef robustness and fragility: a few concluding remarks 247
Case study 6.1: Kanehoe Bay, Hawaii 247
Case study 6.2: The crown-of-thorns starfish: a highly complex biological phenomenon 250

Chapter 7
Cold coasts: permafrost, glaciers, sea ice and fjords 254

Introduction 254
Ice at the coast 256
Types of cold coast 258
Geomorphic change and ecology of cold coasts 262
Human impacts on cold coasts 275
Overview 282
Case Study 7.1: The Exxon Valdez oil spill, Alaska: catastrophe or not? 283
Case study 7.2: The Antarctic coast: exploitation or conservation? 285

Chapter 8
Managing the coast: coping with coastal problems 289

Managing the coastal zone — regaining a holistic perspective on coastal problems 289
Human influences on the coast 289
Sustainable coastal zone use and management 293
The coast of Bangladesh, Bay of Bengal 293
The Mediterranean Sea 300
Conclusions 308

References 312

Index 344

ACKNOWLEDGEMENTS

Writing this book has, at times, seemed to be an enormous labour. We have, however, always remained convinced of the grandeur and exciting variety of the coastal environment, and hope that this book will inspire others to go off and see for themselves, as well as becoming actively involved with understanding and tackling coastal problems. A whole shoal of people have helped to make the process more enjoyable, either by helping or by taking our minds off the whole enterprise. Susan Sampford initiated the project, and the ever-optimistic and cheerful Laura McKelvie steered us towards completion. Jim Hansom, Julian Orford and Denise Reed provided expert advice on earlier drafts of part of the manuscript. Anne Murray and the librarians of the School of Geography gave invaluable bibliographic assistance, and Ian Agnew, Jenny Wyatt and Peter Hayward produced the figures quickly and stylishly. Terry Scoffin, Sandy Tudhope, Colin Woodroffe and, particularly, Jon French have proved excellent companions for Tom in the field; as have Andrew, Amy and Alice Goudie for Heather, in a wide range of field settings. We will both miss the flurry of faxes which have whizzed between Oxford and Cambridge over the past months during the production of this book; although now, perhaps, our students will be able to pin us down more easily.

We are grateful to the following for permission to reproduce copyright material:

Academic Press (London) Ltd for Figs. 2.2, 2.13, 3.4, 3.5, 4.13, 4.14, 5.10b, 6.8; Academic Press (Florida) Ltd for Figs. 1.1, 3.12, 7.7 (and D. Nummedal); Allen and Unwin Inc. for Fig. 3.10; Dr Bob Allison for Fig. 4.4; American Chemical Society for Fig. 7.9; American Meteorological Society for Fig. 2.16; American Geophysical Union for Fig. 8.1; American Shore and Beach Preservation Society for Fig. 4.16; Edward Arnold for Figs. 2.6, 2.9, 2.14, 5.4, 7.1, 7.2, 8.4; Blackie and Sons for Fig. 6.6b; Dr Malcom Bray for Fig. 4.8; Cam-

bridge University Press for Figs. 1.2 (and J. Hansom) and 6.4; Clarendon Press for Fig. 8.5; Coastal Education and Research Foundation for Figs. 2.11, 3.3, 5.6; Dr R. Dolan for Fig. 3.11; Ecological Society of America and G.M. Wellington for Fig. 6.9; Elsevier Science Publishers for Figs. 3.1 (and E. van der Maarel), 3.2 (and A.C. Brown), 3.9 (and R. Silvester), 4.6 (and A.J.C. Malloch), 5.9 (and T. Spencer), 6.2a (and D.J. Barnes), 6.3 (and Y.I. Sorokin), 7.3 (and R.I.L. Smith); Elsevier Science Ltd for Figs. 3.2, 6.7a, 6.7b; Gebrüder Borntraeger for Figs. 4.5 and 4.15; Geographical Journal for Fig. 5.13; Geological Society of America and Dr K.O. Emery for Figs. 4.1 and 4.2; Geological Society for London and J.R.L. Allen for Fig. 5.8; Geoöko Verlag for Fig. 8.2; President of the Gulf Coast Association of Geological Sciences, Dr W.L. Fisher for Fig. 5.16; Institute of British Geographers for Fig. 5.11; International Association of Sedimentologists for Fig. 6.5; Inter-Research (and J.N. Heine) for Fig. 7.4; Secretary of the Intergovernmental Panel on Climatic Change for Fig. 2.17; Mrs D.L. Kinsman for Fig. 2.3; Kluwer Academic for Fig. 3.15; Longman Education for Figs 2.4 and 2.15; Longman Group and University of Chicago Press for Fig. 2.1; Macmillan Ltd, The Mersey Docks and Harbour Company, D.M. McDowell and B.A. O'Connor for Fig. 5.2; Macmillan Magazines for Fig. 7.10; Prentice-Hall Inc. for Figs. 2.7, 2.8, 2.12, 3.6, 3.7, 3.8; Norwegian Institute for Water Research for Fig. 7.8b; Routledge for Fig. 1.5; Royal Society of London for Fig. 4.7; B. Salvat and the Publications Committee, Fifth International Coral Reef Congress for Fig. 6.14b; SEPM (Society for Sedimentary Geology) for Fig. 5.15; Society of Systematic Biologists for Fig. 6.1; SPB Academic for Fig. 3.16; Springer-Verlag GmbH for Figs. 5.5 (and E.A. Shinn), 6.10, 7.8a (and J. Molvaer); Springer-Verlag New York Inc. for Figs. 5.1 and 7.5 (and J.P.M. Syvitski); Swedish Society for Anthropology and Geography for Fig. 5.14; Taylor and Francis Inc., Coastal Management, Vol 18(1), 1990, J.G. Titus for Fig. 3.14. Reproduced with permission. All rights reserved. United Nations Environmental Programme for Fig. 6.13; Universitätsbibliothek der technischen Universität Berlin for Fig. 4.12; University of Hawaii Press and S.V. Smith for Fig. 6.14a; Van Nostrand Reinhold Co. for Fig. 5.3, 8.3; John Wiley and Sons Ltd for Figs. 2.10, 3.13, 4.3, 4.9, 4.11, 5.7, 5.12, 6.2b; Dr C Wilkinson for Fig. 6.15.

Every effort has been made to trace and contact the owners of copyright material but we offer our apologies to any copyright holders whose rights may have been unwittingly infringed.

All plates are by T. Spencer unless otherwise noted.

Heather Viles and Tom Spencer
Oxford, June 1994

Chapter One

THE COASTAL CONTEXT

What are coastal problems?

About 20% of the world's coast is sandy and backed by beach ridges, dunes, or other sandy depositional terrain. Of this, more than 70% has shown net erosion over the past few decade. (Bird, 1985a)

Fifty per cent of the population in the industrialised world lives within one kilometre of a coast. This population will grow at about 1.5 per cent per year during the next decade. (Goldberg, 1994)

On 29 April 1991, a severe tropical cyclone swept up the Bay of Bengal and like many other such storms in the region's turbulent past caused the sea to rise rapidly and destructively . . . fatalities were estimated to be well over 100,000. (Warrick *et al.*, 1993)

These three quotations illustrate growing recent concern about coastal problems, which we define here as natural and/or human-induced events or processes that affect environment and society at the coast. Thus, coastal problems include beach pollution, shoreline erosion, coastal flooding and the reduction of biodiversity in mangroves and marshes. Such problems are not new, of course, and there has long been recognition by scientists and others that the coast is a hazardous and dynamic zone. Reclus (1873), Wheeler (1902), Owens and Case (1908) and Stamp (1939), for example, all made early attempts to consider various aspects of coastal change and human society.

Coastal problems are now recognized as being pressing concerns in many parts of the world. Examples are provided by Wickremathe (1985) who discusses the coastal problems of Sri Lanka, Rodriguez (1981) who reviews coastal environmental stresses in the Caribbean, Dahl (1984) who assesses the problems in Oceania, and Nunn (1994) who looks at the wider issues of coastal problems on oceanic islands around the world. UNEP has also produced many recent reports on the state of marine and coastal environments, including ones on the South Pacific, East Africa (Bryceson *et al.*, 1990) and South Asia (Sen Gupta *et al.*, 1990), which all point to the severity of the problem. Global surveys include those of GESAMP (1990) and Hinrichsen (1990). All these investigations point to an increasingly polluted coastal environment, and one which is becoming ever more densely populated, built-up and prone to natural hazards. A major aim of this book is to elucidate the ecological, geomorphological and, to a lesser degree, societal, setting of such problems with a view to improving the success of coastal management.

Increasingly, coastal problems have been seen within the context of a future, accelerated rise in sea level due to global warming. Many publications have made suggestions as to the likely impacts of such a sea level rise on coasts in many places (Warrick and Farmer, 1990; Holligan and Reiners, 1992; Bird, 1993; Warrick *et al.*, 1993), although huge uncertainties still remain over the magnitude and rate of sea level rise, as well as over how individual coastlines will respond. Regional scale studies have been hampered by such problems, although reviews have been produced for, amongst other places, the British coast (Boorman *et al.*, 1989) and Bangladesh (Mahtab, 1992). Over the past few years there has been a general tendency for predictions of future sea level rise to become less and less dramatic, as illustrated in Chapter 2. One of the major themes of this book is to illustrate that, given such scaled-down predictions, it is the local ecological and geomorphological responses to sea level change which become the key factors.

Coastal problems are one manifestation of environmental problems, or hazards and, as such, have ecological, economic and social dimensions. In simple terms we can regard coastal problems as resulting from 'stresses' on 'coastal systems'. The way that people interact with, and use, the coast has important consequences for the nature of coastal problems. As discussed in more detail in the following sections, there is a huge range of human activities along the world's coastline, but in general terms these tend to produce relatively localized coastal problems, as compared with the large scale problems of tropical forest destruction, degradation of agricultural land and desertification. Natural stresses on coastal systems are, of course, also an important component of many coastal problems. The interaction of human and natural stresses is, however, beginning to produce large scale coastal problems affecting long stretches of coastline. Some 25 per cent of the coastline of the USA is significantly affected by coastal erosion, costing some US$300 million per year in loss of property and building protective structures (Alexander, 1992). Thus, successful coastal zone management is becoming urgently needed in many parts of the world.

This book addresses the causes and manifestations of, and some possible solutions to, a range of coastal problems working from a physical geographical perspective, but one which tries to integrate social and economic dimensions as well. Finding a suitable methodology for investigating the ever more damaging relationship between society and nature is extremely difficult and, as yet, there is no easy solution. Our approach, like others, is imperfect and unbalanced. We aim here to develop a biogeomorphological perspective (Viles, 1988), showing that a solid understanding of the workings of coastal ecological and geomorphological systems is a necessary prerequisite to solving coastal problems. We also acknowledge, however, that coastal management needs equally to tackle the human dimensions of the problems. Our major goal is to point out that coastal problems involve geomorphological, hydrological, ecological and societal phenomena within unique settings (places) and that some sort of holistic approach, however flawed, is needed to tackle such complex problems successfully. We cannot, of course, address all sides of the problem in sufficient and equal detail. In the rest of this chapter we set the scene for the book by amplifying some of the issues introduced above.

The coastal zone: some fundamental characteristics and principles

Coasts are dynamic interface zones involving the meeting of atmosphere, land and sea. Within the coastal zone, major movements of sediments and nutrients are powered by waves, tides and currents (in water and air). These movements shape the coastal profile, producing erosional and depositional landforms. The land is no mere passive canvas, however, as rivers bring sediments, freshwater and nutrients down to the coastal zones (Plate 1.1), and subaerial weathering and erosion also help to shape the coast. As well as providing a base for many human settlements, the coast is home to some of the world's most productive and diverse ecosystems. The organisms in the coastal zone are also active participants, forming reefs and aiding sedimentation, as well as providing an important buffer zone and filtering system. All these components are in a fragile balance controlled by physical and biological processes which can easily be upset by natural or human-induced perturbations. In this context it is useful to identify some fundamental characteristics of the coastal zone.

Coastal hierarchies

Characteristically, the coastal zone is taken to include the area between the tidal limits as well as the continental shelf and coastal plain (Fig. 1.1). Large

Plate 1.1 Contributing to the littoral sediment budget: the Malibu River, from a catchment in the Santa Monica Mountains, enters the Pacific Ocean north of Santa Monica, southern California

scale coastal landforms such as capes and embayments, estuaries, deltas, large dunefields and major coral reef complexes properly belong at this scale. Sandwiched between wholely marine and largely terrestrial environments is the intertidal and immediately supratidal shore zone (Fig. 1.2). This includes beaches, cliffs, tidal and brackish water wetlands and individual reef communities. Below this scale, and superimposed upon these landforms, are — to take the beach environment as an example — features such as longshore bars and troughs, and beach cusps. These forms in turn provide a base for ripples, rills, swash marks and the bioturbation structures of intertidal organisms. These different scales, with smaller scales nested within larger scales, also have characteristic time-scales associated with them (Table 1.1). This book is chiefly concerned with the shore zone scale, within specific coastal zone contexts.

Coastal dynamics

Coastal systems are dynamic over a range of time-scales, from short term fluctuations (over a few weeks or months) to long term changes over thousands of years. Sediment movements within a beach–offshore sediment cell also show dynamism with short term fluctuations (over a few tidal cycles) in the position

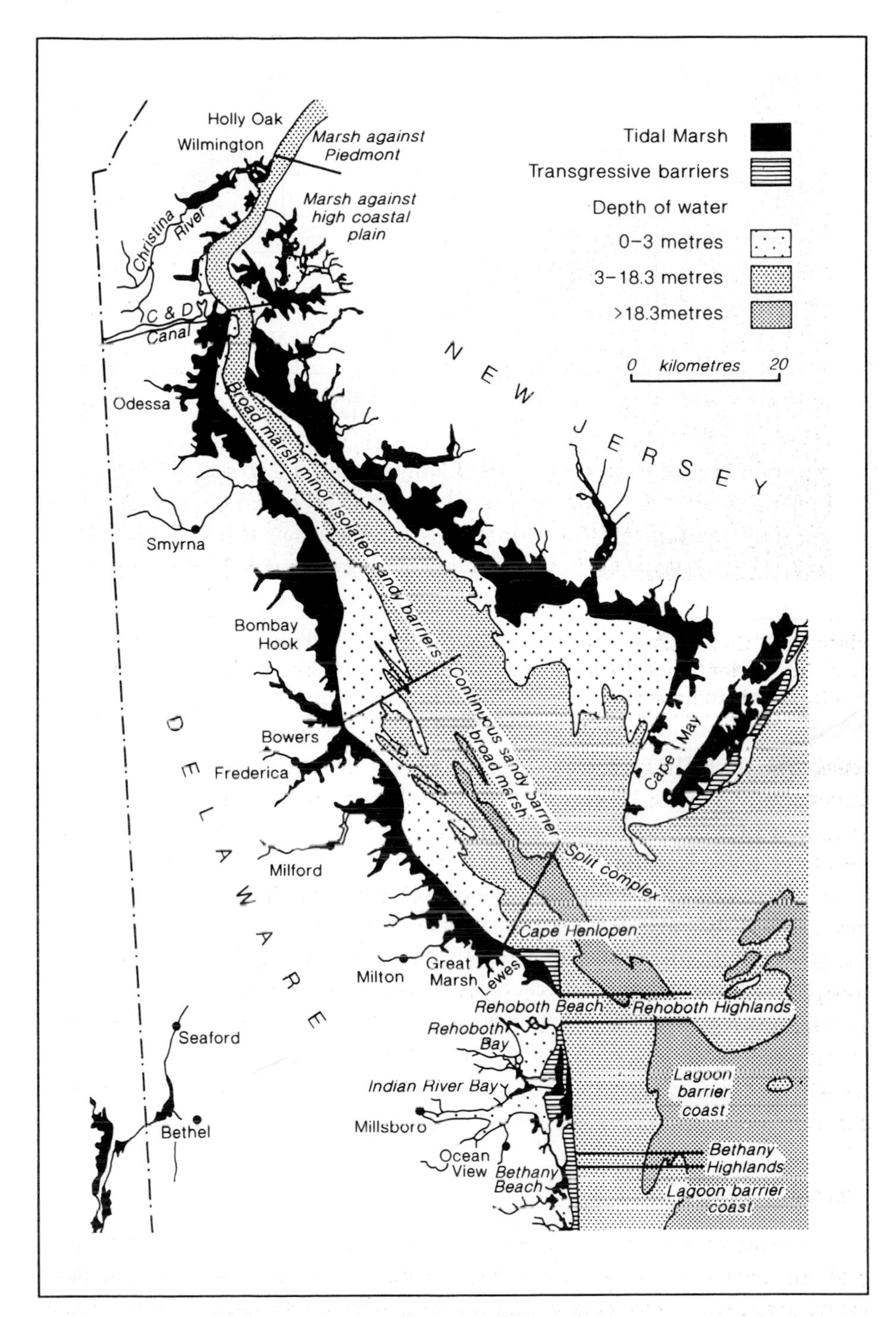

Fig. 1.1 The coastal zone: an example from Delaware, USA

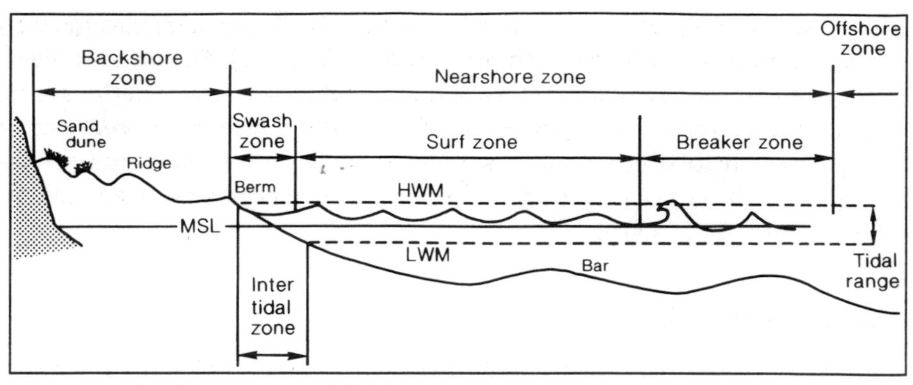

Fig. 1.2 The shore zone. Modified from Hansom (1988)

of berm, cusps and bars, coupled with longer term changes in the whole size and shape of the beach plan and profile. Similar processes characterize other shore zone environments, often in quite complex ways: thus sand dune ecosystems show seasonal changes in productivity, biomass and thus sand-trapping ability, but are also influenced by long term successional changes in species composition. Dynamics, and the range of possible landform changes, must be allowed for in shore management policies: many coastal problems result from attempts to effectively fossilize a particular shoreline configuration or habitat. Such programmes then face expensive remedial action to re-establish quasi-natural patterns of landform change and ecosystem function.

Coastal linkages

Patterns of shoreline erosion and deposition are often indicative of changes in the sedimentary budget along a particular stretch of coastline; in the presence of coastal change it is often useful to isolate sources and sinks of sediments in a particular coastal circulation cell. Such studies clearly identify that up-drift, offshore or inland changes in sediment supply have often quite distinct ramifica-

Table 1.1 Time-scales in the coastal system

Form	*Spacing*	*Timespan*
Cusplet	0–3 m	Minutes–hours
Cusp	3–30 m	Hours–days
Sand wave	100 m–3 km	Weeks–years
Secondary capes	1–100 km	Decades
Primary capes	200 km	Centuries

Source: Dolan *et al.* (1974)

tions elsewhere on the coast and identify the need for shoreline management on an appropriate, often larger than present, scale (Fig. 1.3). Similar arguments might be made for nearshore ecosystem degradation and water pathways to, and along, the coast. Here, however, is the added complication of considering the interactions between landforms, geomorphic processes, community structures and ecological processes: these biogeomorphological aspects are considered in more detail below.

Coastal sensitivity

Coastal systems are subjected to a range of disturbances, both human-induced and natural and there are many approaches to understanding the response to such changes. Sensitivity and resilience are two concepts often employed to explain the response of environmental systems. Sensitivity (also often called resistance) is the degree to which a system undergoes changes because of human or natural stresses, and resilience relates to the speed at which the system recovers from such changes. Combining the two characteristics in a simple 2 × 2 matrix, we can recognize different categories of coastal system which will respond differently to stress (Table 1.2).

As Table 1.2 indicates, hard rock cliffs have low sensitivity and high resilience and will be affected only by the largest stresses. The corollary of this is that such systems are very difficult to manipulate (in terms of road building, for example), which results in a particular type of coastal problem. At the other end of the scale, highly sensitive systems with low resilience, such as easily erodible earth cliffs, suffer from stresses and are very difficult (and expensive) to

Table 1.2 Coastal sensitivity and resilience

Sensitivity	*Resilience*	
	Low	*High*
Low	Fjords (to pollution) (Chapter 7)	Hard rock cliffs (Chapter 4)
High	Cliffs prone to failure (Chapter 4) Mangroves (to pollution) (Chapter 5) Corals (to crown-of-thorns starfish) (Chapter 6)	Beaches (to storm-induced erosion) (Chapters 2 and 3) Dunes (to storms) (Chapter 3) Corals (to storms) (Chapter 6) Shore platforms (to pollution) (Chapter 4)

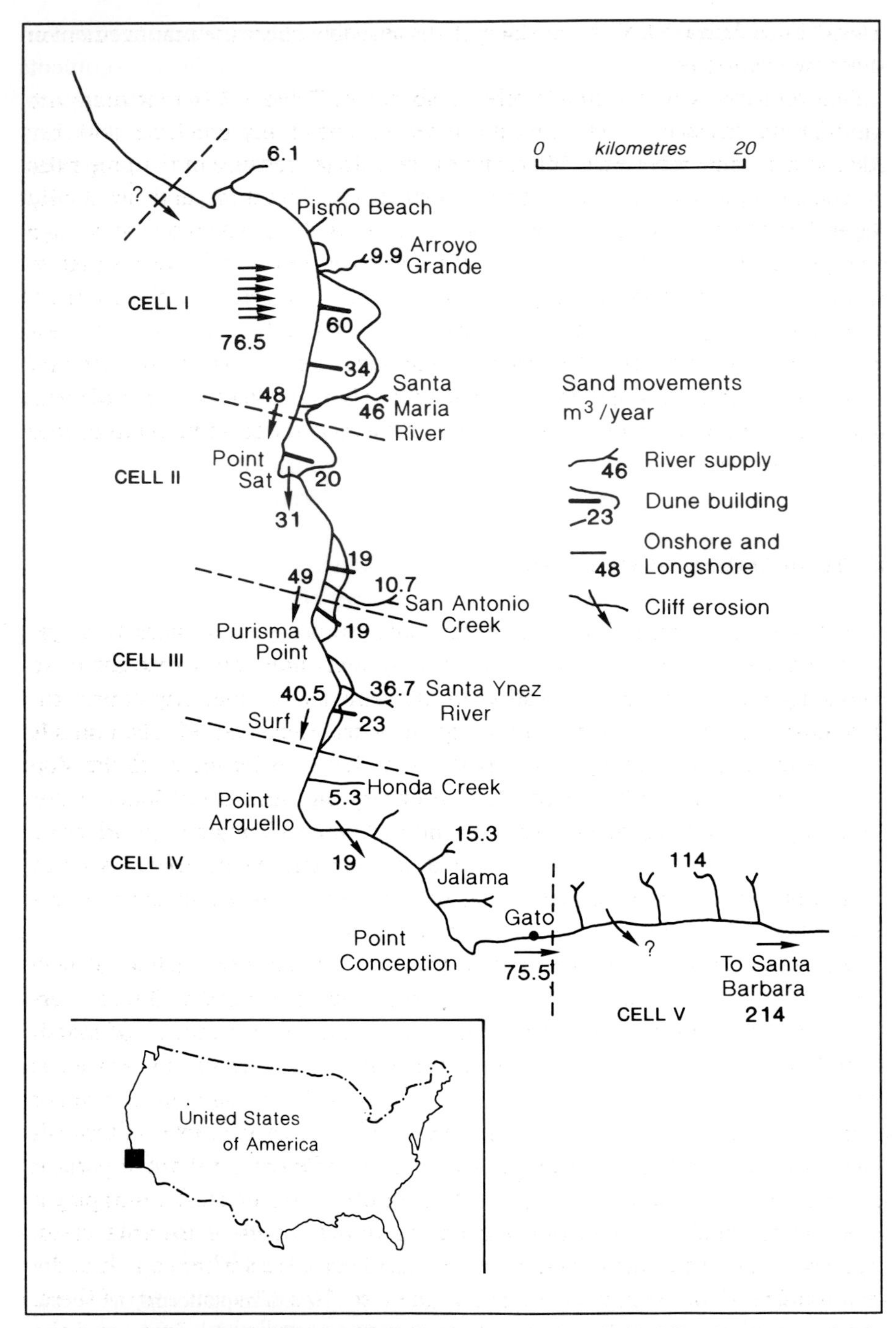

Fig. 1.3 A classic example of a littoral sediment budget: the California coast north of Santa Barbara. After Bowen and Inman (1966)

'mend' once damaged. Most coastal systems probably fit somewhere in between these two extremes.

One problem with the simple scheme shown in Table 1.2 is that there are multifarious stresses acting on most coastal systems at any one time, and that each system may react very differently to each stress, because of the properties of coastal systems discussed above. Furthermore, reaction and sensitivity depend on the time-scales under consideration — a cliff profile may show huge changes on the scale of weeks, for example, and yet show longer term stability. Human activity may affect many parts of the system. Oil pollution of coral reefs, for example, may increase their sensitivity and decrease their resilience to stress from hurricanes and global warming. Time lags can complicate the picture, with, for example, oil stored in mangrove soils being released a number of years after contamination as a result of sediment movements, and going on to pollute nearby coral reefs.

Coastal biogeomorphology

The idea of treating the coast as a series of interconnected and interacting systems is nothing new, although often these systems have been thought of as consisting merely of sediment and water movements, without any consideration of the ramifying influence of the organic world (Fig. 1.4). Plants, animals and microorganisms living in the coastal zone have an intimate relationship with their physical environment. Thus, they rely on mineral nutrients, water for transport and a life support system, and rock and sediments for a substrate to grow upon. As a result of such dependencies, extreme events (such as storm surges, droughts and landslides) have the capacity to destroy or change dramatically the ecological characteristics of the area.

Conversely, changes to coastal ecology, such as the introduction of new hybrid species of *Spartina* to the British marshes, may be echoed by linked alterations in the geomorphological system, such as an encouragement of marsh accretion in the case of *Spartina*. This example shows that organisms, as well as relying upon, and being influenced by, their physical environment, are major agents of change themselves. In many tidal wetlands, for example, salt marsh plants are not merely passive occupants of marsh sediments, but are important in anchoring sediments. Similarly, plants colonizing pioneer sand dunes play a vital role in stabilizing and building up the dunes. In more extreme cases, organisms may dominate the environment, as in coral reefs where much of the solid framework of the reef comes from organic sources. The corollary of these, and many other less well understood interactions between the living and the non-living worlds, is that a change to either the organisms or the geomorphol-

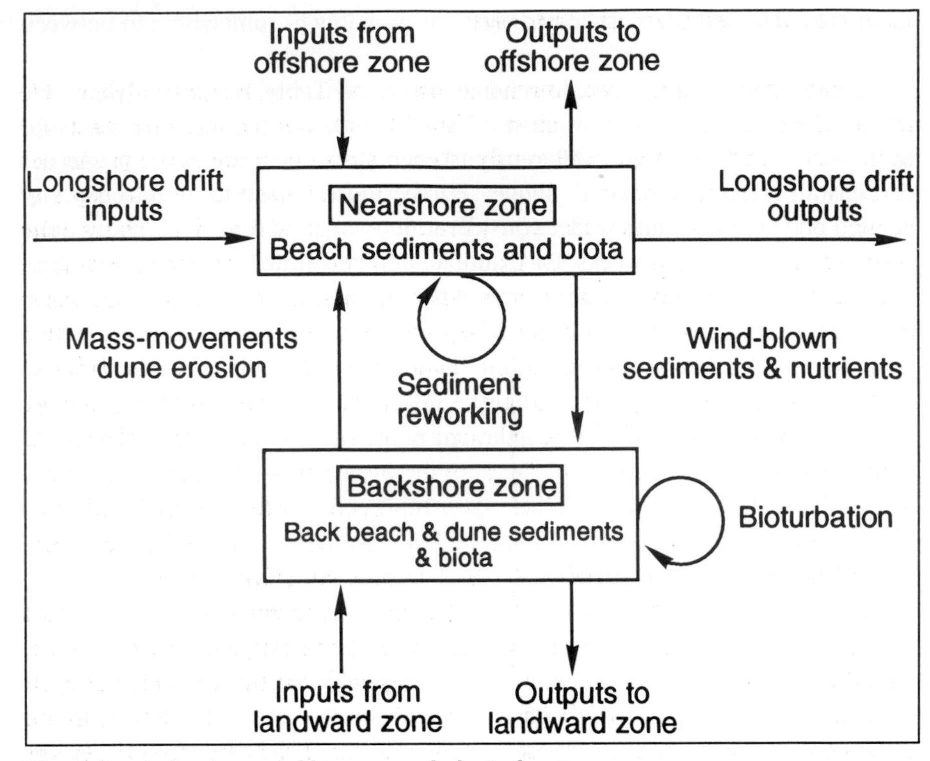

Fig. 1.4 A simple coastal biogeomorphological system

ogy of the coastal zone can have often unexpected repercussions for the whole system.

From a biological perspective, it is clear that different ecosystems function in different ways: thus the strongly detritus-based food webs of seagrass, salt marsh and mangrove swamp systems (Chapter 5) can be compared with the role of grazers within coral reef systems (Chapter 6) and the macro-, meio-faunal and micro-floral communities of the beach environment (Chapter 3). Further complexities arise because these biogeomorphological linkages are often not isolated to one small area, but involve movements and interchanges along the coast, and between the land and offshore zones. Sediment destined for an estuarine mangrove swamp, for example, may originate from many kilometres inland in a river system, whilst organisms, such as fish, which play a vital role in the same ecosystem, may come from a long way offshore. Thus in biogeomorphological systems linkages are not just spatial (i.e. lateral) but also trophic (i.e. ecologically vertical).

Coastal use and management

The coast attracts human settlement because of its beauty, its generally equable and mild climate, the presence often of suitable flat, low-lying land suitable for agriculture, and the use of the sea for shipping and communications. As the coast is a buffer zone between land and sea, it lends itself to certain types of human activity, and not to others. In general, because of the tidal influence the shore zone (and especially the intertidal zone) is not suited for permanent land uses such as agriculture, industry or settlement, apart from those that are based on marine resources, such as types of aquaculture. However, above the active cliff line, or behind the upper beach on low-lying coasts, a much wider range of land use options is available. Reclamation, or 'land claim', as it is probably more accurately described, has enabled the population of many coastal areas to make increasing use of the coastal zone for agriculture, industry and settlement. Singapore, where land reclamation has been practised from 1820, provides a good example. Here, reclamation has increased the land area of the island from 581.5 km^2 in 1960 to 620.5 km^2 in 1986 (Wong, 1992).

Many coasts have long been a focus of settlement as evidenced by the great antiquity of many coastal towns and villages, and the range of archaeological material found in many sites. For example, shell middens (mounds of debris left from human consumption of shellfish) dating back to 4250 BP have been found in Casco Bay, north of Boston harbour, USA (Yesner, 1988) and coastal settlers have been recorded from around 6600 BP in the Hunter Islands, Tasmania (Bowdler, 1988). In most places the coast provides an easy source of food (fish, shellfish and algae), a starting point for oceanic travel, and a place from which to explore inshore areas. Over the centuries these advantages, coupled with such attractions as sea air, beaches and fertile agricultural land on the coastal plain, have produced a massive coastal population. Worldwide, over 60 per cent of the population lives in the coastal zone, and it has been estimated that by AD 2000, 75 per cent of the world's population will live in a 60 km wide strip along the continents' coasts (Lindén, 1990). The distribution of population is very patchy, however, with much of the Arctic coastline sparsely populated. Climatic and topographic influences are undoubtedly important; mountainous coastlines are usually little-settled, whereas fertile low-lying coastal plains, estuaries and deltas have provided the home for many civilizations, such as those based on the Nile and Irrawaddy deltas.

As shown in Table 1.3, all of the major land use options are present in the coastal zone. Within the shore zone, a more limited range is found, dominated by mining, hunting and gathering (primarily fishing), recreation, waste disposal, wave and tidal power generation, and coastal and sea defences. Along the continental shelf, fishing, mining (including oil and gas exploration), ship-

Table 1.3 Human uses of the coastal zone

Zone	*Type of coast*	*Uses*
Subtidal-offshore zone	Continental shelf	Fishing
		Oil exploration
		Mining
		Sand dredging
		Dumping of wastes
		Sewage outfalls
	Coral reefs	Tourism
		Fishing
		Quarrying
	Estuaries	Tidal barrages
		Coastal protection schemes
Intertidal-nearshore zone	Sand and gravel beaches	Recreation
		Sand and gravel mining
		Back beach buildings
		Coastal protection schemes
	Wetlands	Aquaculture in converted ponds
		Oyster beds
		Reclamation
		Grazing
		Reed and timber extraction
		Canals/pipelines
		Nature conservation
	Shore platforms	Seafood hunting/collecting
		Quarrying
Backshore zone	Dunes	Recreation
		Golf courses
		Nature reserves
		Building
		Water extraction
		Army manoeuvres
	Cliffs	Shore protection works
		Building on cliff top
		Mining
		Conservation coasts
Onshore zone	Coastal towns and reclaimed land	Ports and harbours
		Marinas
		Housing
		Industry
		Agriculture
		Nature reserves
		Tourism

ping and waste disposal are all important. Sediment from the continental shelf is often extracted for use as aggregates, or perhaps as beach fill material. An unusual example of such resource exploitation is the dredging of shells to supply Iceland's cement and agricultural lime industries (Earney, 1990). Of course, human activity has also resulted in major changes to coastal zone topography with land reclamation schemes effectively removing areas from the shore zone and enlarging the coastal plain. Most human activity within the shore zone, and on continental shelves, is linked to land uses further inland. Thus, recreation on tourist beaches is linked to hotel and resort developments on the coastal plain.

All these facets of human use of the coast produce specific results in each place, conditioned by the unique circumstances of history and location. Despite the long history of coastal settlement recorded in some areas, other places have only recently been affected. The shores of Antarctica, for example, remain almost pristine and are only intermittently touched by tourists and research scientists. In comparison, the coast of the Netherlands has been the site of human activity for many centuries, with dykes being built along the coast by the tenth century AD (Walker, 1988a). Along the Severn Estuary, England, most of the tidal wetlands had been reclaimed by the end of the Roman period (Allen, 1990a), whilst in China, dykes for salt production and agriculture may date back to over 4000 BC. In many countries, rapid changes in economic development and technology have been associated with major shifts in the nature and severity of the human impact on the coast. Hunting and gathering societies based on fishing produce only minor coastal modifications but may be revolutionized if oil is discovered offshore, or if international tourism develops, resulting in high-technology coastal constructions. Concurrently, increased wealth and changing public perception of the coastal resource may result in some coastal areas becoming 'preserved' and an increasing desire to prevent widespread 'damage' to the coast in general.

Ownership of the coastal zone takes many forms and has ramifications for use and environmental exploitation. Recent legislation has created Exclusive Economic Zones (EEZs) that stretch 200 nautical miles (just over 370 km) from the land (Fig. 1.5). Effectively, this has resulted in coastal countries 'owning' the adjacent coastal shelf and tiny (both in terms of land area and population) oceanic atoll states having jurisdiction over vast areas of deep ocean. Within the shore zone and onshore areas ownership becomes more complicated at the national and regional scale, with all countries having different arrangements. In England and Wales, for example, the National Trust (a national charitable body) owns 760 km of the coastline. Individual owners also own properties above the high tide limit. According to Alexander (1992) almost 70 per cent of the shorefront of the USA (excluding Alaska) is privately owned. Much of the

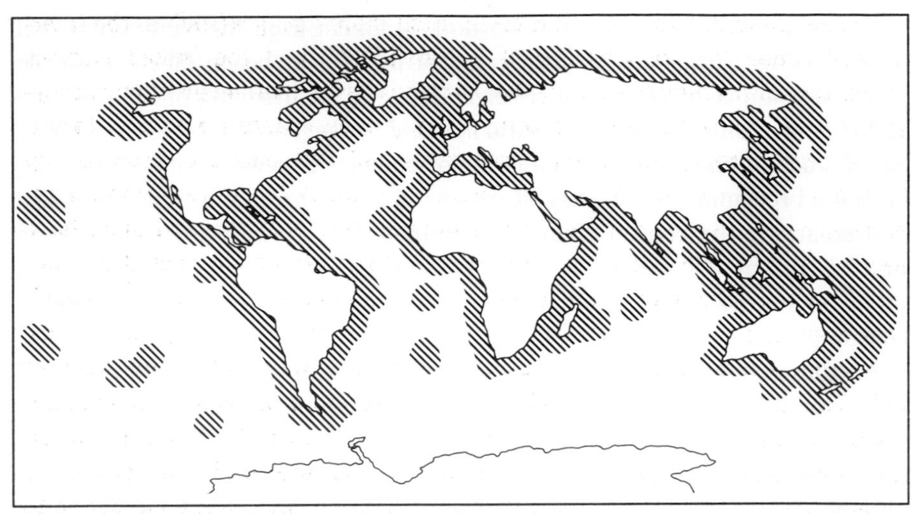

Fig. 1.5 Worldwide 200 nautical mile exclusive economic zones. After Earney (1990)

intertidal zone around the British coast is Crown Land or is owned by the Duchy of Lancaster.

Some of the land within the coastal zone is thus a private resource; other areas are treated as common land. Some facets of the coastal environment, such as seawater, can be regarded as public goods or 'common property'. In many countries complex tenure arrangements have been developed, as in Japan where coastal fishermen have legally 'guaranteed' access to, and ownership of, marine life in their village waters (Ruddle, 1989). These complicated patterns of resource ownership mean that multiple uses of many coastal environments are possible, with concomitant conflicts. With increasing international linkages, many small coastal areas are 'used' by people from a wide range of places. Thus, a beach in the Caribbean may be used for sand extraction by the local building trade, for recreation by tourists from rich, industrialized countries and for sewage disposal from nearby hotels and houses. Sometimes conflicts arise between different users of coastal land and water. In New York harbour, for instance, dredging of aggregates is an important industry, but one which has become contentious in recent years. Between 1966 and 1973 an average of 6.7 million tonnes of aggregate was removed each year. Environmental groups have become concerned about the possible impact on fish stocks and beach profiles (Earney, 1990).

The multiple uses of coastal areas require management, usually by national or local government agencies, who are responsible for protecting coastal settlements, permitting certain uses and proscribing others. In 1972 in the USA, the Federal Government passed the Coastal Zone Management Act, which provides for the development of coastal planning and management schemes by state authorities (Marsh, 1983). In some cases, international bodies (such as the International Union for Conservation of Nature (IUCN) and the United Nations Environment Programme (UNEP) play a role in coastal management by surveying the area and setting up and administering coastal nature reserves. Examples of coastal zone management strategies and problems are discussed by Hansom (1986) for the UK, and by Katz (1989) for Belize. Coastal zone management plans usually involve some notion of sustainable use, and there have been many studies of the range of factors that need to be considered by integrated coastal zone management plans. Types of coastal protected landscapes range from strictly preserved nature reserves, to multiple-use management areas. IUCN guidelines for coastal management to include conservation are given in Table 1.4. There has been a gradual move away from management schemes which consider the coastal environment as either a hazard (in terms of erosion and coastal flooding, for example) or a resource to be plundered, towards a more dynamic view of the coast as providing a challenge. This is not a challenge to be overcome in the sense of attempting to control the environment, but rather a challenge to be met with imaginative and flexible responses which must be based on firm understanding of the workings and dynamism (including complex responses, time lags and linkages) of the coastal environment.

Approaches to the study and solution of coastal problems

The acknowledgement that many coastal areas are hazardous, threatened and yet valuable environments has led to an increasing need for understanding and solving coastal problems within a general context of coastal management. A crucial, but often very difficult, first task is to understand the natural functioning of the coastal environment under study, plus its interrelationships with neighbouring systems. Attempts need to be made to understand the functioning of sediment circulations and ecosystem processes, such as nutrient cycling. Furthermore, an inventory of the species present and the fundamental characteristics of the area needs to be compiled. The comparison of old and new maps has been used intensively to show the nature of coastal change, and repeated air photography and satellite imagery have also proved invaluable for the detection and measurement of coastal morphological and vegetational changes (El-Ashry, 1977; Frihy, 1988; Leatherman, 1993). Computer technology has increased the power of surveying, leading to the production of digital terrain models (e.g. Kidson *et al.*, 1989 who have carried out repeated surveying at Braunton Burrows dunes, Devon, England using such techniques). Once data have also been collected on the nature of the threat to the area and the human dimensions of the problem, an analysis can be undertaken.

Table 1.4 IUCN guidelines for coastal zone and wetlands policy

Priorities
1. Protection of significant conservation values
2. Maintenance and restoration of the essential character and functioning of each environment
3. Preservation of estuaries, predominantly unmodified islands, reefs, coastal wetlands, lakes, lagoons, ponds and dunes
4. Restoration of degraded water quality
5. Prevention of any new discharges of untreated waste into water
6. Recognition of interests of indigenous people
7. Maintenance and improvement of public access, and of recreation opportunities which would neither modify the environment nor adversely affect the enjoyment of other users
8. Prevention of the alienation of foreshore, seabed, and public lands immediately adjacent to the foreshore
Planning requirements
1. Coastal developments conform to regional and national government policy
2. Consents after environmental impact assessments for: reclamations, marinas, jetties, breakwaters etc.
3. Control or prohibition of removal of sand or reef material, disturbance of coastal vegetation and seaweed, and disturbance of nesting sites and migration paths of species such as turtles and penguins
4. Constraints on siting residential, tourist and industrial enterprises on or near the coast, including controls on discharge of effluent and water extraction
5. Guarantee of public access
6. Management of wetland catchments so that the complex relationships between wetland and surrounding ecosystems are taken into account
7. Protection of specific areas of the coastal marine environment for their natural values and providing integral protection of the land/sea interface

Source: modified from Lucas (1992)

Predictions based on computer modelling, often in association with Geographical Information Systems (GIS), are increasingly attractive methods of presenting a number of different solutions. Thus, Shennan (1993) illustrates the potential of GIS to help decision-makers assess the future threats and options for two British lowlands, the Tees estuary and the Fenland, related to a future sea level rise. Such modelling exercises help display alternative scenarios, but are heavily reliant on the accuracy of the data fed in and on the model specifications themselves. GIS-style applications also assume that coastal landforms are passive, so that sea level rise will induce flooding and loss of land up to a certain contour level. As we show in the rest of this book, it is unrealistic and unhelpful to consider the coast in such a passive way. Increasing sea levels will produce geomorphic changes which will complicate any simple picture. Assuming that a suitable model of the coast can be built, assessment of which solution to adopt requires a further dimension to be considered, i.e. economic factors. Also, any solution must be tailor-made to the social values of the place under consideration. Different responses would be applicable to identical problems in an area of outstanding natural beauty near a large urban agglomeration in a developed country, than in an area exploited for oil reserves in a developing economy.

Solutions to coastal problems, which are in fact often applied without highly detailed prior surveys and studies, may involve high-tech engineering, so-called 'soft engineering', or 'ecological' strategies, and may, of course, be more or less successful! Assessing the success or otherwise of a solution to a coastal problem is a further vital part of the continual process of evaluation and re-evaluation. For example, after engineering works to strengthen the sea wall and to armour adjacent areas of the coast at West Bay, Dorset, England, several of the Portland Stone armour blocks were selected for long term monitoring, using non-corrodible pins grouted into the rock as fixed reference points. Using this technique, a cumulative erosion of up to *c.* 50 mm depth of material was recorded over the period 1983–4 (Clark, 1988), but subsequent high erosion has led to the replacement of many blocks.

Types of coastal problem

The diversity of coastal problems is remarkable and several classifications could be attempted. We may distinguish, for example, between those problems primarily caused by natural stresses and those involving a significant human input. At a finer level of resolution, we could make distinctions between the type of human interference involved. Conversely, we could organize problems according to their scale and severity — separating local, minor problems from those seriously affecting large tracts of coast over a long time span. We could also classify problems in terms of the more detailed nature of their impact upon the coastline, i.e. acceleration or deceleration of naturally occurring processes, or interference with them in general, introduction of new processes or inputs, or the introduction of new landforms, organisms or chemical compounds. All such classifications are of necessity inadequate, although valuable, and many coastal problems are intimately related to others because of the integrated nature of coastal biogeomorphological systems. Thus, naturally high rates of erosion on coastal cliffs may lead to cliff protection strategies, which themselves produce accelerated erosion down-drift.

In this book we have followed the approach of trying to divide coastal problems according to the particular part of the coastal environment which they affect, e.g. problems faced by dunes, beaches and coral reefs. Within this broad classification, we recognize that there are different types of problem affecting primarily the biological, chemical and physical characteristics of the environment, which in many cases act in combination. Furthermore, we realize that the coastal problems of many parts of the coastal environment (e.g. beaches, dunes and wetlands) are linked in particular environmental settings, such as barrier island coasts. No single book can cover all coastal problems, and we have deliberately tried to be eclectic in our coverage, aiming to show diversity.

How this book is structured

Bearing in mind the points discussed in this chapter, we focus in Chapter 2 on the fundamentals and dynamics of coastal systems and their relation to changing sea level and sediment supply. Subsequent chapters detail the nature and distribution of different coastal environments, based on an adaptation of Inman and Nordstrom's (1971) coastal classification, followed by an assessment of the major pressures, processes, problems and potential solutions. As one of the overall aims of the book is to stress how an understanding of coastal problems can only be gained through a detailed understanding of the workings of coastal biogeomorphological systems, we feel that it is vital to provide detail on the key characteristics of each system. Each chapter uses two or more case studies to exemplify the complexities involved, and to show that there are no easy solutions. Chapter 3 deals with beaches and dunes, Chapter 4 with rocky coastlines — cliffs and platforms, Chapter 5 with coastal wetlands, Chapter 6 with coral reefs and Chapter 7 with glaciated and high latitude coasts. Chapter 8 illustrates the linked nature of many coastal problems, and the extra complexities faced by coastal management schemes which relate to economic and social characteristics, by consideration of two case studies of sub-continental areas. Thus, the thread of our argument is that you must understand the complex, physical basis of coastal problems in order to attempt solutions (Chapters 2 to 7), but that such solutions will fail unless you also can grapple with the equally complex socio-economic constraints on human interactions with the coastal environment (Chapter 8).

Chapter Two

HOW COASTS WORK

Introduction

From the dramatic, precipitous cliffs of the Banks Peninsula, South Island, New Zealand to the palm-fringed, gleaming white, coral sands of a Seychelles beach there is an almost infinite variety of coastal scenery. In any one place the coastal zone consists of the same basic components (water, air, sediments, rocks and organisms), but how they are structured depends on a range of factors including geology, climate and oceanographic regime. Contemporary processes interact with landforms produced by past changes in sea level, sediment supply and process regime. Many North American and north-west European coasts, for example, show strong legacies, in terms of sea level change and recent sediment supply, from the repeated advance and retreat of large ice sheets over the last 1–2 million years.

Many attempts have been made to classify and explain the diversity of coastal environments using genetic or descriptive criteria, or a mixture of the two. Thus Johnson (1919) split coasts into submergent, emergent, neutral and compound, depending on their recent history of relative sea level, Shepard (1963) divided coasts according to the relative importance of coastal and non-coastal processes, recognizing primary coasts (virtually unmodified by coastal processes) and secondary forms, and Davies (1964, 1972) split the world's coastline up according to wave activity leading to a division into high latitude storm wave coasts, low latitude swell wave coasts and low energy coasts. Clearly all these different approaches are of value. Here we recognize the importance of variations in geological structure, sea level history and biogeography in providing contexts within which both contemporary ecological and geomorphological processes act and future environmental changes will be placed.

Plate tectonics and coastal type

Inman and Nordstrom (1971) produced an important classification based primarily on plate tectonics. They define three major coastal scales as shown in Table 2.1. The controls operative at the first and second order provide a classification (tectonic and morphological) for the whole coastal zone. The controls operating at the third order help classify features within the shore zone. Coastal type at the large scale, therefore, is determined by the position of the area relative to plate margins, leading to a division of the world's coastline into collision, trailing-edge and marginal sea coasts (Fig. 2.1). Collision, or active margin, coasts are typically characterized by the delivery of relatively coarse sediments from mountainous catchments to a coastal zone characterized by a narrow continental shelf and deep water not far offshore. Trailing edge, or passive margin, coasts by comparison are fed by large rivers often draining enormous drainage basins, thus contributing large volumes of fine sediments to wide, low-angle continental shelves. Most of the world's major deltas are associated with trailing-edge coasts and they are also characterized by extensive barrier island development. Collision coasts, as in the Mediterranean and south-east Asia, often support only rocky shorelines, with poor beach or reef development, and exhibit sea level histories complicated by regional tectonic movements. These controls help provide a basis for the second order of Inman and Nordstrom's (1971) classification (Table 2.2).

Table 2.1 The three major coastal scales

Order	*Dimensions*		*Controls*	*Results*
1st order	Length Width Height range	*c.* 1000 km *c.* 100 km *c.* 10 km	Plate tectonics	Coastal plain and continental shelf
2nd order	Length Width Height range	*c.* 100 km *c.* 10 km *c.* 1 km	Erosion and deposition modifying 1st order features	Deltas, coastal dunefields, estuaries
3rd order	Length Width Height range	1–100 km 10m–1 km ?	Wave action and sediment size	Beaches longshore bars, mudflats

Note: 1st and 2nd order factors define the coastal zone, 3rd order factors define the shore zone
Source: Inman and Nordstrom (1971)

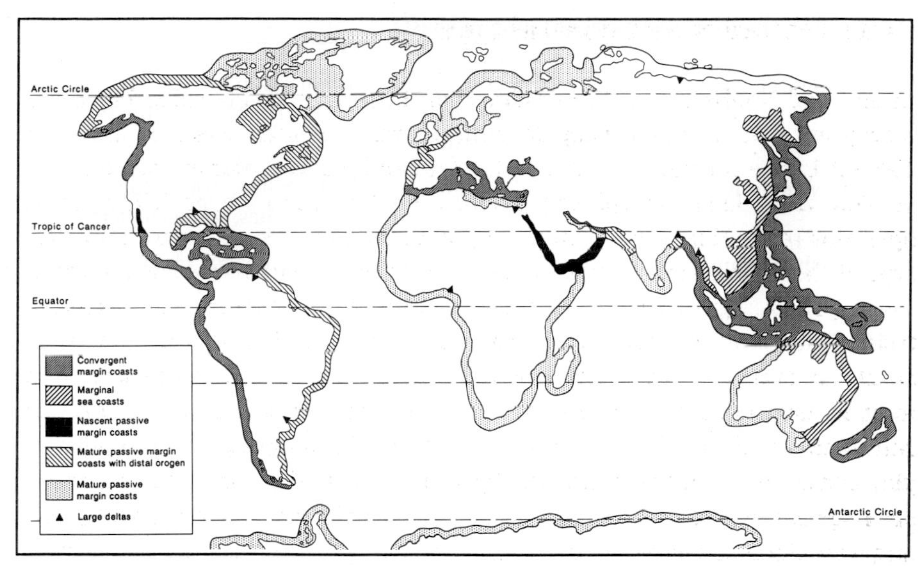

Fig. 2.1 Tectonic setting of the world's coastlines. After Summerfield (1991)

The coastal setting: Late Quaternary sea level changes

Over the past 18 000 or so years, since the last glacial maximum, changes in the loading of ice on the continents (glacio-isostasy) and alterations in the volume of water in the ocean basins (eustasy) (which itself produces changes in the loading on the sea floor, or hydro-isostasy) have produced a whole suite of sea level responses along different coasts. Within glaciated areas, the presence of a large ice sheet causes depression of the land surface, coupled with a compensatory forebulge in the crust some distance away. This occurs because of the nature of the earth's crust, which flexes in response to loading stress. Once the ice retreats the land underneath begins to rebound, and the forebulge starts to collapse. At the same time, eustatic sea levels start to rise once water is released from the ice sheet. In general, the isostatic changes take a longer time than the eustatic ones, as the earth's crust reacts quite slowly to changes in loading. These changes produce a relatively complex series of coastal sea level profiles in different areas, depending on the proximity of the area to an ice sheet.

By assuming that the earth's structure can be characterized by a thin and rigid lithosphere, with some ability to flex, overlying a more fluid, low long-term strength asthenosphere — the 'elastic plate' model — and with a knowledge of the distribution of ice at the glacial maximum and the subsequent pattern of meltwater contributions from the northern and southern hemisphere ice masses, it is possible to model sea level change over the 10 000 years of the

Table 2.2 Morphological classification of coasts

Coast type	*Characteristics*
Mountainous coast	Shelf < 50 km wide, coastal mountains > 300 m high, rocky shore zone with pocket beaches. Mainly on collision coasts
Narrow-shelf hilly coast	Shelf < 50 km wide, coastal hills *c.* 300 m high or less, occasional headlands and beaches, some barriers
Narrow-shelf plains coast	Shelf < 50 km wide, low-lying coastal plains, barrier beachs, occasional low cliffs
Wide-shelf plains coast	Shelf > 50 km, low-lying coastal plains and wide shore zone, often with barrier beaches
Wide-shelf hilly coast	Shelf > 50 km wide, coastal hills *c.* 300 m or less, barrier beaches and occasional headlands and cliffs
Deltaic coast	Sediment deposited where river enters sea; low-lying coastal bulge
Reef coast	Organic origin, resistant; fringing or barrier type
Glaciated coast	Coastal features dominated by erosional effects of glaciers, precipitous cliffs and fjords common

Source: Inman and Nordstrom (1971)

postglacial period (Clark *et al.*, 1978; Fig. 2.2). The reality of individual sea level histories is more complex than Fig. 2.2 suggests, as the earth's crust may not react in such a smooth way. Areas once depressed by ice loading tend to show emergent coastal features, such as the staircases of raised beaches characteristic of Hudson Bay, Canada, as relative sea level has declined over the past 15 000 years (I in Fig. 2.2). Coasts which were on a forebulge adjacent to ice loaded areas show submergent trends (II) as relative sea level has risen because of the collapse of the forebulge. A complex suite of profiles is found in the transitional zone between I and II where both submergent and emergent trends exist, with the upper curves representing an area like Scandinavia, and the lower ones areas like the Netherlands and eastern England. Most other areas show either a general submergent trend, where eustatic changes dominate well away from ice sheets, or submergence followed by emergence, where hydro-isostatic depression of the ocean floor dominates, e.g. the south Pacific (zone V).

The importance of sea level history for the study of today's coastal problems is that the length of time that sea level has been at or around its present level affects the functioning of the coastal system. Thus, for example, coastlines in Australia have experienced coastal processes at the present level for 6000 years whereas on the Florida, USA coastline, present sea level has only been reached in the last 1000 years. Some coastlines are still being subjected to isostatic uplift

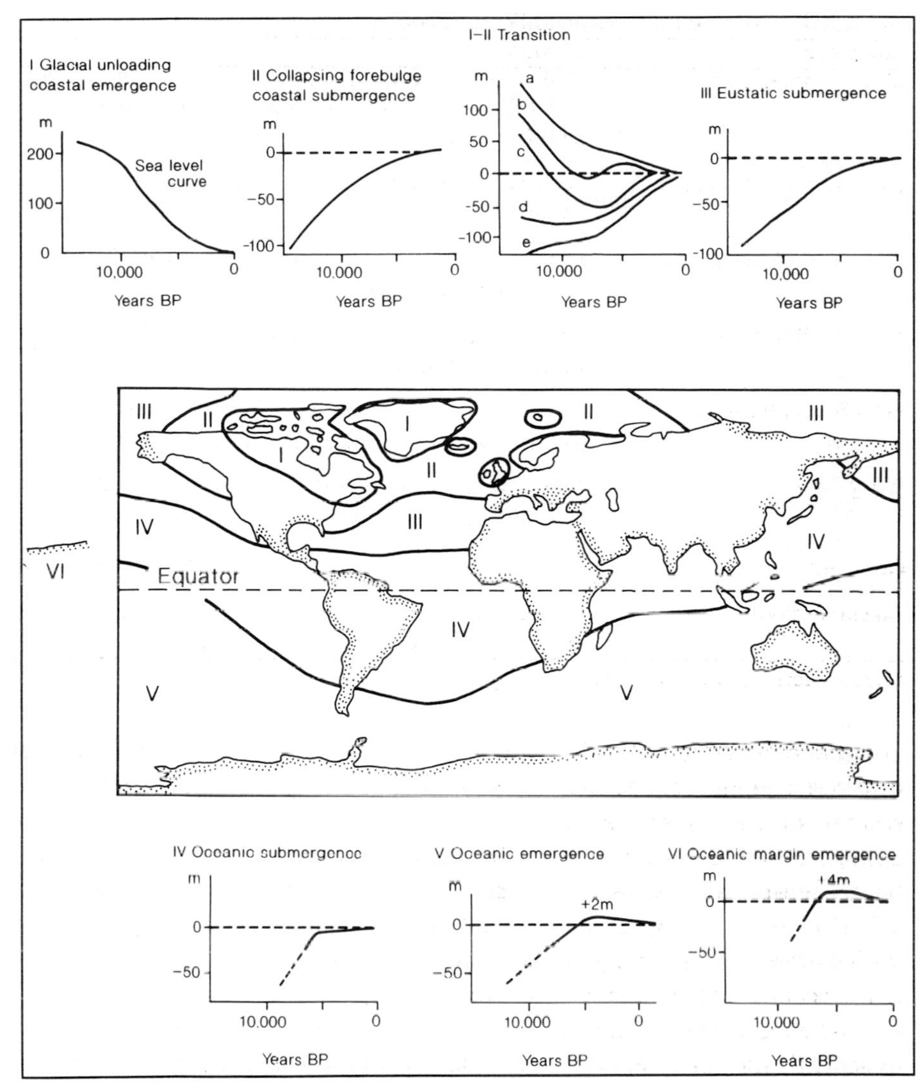

Fig. 2.2 Probable variations in sea level from 15000 years BP for five areas of the world. After Clark *et al.* (1978)

and sea level fall; others are undergoing isostatic adjustments that lead to sea level rise. These continuing movements are important in the context of future sea level rise: isostatic uplift will offset some of the expected eustatic component whereas isostatic downwarping will add to the ocean warming sea level effect. Finally, sea level history is important in the context of sediment supply at the modern coast. In areas such as the southern North Sea basin where there has been a general submergent trend over the last 10000 years, sediment sources such as glacially derived sands and gravels available in the early stages of the postglacial transgression are now below the wave base. Thus, many coastal landforms may be largely fossil features dominated by sediment reworking and clearly vulnerable to human activities which interfere with sediment redistribution processes. By comparison, in Scandinavia where sea level has fallen over the same period, new sediment sources become available as sea level drops.

Wave processes in the shore zone

Tectonics and sea level history provide the contexts for the operation of contemporary processes. Table 2.3 shows a modification of the classification of shore zone features proposed by Inman and Nordstrom, which divides them in terms of processes. According to calculations presented by Inman and Nordstrom (1971) almost 45 per cent of the world's coastline is wave-eroded, 36 per cent glaciated and a mere 1 per cent is dominated by wind deposition. Such a process-based classification includes the effects of latitude, as reef building, glaciation and other such processes are primarily climatically controlled; these azonal processes are dealt with in Chapters 6 and 7, respectively.

Almost all nearshore erosion, sediment transport and deposition can be explained by the action of different types of waves; these can be classified according to their frequency characteristics (Fig. 2.3). Waves that are caused by periodic forces — such as the effect of the sun and moon in producing tides — have frequencies coinciding with the causative forces (and are dealt with later in this chapter). Most waves, however, result from non-periodic disturbances of the water surface which displace water particles from their equilibrium position. Regaining that position requires a restoring force; for waves of geomorphological significance, gravity is the main force acting to return seawater to its

Table 2.3 Processes affecting shore zone

Erosional effects	Wave erosion
	Glacial erosion
	Water erosion at previous ocean low-stand
	Wind erosion
	Biological erosion
Depositional effects	Wave deposition
	Glacial deposition
	Fluvial deposition
	Wind deposition
Organic accretion	Coral and algal reefs
	Mangroves and salt marshes

Source: Inman and Nordstrom (1971)

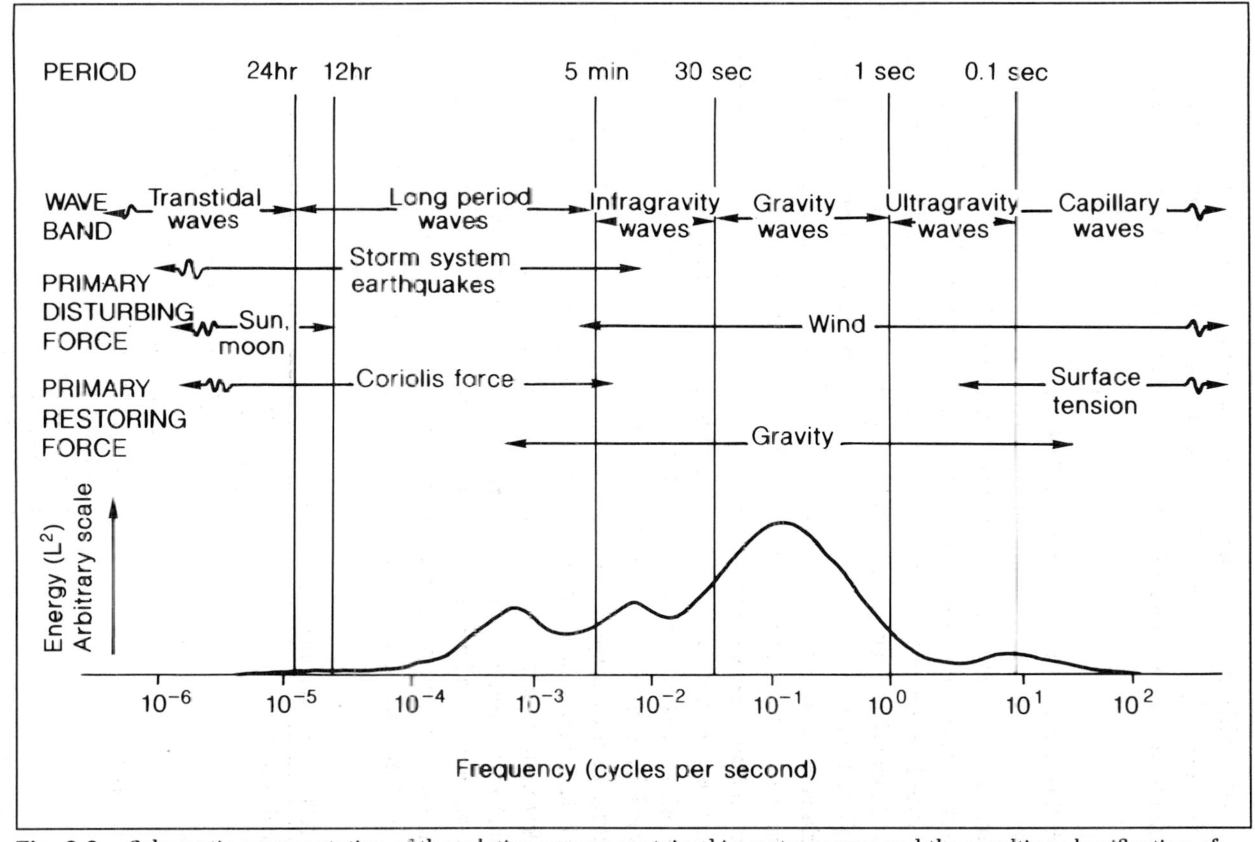

Fig. 2.3 Schematic representation of the relative energy contained in water waves and the resulting classification of ocean waves. After Kinsman (1965)

equilibrium position. Such gravity waves, or wind waves, contain a very large proportion of the total energy present across the frequency spectrum (Fig. 2.3).

Gravity wave generation

The degree of energy transfer from the atmosphere to the ocean surface, and thus the degree of wave development, is dependent upon three factors: windspeed, wind duration and the distance (or fetch) over which the wind blows. Waves may be duration- or fetch-limited: thus maximum wave heights in the eastern North Atlantic are 35 m but only 10 m in the English Channel (Hardisty, 1994). Numerous wave forecasting methods have been devised to predict wind wave characteristics from meteorological data; they are well described by Komar (1976). Waves formed in a storm area form a sea of steep, short-lived and confused crests and troughs. However, as waves move out of the area of generation, they become sorted by wave period, with the highest period waves moving most rapidly. This sorting process also narrows the wave spectrum, yielding swell waves of more regular wave height, length and period. Such swell waves may travel over huge ocean distances with little energy loss before making landfall; thus waves breaking on the Alaskan coastline may have been generated by the great storms that characterize the southern oceans around Antarctica (Snodgrass *et al.*, 1966). Elsewhere, such as in the Doldrums near the Equator, winds are light and variable in direction and yield only small waves. A knowledge of atmospheric circulation systems, of patterns of storminess under different climatic regimes and an understanding of the movement of ocean swells allows a global map of world wave environments to be constructed (Fig. 2.4).

Incident waves and their transformation in shallow water

Although waves driven by the wind and arriving at a coast (incident waves) contain a spectrum of wave periods and a variety of wave heights, it is useful to start with simple wave forms which can be described by reference to a few basic parameters (Fig. 2.5): height (H), wavelength (L), period (T, the time taken for a wave to travel one wavelength) or frequency ($f=1/T$), and wavespeed or wave celerity (C). Relationships can be established between these parameters (Komar, 1976) as follows:

(eq. 1)
$$L = gT^2/2\pi$$
$$(L = 1.56T^2)$$

and

(eq. 2)
$$C = L / T$$
$$(C = 1.56T)$$

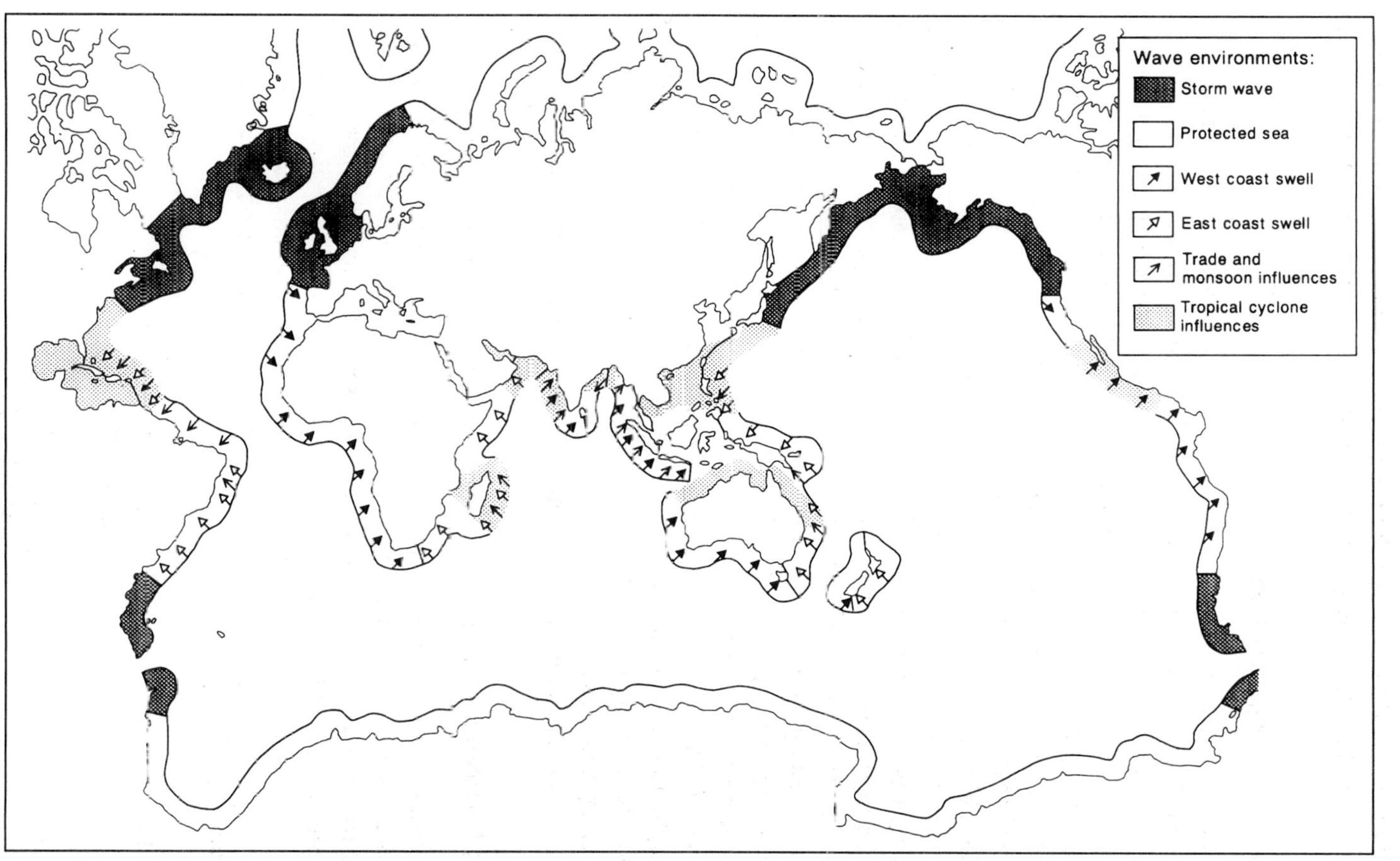

Fig. 2.4 World wave environments. Modified from Davies (1972)

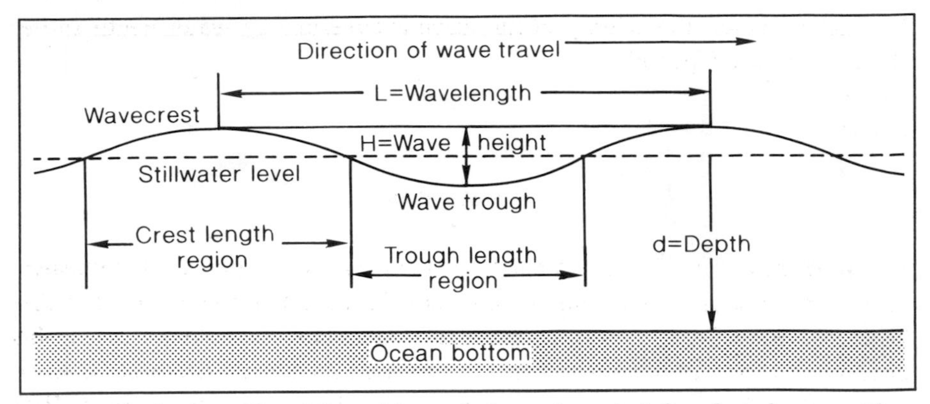

Fig. 2.5 Schematic illustration of the main wave characteristics of wind waves. After Coastal Engineering Research Center (1984)

In practice, the speed of the wave train is less than the speed of individual waves and thus:

$$\text{Group wave celerity } (Cg) = Cn$$

where:

$$n = 0.5 \text{ in deep water}$$
$$n = 1.0 \text{ in very shallow water}$$

These relationships apply in deep water, or water depths greater than one quarter of the wavelength. However, when waves enter shallow water (defined as water depths less than one twentieth of a wavelength) a series of transformations take place. These are best envisaged by thinking about wave energy in the nearshore zone. A fundamental relationship is that between wave height and wave energy:

(eq. 3) $$\text{Wave energy} = E = \tfrac{1}{8} \rho g H^2$$

where:

ρ = water density
g = acceleration due to gravity
H = wave height

The rate at which this energy is transferred across the water surface, the energy flux or wave power, is given by:

(eq. 4) $$\text{Wave power} = P = ECn$$

In shallow water, the celerity of the wave is governed by water depth rather than by the wave period:

(eq. 5) $$C = \sqrt{gh}$$

where:

h = water depth

As the energy flux, P, must remain constant and as Cn generally decreases, then wave energy, E, and thus wave height, H, must increase as the shoreline is approached. Furthermore, as wave period remains constant during this process, a decrease in wavelength, L, must accompany the decrease in wave celerity (Fig. 2.6). Thus as waves approach the shore wave steepness (= H/L) increases. Wave crests become narrower and more peaked with flatter intervening troughs.

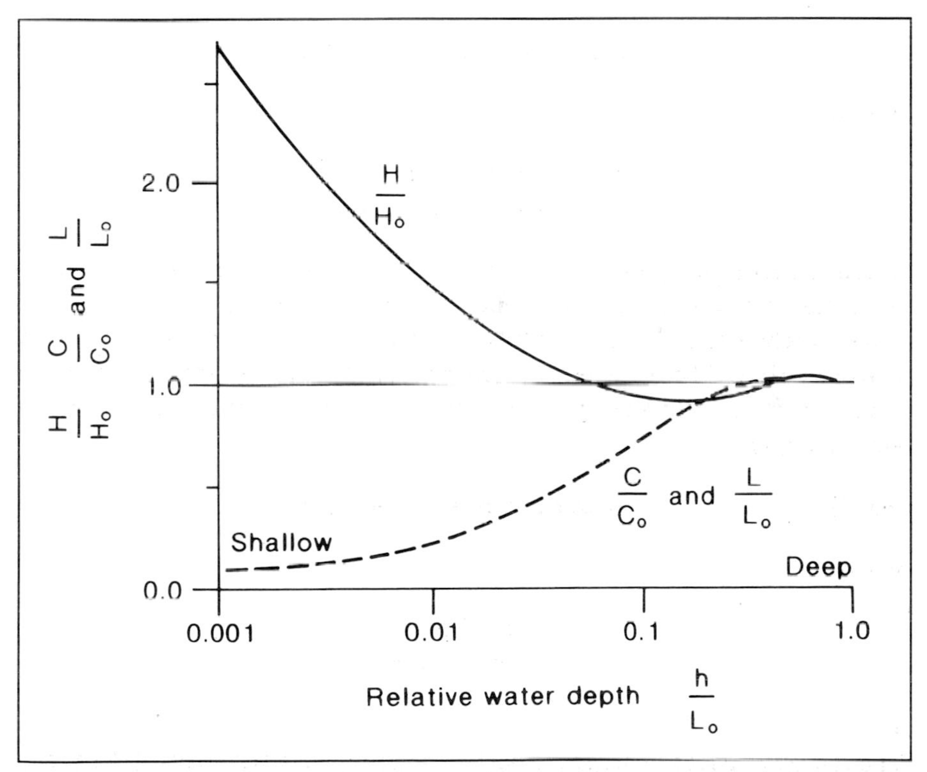

Fig. 2.6 Wave shoaling transformations, relative to deep water values (subscript 'o'). Note that wave height increases, wave length and wave celerity decrease, with shallowing. Wave period remains constant throughout. After Pethick (1984)

Incident wave refraction

Wave heights also change as a result of wave refraction; in shallow water wave crests bend in an attempt to run parallel to the depth contours.

(eq. 6) $$\sin \alpha = (C/C_0) \sin \alpha_0$$

where:

α = the angle between the wave crest and the depth contour
C = wave celerity
C_0 and $\sin \alpha_0$ = deep water values for C and $\sin \alpha$ respectively

Thus the degree of refraction is related to the initial angle of deep water wave approach and to the change in wave celerity. As wave celerity decreases in shallow water, the angle between the wave crest and the bottom contour must decrease landward (Fig. 2.7).

What happens to wave energy and wave height with wave refraction? Wave rays are lines drawn normal to the wave crest; in a simple case of parallel offshore contours and oblique wave approach, the spacing between the rays increases as refraction takes place (Fig. 2.7). A term to account for wave refraction can be added to the standard energy flux relationship:

(eq. 7) $$P = ECnS$$

where:

S = wave ray spacing

As spacing increases in shallow water and C and n are not affected by refraction, wave energy must decline. Thus, wave heights must be lower in the presence of the process of wave refraction. Convergence of wave rays as a result of shallow water around rocky headlands indicates that wave heights increase in these localities. Conversely, wave divergence over deep water submarine canyons, decreases wave heights at canyon heads (Fig. 2.8). These variations in wave height translate into variations in the height of breaking waves and these variations in turn drive alongshore current flows and circulation systems (see below).

The breaking wave

Eventually waves in progressively shallower water over-steepen, become unstable and break. The point of breaking is a function of wave characteristics and water depth and can be simply defined by the gamma ratio:

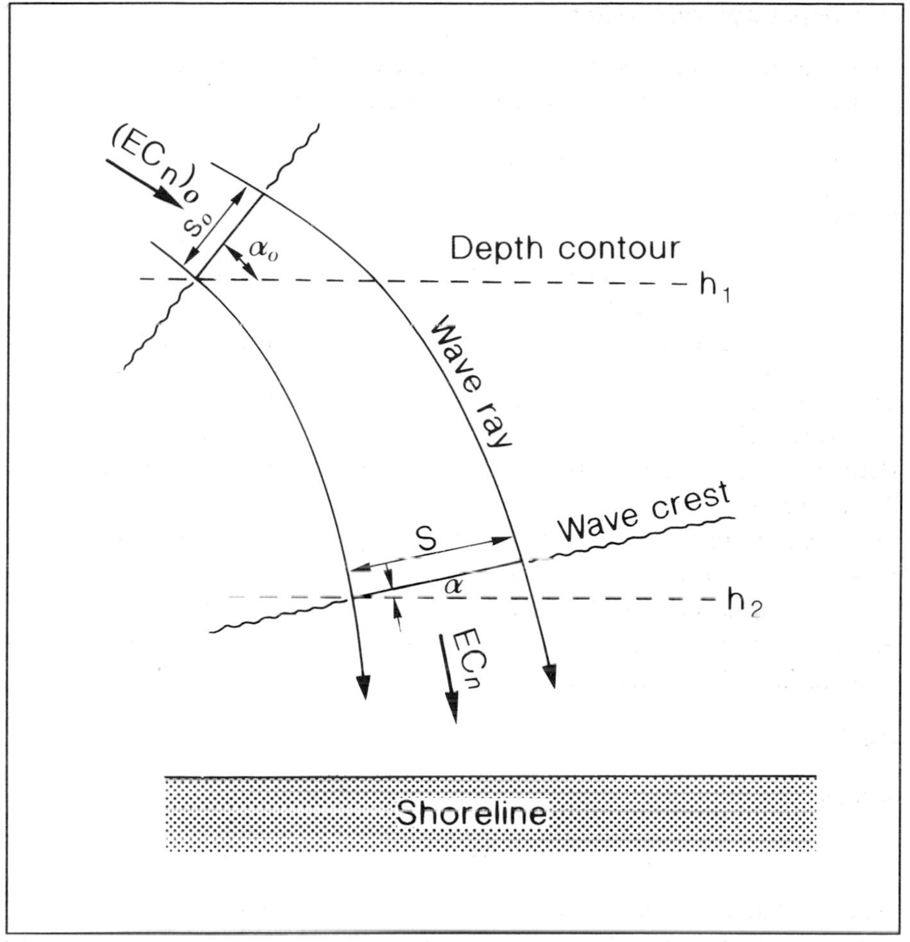

Fig. 2.7 Relationship between wave approach angle (α), water depth (h) and wave crest length (S). Subscript 'o' denotes deep water values. Note spreading of orthogonal wave rays in shallower water (see text for details). After Komar (1976)

(eq. 8) $$\gamma = \frac{H_b}{h_b}$$

where:

subscript 'b' denotes 'at the wave break point'.

Thus, low waves run into shallower water than high waves before breaking and as a wide range of wave heights arrive at the coast at any one time then not all waves break at exactly the same point. However, the breakpoint is not only determined by wave height; the gamma ratio typically ranges in value from 0.6 for 'flat' beaches to 1.2 for 'steep' beaches. For a given wave height,

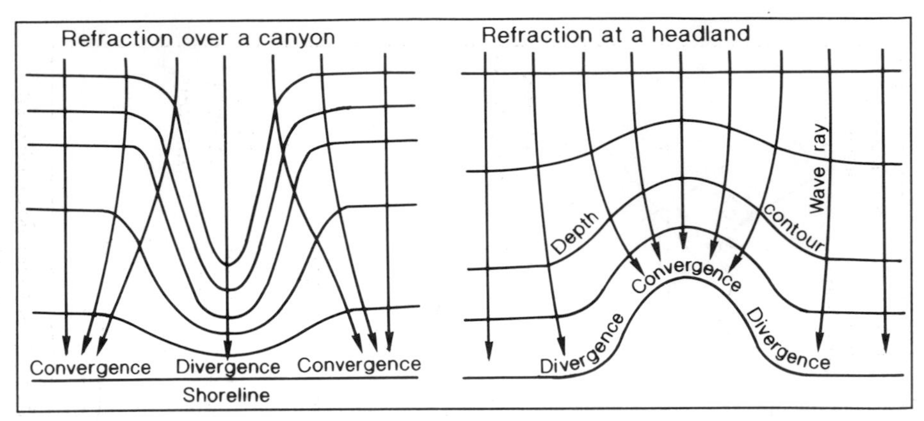

Fig. 2.8 Patterns of wave convergence and divergence, as indicated by wave rays, with complex bottom topography. After Komar (1976)

therefore, breaking will occur further inshore on a steep beach than on a low angle beach. These controls were formalized from laboratory experiments by Galvin (1968, 1972) into a breaker coefficient which for shallow water takes the form:

(eq. 9)
$$B_b = \frac{H_b}{g\beta T^2}$$

where:

β = beach slope

This coefficient also usefully relates breaking to the type of breaking wave (Fig. 2.9).

Swash, backwash and the beach profile: on to beach morphodynamics

Water movement and sediment transport under breakers of different types and subsequent breaker type-specific processes in the swash/backwash zone in part translate into changes in the beach profile. Within the swash/backwash zone, a useful measure is Kemp's (1975) phase difference, P, the ratio between wave run-up (t, the time from the moment of wave break until the furthest point water is seen to move up the beach) and wave period (T). For surging breakers $P \leqslant 0.5$ whereas for spilling waves $P \geqslant 1.0$. The explanation for these differences is clear from careful study of swash and backwash velocities at different locations on the beach. Long-period surging waves (Plate 2.1), characterized by a short burst of high onshore velocities at time of breaking (Fig. 2.9), introduce well-spaced, discrete swash packets on to the beach, which in a simple and

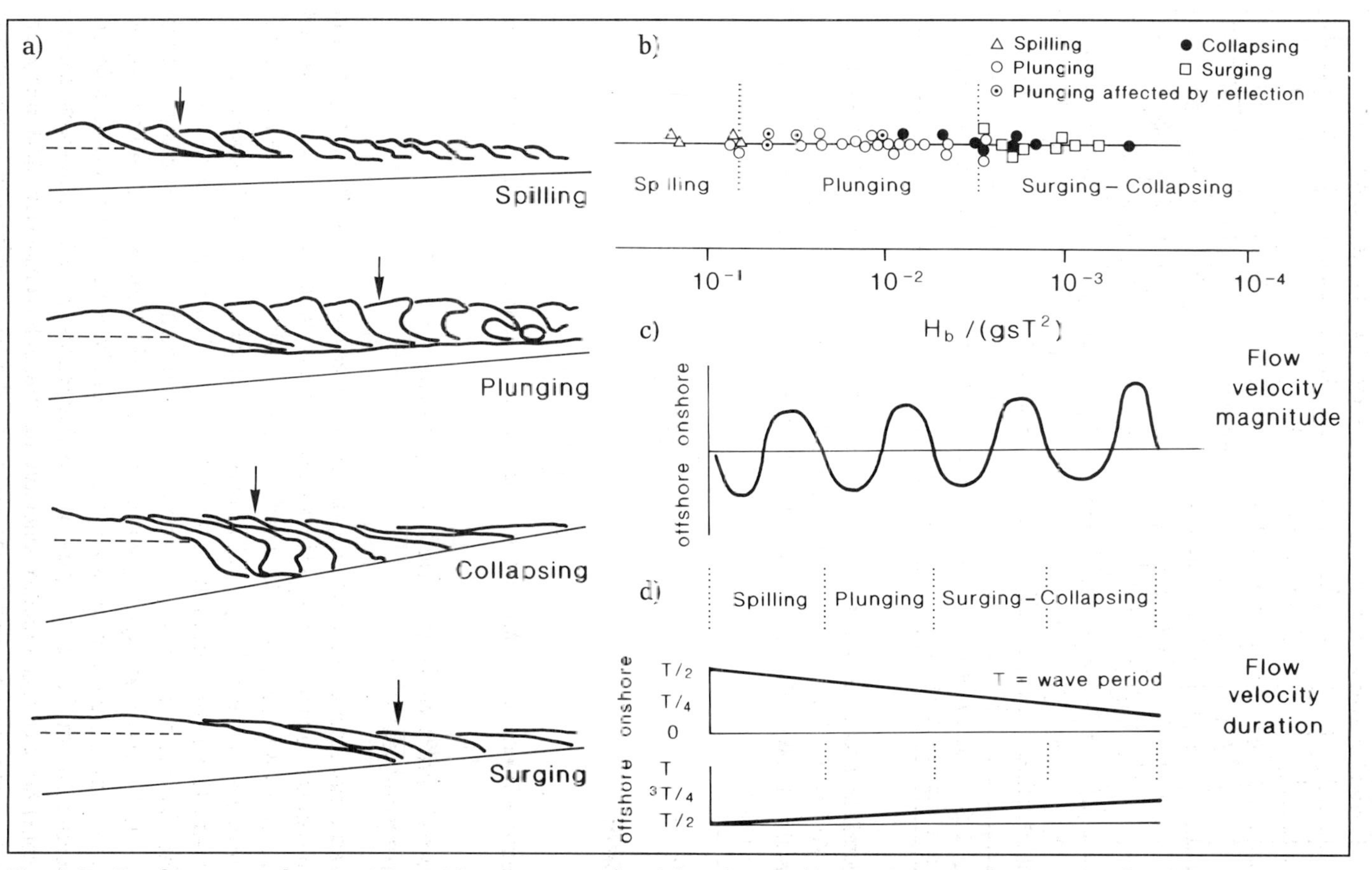

Fig. 2.9 Breaking wave characteristics. a) Breaker types after Galvin (1968), b) according to Galvin's (1968) breaker coefficient (equation (9)), c) and d) changing onshore/offshore flow asymmetry with different breaker types. After Pethick (1984)

symmetrical pattern, decelerate, reverse and return as backwash before the arrival of the next wave (Fig. 2.10a). By comparison, in the case of shorter period spilling and plunging waves, water near the breakpoint is already moving seaward before water further up the beach has completed its swash phase; as these upper beach flows form the backwash they are overridden by the next uprush (Fig. 2.10b). As a first approximation, it is not difficult to see how surging waves, particularly where percolation of swash volumes is high, as on beaches composed of coarse sediments, might be associated with the build-up of the beach profile, whereas the high volumes of water on the beach and in the backwash under spilling wave conditions might be conducive to the combing of material from the beach and the flattening of the beach profile.

Many authors have suggested links between breaker types and beach morphology, to give more meaning to the typical division of beach states into 'winter' vs. 'summer' or 'storm' vs. 'swell' profiles. The most successful of these parameters has been the surf scaling factor and its variants. One derivative, for example, has the form:

(eq. 10)
$$\text{Surf scaling factor} = \epsilon = \frac{H_b\,\omega_i^2}{g \tan \beta}$$

where:

$$\omega_i = 2\rho/T$$

The surf scaling factor introduces two important terms of beach description. At low values ($\epsilon \leqslant 2.5$) the beach is known as 'reflective' (Plate 2.2), characterized by surging breakers and a high percentage of incident energy being reflected from the beach face. At high values ($\epsilon \geqslant 33.0$) the beach is 'dissipative' (Plate 2.3), where wave energy from spilling breakers is lost across across a wide, flat beach (Table 2.4).

Masselink and Short (1993) have extended this work to include higher tidal situations by introducing a new parameter, i.e. the relative tidal range (*RTR*) which is defined as:

(eq. 11)
$$RTR = \frac{TR}{H_b}$$

where:

TR = tidal range (m)
H_b = breaker height (m)

Large tidal ranges have several effects on beaches as they retard the rate at which sediment transport and morphological changes occur, increase the

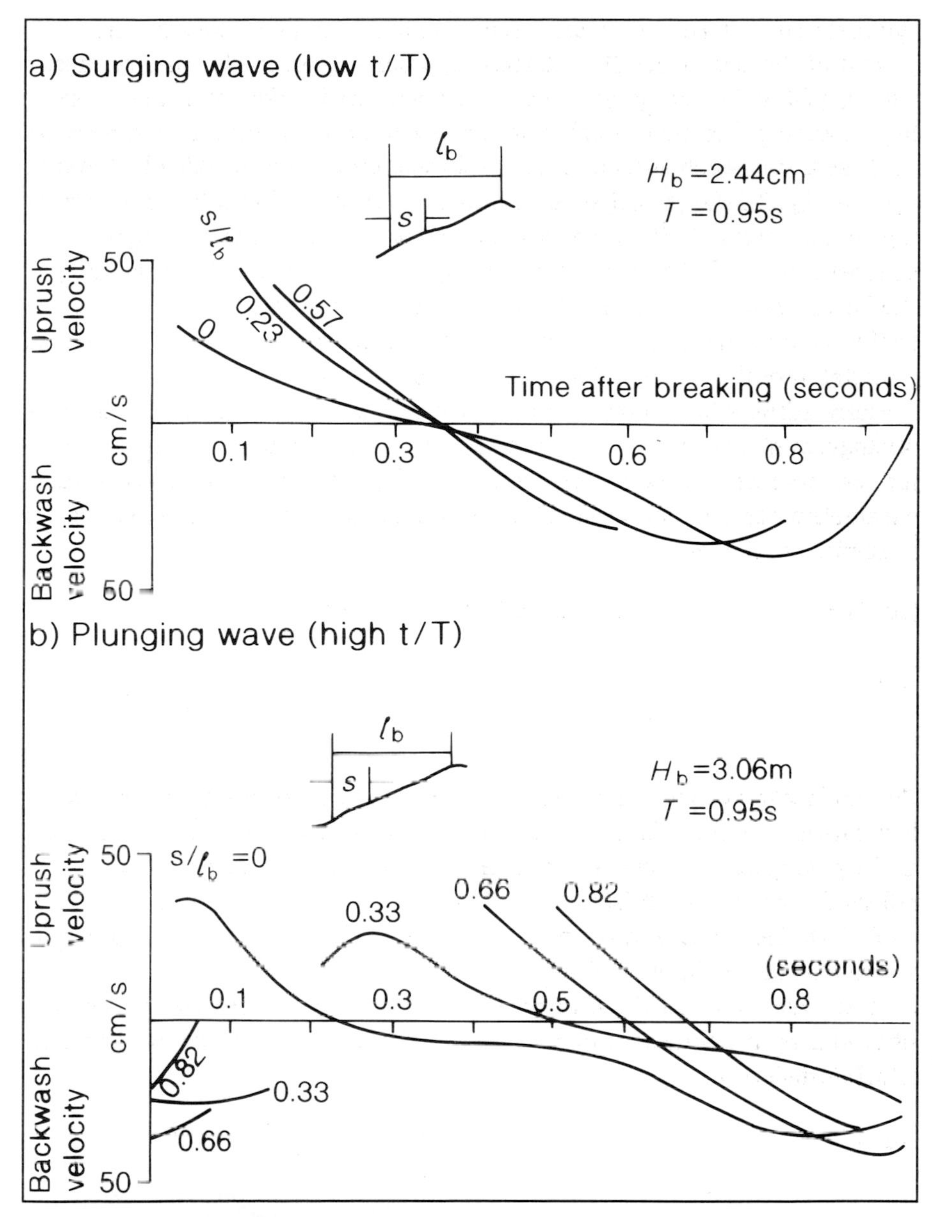

Fig. 2.10 Swash and backwash characteristics by wave type. s = point of wave break, l_b = point of maximum uprush, low values of s/l_b near the breakpoint, high values near top of the swash. a) Surging wave (low t/T): discrete swash event with velocity zero at all points on the beach simultaneously + swash/backwash event completed before next wave arrives. b) Plunging wave (high t/T): water near breaker begins to move seaward before swash completed + backwash not complete before next wave breaks. After Kemp (1975)

Plate 2.1 Surging waves on a tropical beach of volcanic sand: Baie du Contrôleur, Nuku Hiva, Marquesas, east-central Pacific Ocean

Plate 2.2 Reflective beach profile: Lulworth Cove, Dorset coast, England. Note coarse sediments and presence of beach cusps

Table 2.4 Dissipative–reflective beach characteristics

Characteristic	*Reflective*	*Intermediate*	*Dissipative*
Wave breaker height (m)	< 1	1–2.5	> 2.5
Surf scaling	0.1–2.5	2.5–20	20–200
Approx. surf zone width (m)	< 10	10–100	100–>1000
Breaker type	Plunging–collapsing	Spilling–plunging	Spilling
Angle of breaker approach to shore	Strong oblique	Slightly oblique to normal	Shore normal
Edge waves period (s)	< 15	5–30	30–70
Swash period (s)	5–10	10–30	Up to 60
Wave driven currents	Shore parallel	Meandering–helical	Rip cells–circulatory
Rip currents	Absent–weak	Mixed	Strong, persistent, stationary
Nearshore bars	Absent	1 or 2, crescentic	Multiple
Beach slope (°)	> 3°	3–1°	< 1°
Beach profile	Concave	Shallow concave	Rectilinear
Beach platform	Swash cusps	Mixed	Surf cusps
Common sediment size	Coarse sand, gravel	Medium sand	Silt to fine sand
Alongshore grading	Common	Partly	Rare
Sediment transport dominant directions	Alongshore	Mixed	Onshore–offshore
Dunes	Usually small	Intermediate	Usually large
Sand storage	On beach	Shifts between surf zone and beach	In surf zone
Filtered volume (volume of sea water flushed daily through intertidal sand)	Large	Intermediate	Small
Residence time (time filtered volume takes to percolate)	c. 6 h	6–24 h	c. 24 h
Surf-zone diatoms	None	Variable	Rich
Intertidal fauna	Poor	Variable	Rich

Source: compiled from Short and Hesp (1982), Carter (1988) and Brown and McLachlan (1990)

importance of shoaling wave processes thus reducing beach gradients, inhibit bar formation, affect rip current flows at different states of the tide, and encourage the dominance of shore parallel currents seaward of the lower intertidal zone. A study of beaches in central Queensland produced a new conceptual model of beach morphology based on Ω, a variant of the surf scaling factor, and *RTR* (Fig. 2.11). The eight different beach states recognized in this model each contain a different array of small scale beach features (e.g. bars) as well as different dissipative or reflective characteristics (Plate 2.4).

Much of the temporal variation in beach state is related to storm events when wave heights are often much larger than normal. Along the micro-tidal Ebro Delta coast of Spain, for example, there are nine major longshore bar and trough systems which move both across and alongshore, with the main mor-

Plate 2.3 Dissipative beach profile: spilling waves at the most eastern extremity of Australia: Byron Bay, New South Wales

phological changes occuring in storm periods (Guillén and Palanques, 1993). Recent, detailed surveys of beach profile changes as well as measurements of waves, currents and suspended sediment concentrations over storm events have shown how beaches respond. Lins (1985), for example, uses principal components analysis to show that most variations in profile of beaches on the Outer Banks of North Carolina, USA, occur on time-scales of more than 4 months, and that individual storm events contribute relatively little. In the tropical Nigerian environment, studies at Forcados Beach on the north-western flank of the Niger Delta have shown clear seasonal trends in erosion, accretion and beach profiles — with higher berms and concentrated erosion and accretion periods in the wet season (Oyegun, 1991). Thus, the most rapid changes in beach state do not necessarily accompany the most rapid changes in beach height and it is possible, at wave energy levels capable of driving morphological change, for fast changes in beach state to take place when waves remain relatively constant in terms of height and period. These findings support recent arguments that changes in sediment transport are driven not simply by incident waves but also by infragravity, or long, waves with periods of 20–200 s (Fig. 2.3). These waves form a complex family of water level oscillations trapped within the breaker zone; they include standing infragravity waves parallel to

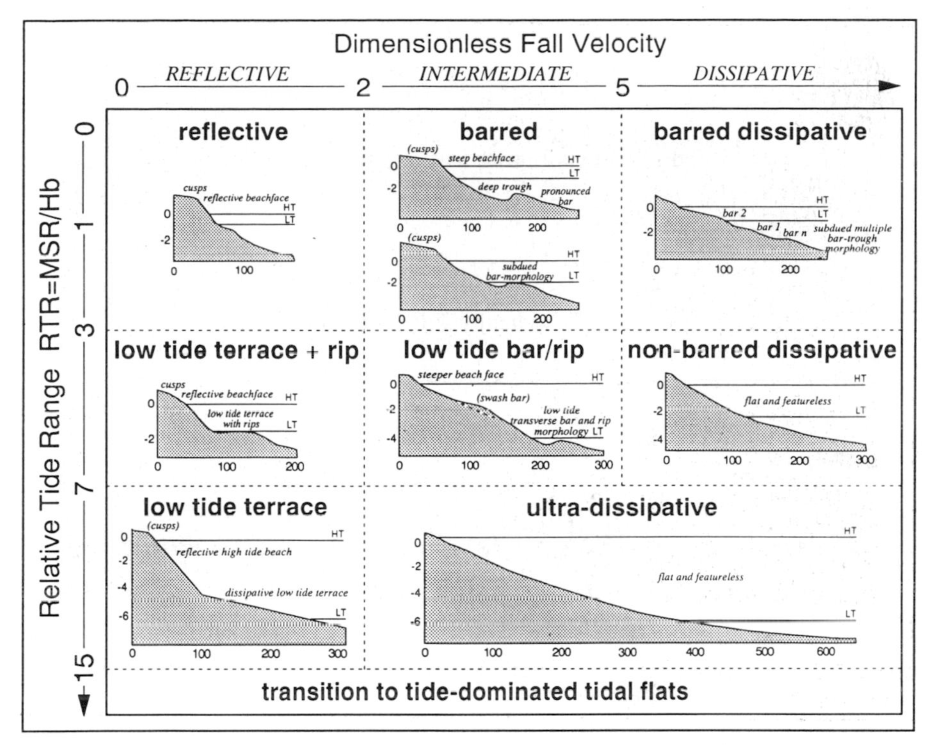

Fig. 2.11 Conceptual model of beach morphodynamics related to dimensionless fall velocity and relative tide range. After Masselink and Short (1993)

the shoreline and shore-normal edge waves (Hardisty, 1994). The interaction of different types of infragravity waves with incident waves leads to longshore variations in breaker height which, like wave refraction processes, then power cell circulation systems at a range of scales. These processes result in beach cusps and rhythmical sand wave topography at small and large shore zone scales, respectively. Return flows in cell circulation systems are associated with concentrated flow, or rip currents, in regions of low breaker height (Fig. 2.12). Offshore sediment transport associated with rip currents may lead to the growth of erosional embayments on the beach and in some cases their continued development into backshore environments leads to severe erosion problems, as on the barrier coastlines of the Atlantic coast (Dolan, 1971) and the Pacific Ocean coast spits (Komar, 1983a, b) of the USA.

More complex circulation systems may result, with the addition of longshore current flows generated within the breaker zone from waves arriving obliquely at a coastline. Measurements of alongshore sand transport rates have been shown to be related to the longshore component of wave power:

(eq. 12) $$P_L = (ECn)_b \sin \alpha_b \cos \alpha_b$$

in the form:

(eq. 13) $$\text{Sand transport rate } (m^3\ d^{-1}) = Q_L = k_L\ P_L f$$

where:

k_L = 0.06–6.2, typically < 0.5
f = 0.54–1.0, typically 1.0 (Hardisty, 1990)

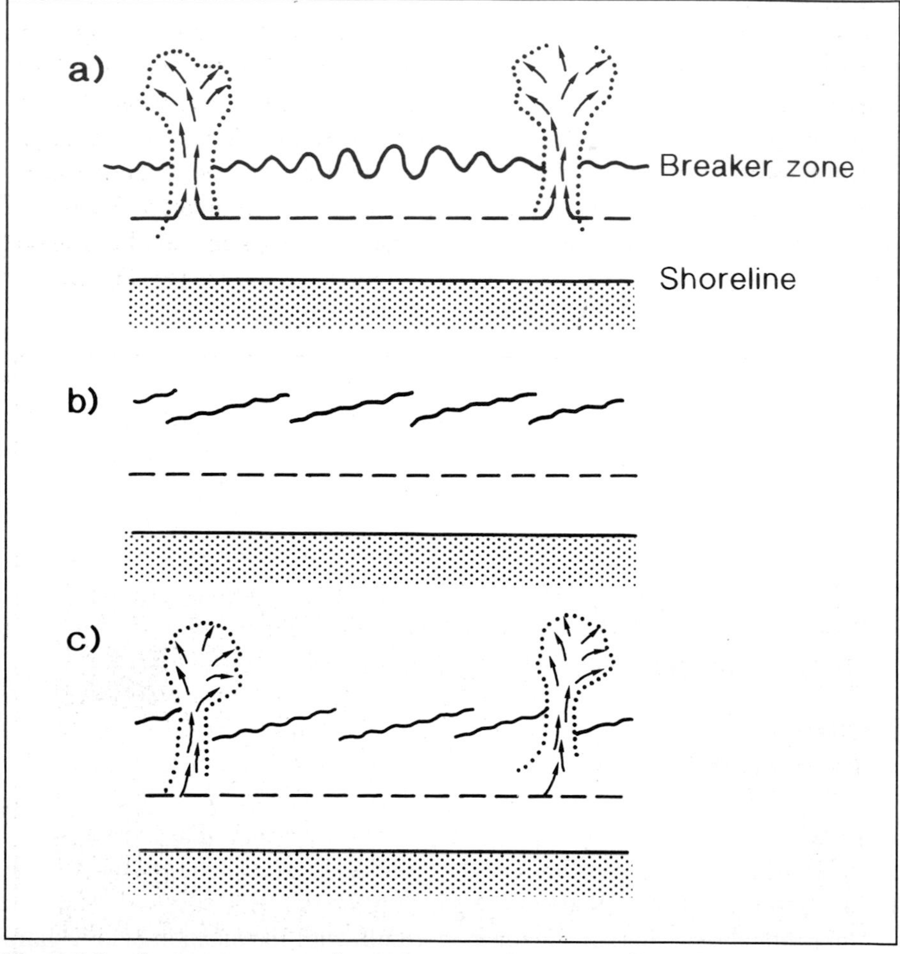

Fig. 2.12 Current systems and nearshore circulation. a) Cell circulation under shore normal wave approach. b) Longshore drift with oblique wave approach. c) Mixed system. Size of arrow indicates strength of current flow. After Komar (1976)

The importance of longshore sediment transport shows that beach dynamics must be understood in terms of the beach as a volume not as a profile and that up-drift processes may override onshore–offshore dynamics. The implications for coastal management are discussed in Chapter 3.

The influence of tides

Tides are a vital component of coastal dynamics, producing important currents and sediment movements, as well as influencing the zonation of coastal organisms, landforms and weathering processes. Tidal processes are particularly important in areas where wave energies are relatively low, such as lagoons, tidal bays and estuaries. Astronomical tides are related to the motion of the earth–moon–sun system, and have a range of periodicities. Explained simply, the gravitational forces of both the sun and the moon produce tides, although the sun has only half the tide-producing effect of the moon. The attractive force of the moon, coupled with the centrifugal force due to the rotation of the earth-moon system, produces a tidal bulge on the moon-facing side and also one on the opposite side of the earth, *c.* 35 cm above average sea level (see Fig. 2.13).

Plate 2.4 Intermediate beach state: longshore bars and troughs, Ainsdale coast, north-west England

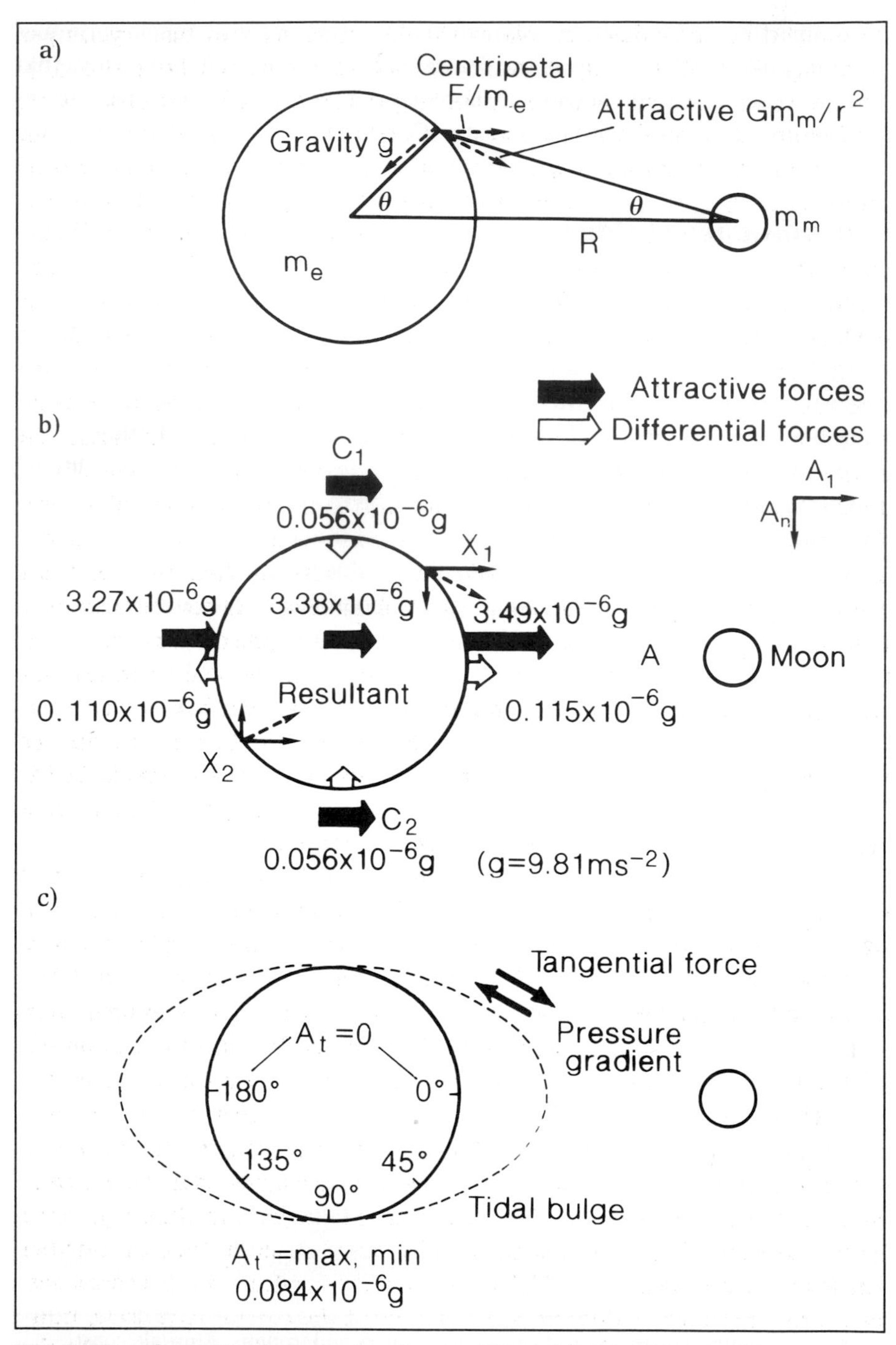

Fig. 2.13 The main tide-generating forces. a) Forces due to the mass of the moon acting on the earth. b) Variations in tidal forces around the earth. c) The tidal bulge resulting from the forces shown in (b). After Carter (1988)

As the earth rotates daily, all places should experience two tidal cycles per lunar day (24 h 50.47 min) of equal amplitude (bulge–trough–bulge–trough). This would only be true, however, if the bulges were centred symmetrically on the Equator at all times. In fact, the moon's orbital plane is tilted at *c.* 5° from the earth's and symmetry is achieved only twice a month as the moon passes through the Equator in the course of its passage from up to 28.5° latitude North to 28.5° latitude South. The tidal bulges follow these changes and different parts of the earth experience one of three kinds of tides: at high latitudes, the tide of only one hemisphere is dominant and tides are diurnal (one tide per day); at the Equator two tides per day are experienced (the semi-diurnal regime); in between, two tides are also experienced, but one is stronger than the other (the mixed regime). The sun also influences the tidal bulges in the same way as the moon, but more weakly because of its greater distance from the earth; the effect is 46 per cent weaker than that of the moon and produces a bulge of only 16 cm. When the earth, moon and sun are all aligned (at new and full moon), tidal ranges are 20 per cent greater than average, whereas when the earth–moon and earth–sun axes are at right angles to one another, and the sun's effect partly cancels out the moon's effect, tidal ranges are 20 per cent less than average. These differences thus give rise to the 29 day cycle of spring and neap tides, respectively. Furthermore, as the orbits of the moon and earth are not spherical but elliptical, the relative distances between the earth, moon and sun vary with the time of the month and the time of the year, introducing further tidal variations which are explained in more detail elsewhere (Pethick, 1984; Carter, 1988). Tides also vary on longer timespans, such as the 18.6 year nodal cycle, which is controlled by variations in the lunar orbit.

The tidal bulges move through the oceans as tidal waves that are reflected from land masses and affected by bottom friction in shallow seas. The tidal wave is deflected by the Coriolis force created by the earth's spin and thus rotates around the point of zero tide known as the amphidromic point. The time-course and character of the rotating tide can be plotted by use of co-tidal lines; all points along a co-tidal line experience high tide at the same time, although tidal range increases along a co-tidal line with distance from the amphidromic point. Ocean tides beating at the entrances to enclosed seas and bays are often reflected back to the entrance where they then interact with another incoming tide. This process may happen several times and thus allow tidal resonance to build very large tidal ranges. The classic example is the Bay of Fundy, Nova Scotia, Canada where the maximum tidal range reaches 15.4 m. Further details on the modification of tidal waves, and the implications for coastal systems, are considered in Chapter 5. Remaining at the coastal zone scale, however, it is clear that tidal patterns can be very complex. Thus, for example, the interactions and modifications of ocean tides as they move through the North

Sea basin result in the presence of three amphidromic points. Their positions, and the varying lengths of the co-tidal lines associated with them, show that tidal characteristics change over quite short distances along the coastline of north-west Europe (Pethick 1984; Fig. 2.14).

The tidal ranges in the North Sea vary from mesotidal to macrotidal. In fact, three tidal ranges can be recognized globally. Davies (1964) divides spring tidal ranges (i.e. the mean difference between spring high and neap low tides) into: microtidal (< 2 m), mesotidal (2–4 m) and macrotidal (> 4 m). Microtidal regimes are found on open coasts where the tidal wave is dominantly reflected, or inland seas, such as the Mediterranean, where there is only low tidal energy. Macrotidal conditions occur where the tidal wave becomes dissipated across wide continental shelf slopes, or confined in estuaries and gulfs. Thus, tidal range should increase as the width of the continental shelf increases. Most of the world's coasts are macrotidal or mesotidal environments (*see* Fig. 2.15).

Tidal range is an important control on coastal ecology and geomorphology, determining the width of coast subjected to alternate wetting and drying and the impact of waves. Rocky shores along Mediterranean microtidal coasts, for example, are characterized by a narrow band of intertidal organisms, whereas similar shores along the macrotidal Bristol Channel display a wide area with clear zonation of forms. On beaches, tidal range has an important influence on profile development, as explained more fully in Chapter 3. On macrotidal coasts, wave activity occurs over a large area across the tidal cycle, whereas it is highly concentrated within a specific band on microtidal coasts.

Coastal ecology

The organisms found at the coast reflect the unique characteristics of this zone as a meeting point of marine and terrestrial systems. In the intertidal and subtidal zones mainly marine species are found, with terrestrial species becoming increasingly dominant in supratidal environments such as dunes and cliff tops. The coastal zones (down to 30 m below sea level) are vital to the ecology of the oceans, as the productivity of these areas forms a large portion of total marine productivity. Many coastal ecosystems are highly productive and diverse. Coral reef productivities are estimated (by changes in oxygen tension in water as it flows over the reef) at 1500 to 12 000 g C m^{-2} a^{-1} and the productivity of *Spartina* marshes in southern USA is estimated at over 1600 g C m^{-2} a^{-1}. Many studies have been made of coastal ecosystems in terms of their structure, functioning and change over time and these provide a vital precursor to any studies of the ecological aspects of coastal problems.

Within the littoral zone benthic (i.e. sediment dwelling) and pelagic (i.e.

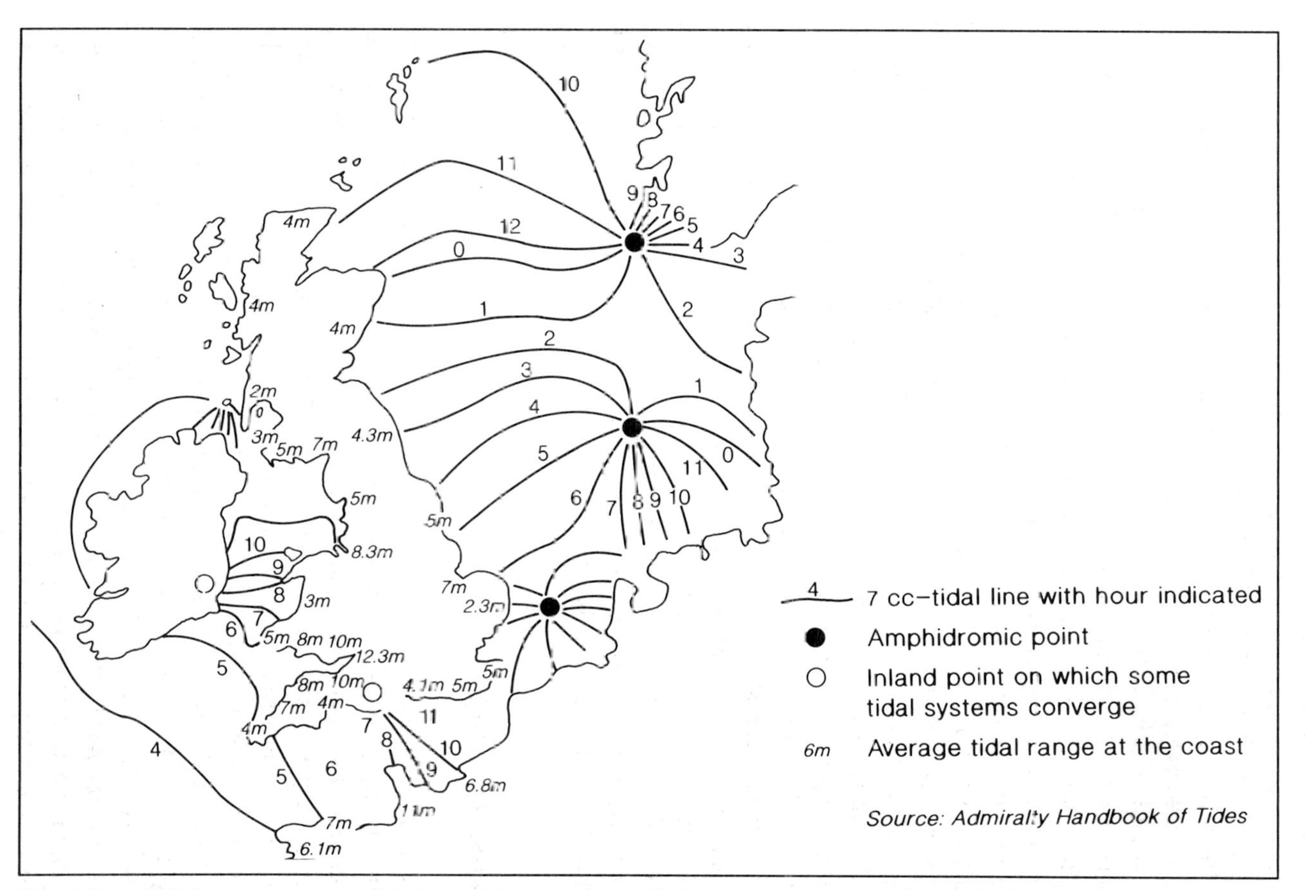

Fig. 2.14 Tidal systems around Britain and the associated tidal range at the coast. After Pethick (1984)

Fig. 2.15 World tidal range environments. Values are in metres at spring tides. After Davies (1972)

within the water column) organisms are highly interlinked, and both benthic organisms and phytoplankton contribute to net primary productivity here. Large, multicellular macroscopic algae, for example, may be very fast growing and cover large areas of the intertidal zone. In general, coastal zones are fertile, nutrient-rich environments, with the nutrients brought in from rivers and the upwelling of coastal waters. There are key latitudinal variations in coastal ecology, with the rocky intertidal zone in both the tropics and the Arctic, for example, being relatively impoverished because of extreme stresses from high air temperatures and ice action, respectively. Conversely, subtidal seagrass beds show increasing biomass and productivity on a pole to tropics gradient, with productivity increasing from *c.* 500 g C m^{-2} a^{-1} in temperate areas, to over 800 g C m^{-2} a^{-1} in the tropics. Superimposed upon this simple latitudinal pattern, however, are variations in productivity and diversity of coastal ecosystems encouraged by differences in nutrient availability and substrate types. Thus, for example, protected, tidal-dominated, muddy coasts in estuaries characteristically support large populations of marsh or mangrove plants, whereas wave-dominated beach systems support a relatively impoverished community of benthic organisms. Coastal zone ecosystems, unlike many terrestrial systems, also contain large numbers of relatively simple life-forms, such as prokaryotic bacteria and blue-green algae (cyanobacteria) and algae and protozoa (eukaryotes), which live as phytoplankton and microbenthos. In mangrove swamps and marshes, for example, Barnes and Hughes (1988) suggest that microalgae produce more biomass used for food by crustaceans and other consumers than do higher plant species.

Ecological interrelations

There are many different pathways for energy, sediment and nutrients within coastal ecosystems which illustrate the interdependence of the geomorphological and ecological systems. Benthic organisms, for example, live on and in coastal zone sediments. The subsurface zone is inhospitable, because of anaerobic conditions fostered by the decay of organic matter. Thus, many benthic fauna have to create burrows in order to live within anoxic sediments. This lifestyle creates bioturbation of the sediments, as well as indirect effects on the size and composition of sediments. Faecal pellets produced by benthic fauna, for example, act to produce larger particle-sized sediments which will have different entrainment characteristics. Benthic organisms thus play a direct role in sediment dynamics. Benthic fauna consume microscopic benthic algae, such as the interstitial flora in sandy beaches, as well as phytoplankton and detritus from the water column. There are many ecological interactions between different benthic organisms as they compete for space and resources and become involved in predation. Thus, Barnes and Hughes (1988) note that many deposit feeders discourage the growth of suspension feeders, through the disturbance of the sediment by bioturbation and the production of faecal pellets.

Many benthic organisms also provide shelter and support for other organisms. Sublittoral kelp forests, for example, which extend down as far as light penetration permits photosynthesis by the kelp species, provide three-dimensional structures, analagous to coral reefs, which support a diverse fauna. The now-extinct giant Stella's sea cow, for example, browsed on kelp, and sea otters (*Enhydra lutris*) browse on sea urchins (*Strongylocentrotus purpuratus*) here, which themselves graze on the kelp. The majority of the biomass of the kelp species goes into the detrital food chain, however.

Such interactions produce dynamic associations of organisms closely interrelated with their geomorphological environment. On rocky intertidal surfaces (Chapter 4), for example, predation by grazing species helps to maintain a high species diversity, by preventing any one species gaining competitive advantage and dominating the system. There are also important ecological interactions between many coastal communities, with some marshes and swamps (Chapter 5), for example, exporting nutrients to help fuel the continental shelf ecosystems. In other cases, however, such systems act more as a sink for sediments and nutrients, depending upon their position within coastal circulations.

Coastal communities are subjected to a range of natural disturbances which act as important controls on their productivity and species diversity, such as storm surges, tectonic events and invasions of species. Some of these disturbances are what we refer to in this book as coastal problems; others provide an important control on the ecosystem and may condition its response to coastal problems in the future. Thus, Case study 6.1 discusses *Acanthaster* predation on coral reefs, which is a biological disturbance seen to pose problems for many reefs.

Water on land at the coast: seaspray, fluvial inputs and groundwater

Although waves and tides provide the major controls on most coastal processes, other forms of water are important in coastal ecology and geomorphology. Ice is dealt with more fully in Chapter 7, but in this section we review the importance of seaspray, river water and groundwater in affecting the functioning of coastal biogeomorphological systems.

Seawater acts as an important agent of weathering, as well as a pathway for nutrient transport inland. The unique geochemistry of seawater (Table 2.5)

Table 2.5 The geochemistry of seawater

Element	*Average contents* (ppm)
Chlorine	19 000
Sodium	10 500
Magnesium	1300
Sulphur	904
Potassium	380
Calcium	400
Silicon	2.9
Bromine	65
Strontium	8
Aluminium	0.001
Boron	4.5
Fluorine	1.3
Phosphorus	0.09
Iron	0.003
Manganese	0.002
Zinc	0.002

Source: Modified from Colinvaux (1986)

controls its effects as a weathering agent, as well as its significance as a source of nutrients. The power of seawater to dissolve limestone within the coastal zone has been debated and is probably minimal (Chapter 4), but it is an important agent of coastal salt weathering, which affects coastal rocks, buildings and engineering structures (such as coastal defences). Salt weathering involves a complex series of weathering processes, both chemical and mechanical (Goudie, 1985). In order to be effective, salt from seawater must be delivered in waves, splash or spray. Laboratory simulations have shown vividly the effectiveness of marine airborne salt as a weathering agent (e.g. Auger, 1988). Salt spray deposition can be traced many kilometres inland in some cases, thus expanding the influence of the coastal zone on weathering and vegetation characteristics. For example, James *et al.* (1986) found high concentrations of calcium, magnesium, sodium and potassium in rain falling on the Ainsdale sand dunes, Merseyside, UK which acts as an important input of nutrients there.

Where rivers enter the sea there is an important input of freshwater, sediments and nutrients to the coastal zone. Globally, it is estimated that around 15 million t. of dissolved nitrogen and 1 million t. of dissolved phosphorus enter the ocean from rivers naturally each year (World Resources Institute, 1992), although human activities have probably increased these figures by 50–200 per cent. The major consequences of fluvial inputs for the development of deltas and estuaries are dealt with more fully in Chapter 5, but it is clear that fluvial inputs also provide a major source of nutrients for many coastal zone ecosystems, as well as providing a suite of habitats from fresh through brackish to saltwater environments which permit colonization of a range of species. Later sections show manipulation of both river water quantity and quality to be major ways in which human activity can influence coastal processes.

At the coast, freshwater and saltwater meet within aquifers. The Ghyben–Herzberg principle describes simply the nature of the freshwater–saltwater interface, showing that for each metre of freshwater head above sea level the interface becomes depressed 40 m below sea level. In reality, however, the situation is more complex as the water bodies are in motion. A zone of dispersion is produced as saltwater penetrates inland, mixes with freshwater and creates a seawards discharge of brackish water. As coastal aquifers are frequently tapped for drinking water, the exact position and nature of the freshwater–saltwater interface is of great concern.

Several studies have revealed the spatial complexity of the chemical composition of many coastal groundwater bodies. In Long Island, New York, for example, hydrogeological investigations revealed three wedgelike extensions of salty groundwater landward of barrier beaches and Jamaica Bay. However, in the 1960s it was felt that no serious salt contamination of wells would occur for at least 20 years (Lusczynski and Swarzenski, 1966), although there have been worries about saline intrusion since then. In Bahrain, there are several areas of saline coastal groundwater intrusions, which have lead to the death of many hectares of palm plantations on Sitrah Island and in the Sha'ban al Kharda valley. In the arid Bahrain climate, salty groundwater is brought to the surface by capillary rise and may cause severe salt damage to building foundations (Doornkamp *et al.*, 1980). On the Dungeness shingle foreland in southern England, surface electrical resistivity studies have shown the presence of saline groundwater (with chloride ion concentrations of up to 3000 mg l^{-1}, compared with background levels of 40 mg l^{-1}) under the freshwater lens. The thickness of the freshwater lens was found to be less than previously thought and water extraction from it may be problematical in the future (Oteri, 1983).

Sea level changes, as well as pumping of the groundwater itself, can cause great changes in the groundwater. In the Charleston area of South Carolina the Santee Limestone–Black Mingo aquifers have not been used since *c.* 1950 because of saltwater intrusion (Kana *et al.*, 1984). Modelling studies of this area indicate that future sea level rise associated with global warming will lead to further saltwater intrusion to the other, still-utilized aquifers. However, Kana *et al.* (1984) suggest that these changes will be small compared with the ongoing effect of groundwater extraction.

Physical disturbances to coastal biogeomorphology: storm surges and large scale pressure changes

Storms and storm surges

Meteorological conditions can alter the quasi-regular oscillations of water level produced by astronomical tides. Storm surges, for example, occur where low barometric pressure and strong winds act to raise tide levels higher than predicted. When a strong depression occurs over the sea, the low pressure and strong winds created affect the sea surface. Falling barometric pressure acts to draw up the water surface producing approximately 1 cm rise for every 1 mb drop in pressure. Strong, onshore winds transfer energy to the sea surface. Surges can produce catastrophic flooding and large scale sediment movements, and may have a big impact on coastal geomorphology.

On the Atlantic coast of the USA in March 1989 a major storm (one of the three biggest in 45 years in terms of erosion caused) generated deep water waves of 1–5 m or above for 115 hours. Another storm here, in late October 1991, was the most powerful for 50 years (Dolan and Davis, 1992). In Europe, a notable, damaging storm was the 1953 event that affected much of the North Sea coast of England and Holland, and led to widespread flooding and loss of life (Grieve, 1959). According to Dolan and Davis (1992) the Ash Wednesday storm on the Atlantic coast of the USA in 1962 caused the loss of *c.* US$300 million in property. It was so severe because it occurred during spring tides and its duration spanned five semi-diurnal tidal cycles.

Such pressure-caused changes may show a seasonal distribution, and Bird (1993) notes that in the South China Sea, mean sea level is *c.* 40 cm higher in November to March, during the north-east monsoon, than during the south-west monsoon. Several other causes may be invoked for changes in sea level over the seasonal to annual time-scale, including changes in sea surface temperatures, onshore wind stress components, rainfall, salinity currents, tectonic movements, mixing of surface and deep ocean waters, and changes in river discharge. In the east China Sea, changes in the large volume of freshwater discharge coming in, mainly from China ($> 1.2 \times 10^{12}$ m^3 a^{-1}) as well as fluctuations in the position of the Kuroshio current, produce important sea level changes (Bryant, 1987).

El Niño Southern Oscillation (ENSO) events

Changes in atmospheric pressure over areas of the globe can lead to changes in sea level (as higher pressures push down and depress the sea surface and lower pressures release the surface allowing it to rise) which, coupled with linked changes in sea surface temperatures, influence coastal processes over a period of a few months to a few years. The most famous atmospheric pressure change is the El Niño Southern Oscillation (ENSO) which occurs every 2–7 years and affects the tropical atmospheric and oceanic circulations in the Pacific and Indian Oceans. ENSO events are discussed in greater detail in climatology texts such as Henderson-Sellers and Robinson (1986) and Philander (1990), but the basic features are described below.

The usual atmospheric circulation system, as shown in Fig. 2.16, is characterized by high pressure over the south-eastern Pacific Ocean and low pressure at the western Pacific margin. This pressure difference, measured between Tahiti (high pressure) and Darwin, North Australia (low pressure) is called the Southern Oscillation. It generates the north-easterly and south-easterly trade winds that blow across the Pacific Ocean, picking up moisture before they rise from the warm water pool surrounding Papua New Guinea, Indonesia and the Philippines, resulting in high rainfall totals. High level return flow completes what is known as the Walker Circulation. Oceanographic processes are associated with this atmospheric circulation system. Easterly wind flow generates upwelling of cold but nutrient-enriched deep ocean water along the coasts of California and, particularly, Peru and Ecuador in South America and this then moves out as a cold water plume along the Equator. As a result, coastal waters are highly productive but coastal margins and islands are typified by aridity. The easterly wind stress over the Pacific Ocean also generates a slope on the sea surface: on average, the western Pacific is 45 cm higher than the eastern Pacific Ocean.

The start of an ENSO event is seen in a rapid fall in the Southern Oscillation; as the pressure difference falls so the trade winds slacken and this allows warm water to 'slosh back' from the western Pacific. This warm water caps the nutrient-rich upwelling water in the eastern Pacific and results in a catastrophic collapse of the fisheries of the South American coasts and of bird populations on equatorial Pacific islands. Furthermore, this warm water causes the centre of convective activity to shift towards the cental and eastern Pacific; thus normally arid coasts and islands are subjected to heavy rainfall whilst northern Australia and the margins of south-east Asia experience abnormal droughts. The movement of warm water from the western Pacific is accompanied by sea level changes which can be traced through tide gauge records and GEOSAT altimetry (Miller *et al.*, 1988) along the Equator. On reaching the eastern Pacific the sea level high splits north and south along the west coast of North and South America. It becomes trapped along the shore here because of refraction over the continental shelf and the effects of the Coriolis force. This produces a high sea level along the west coast of the Americas.

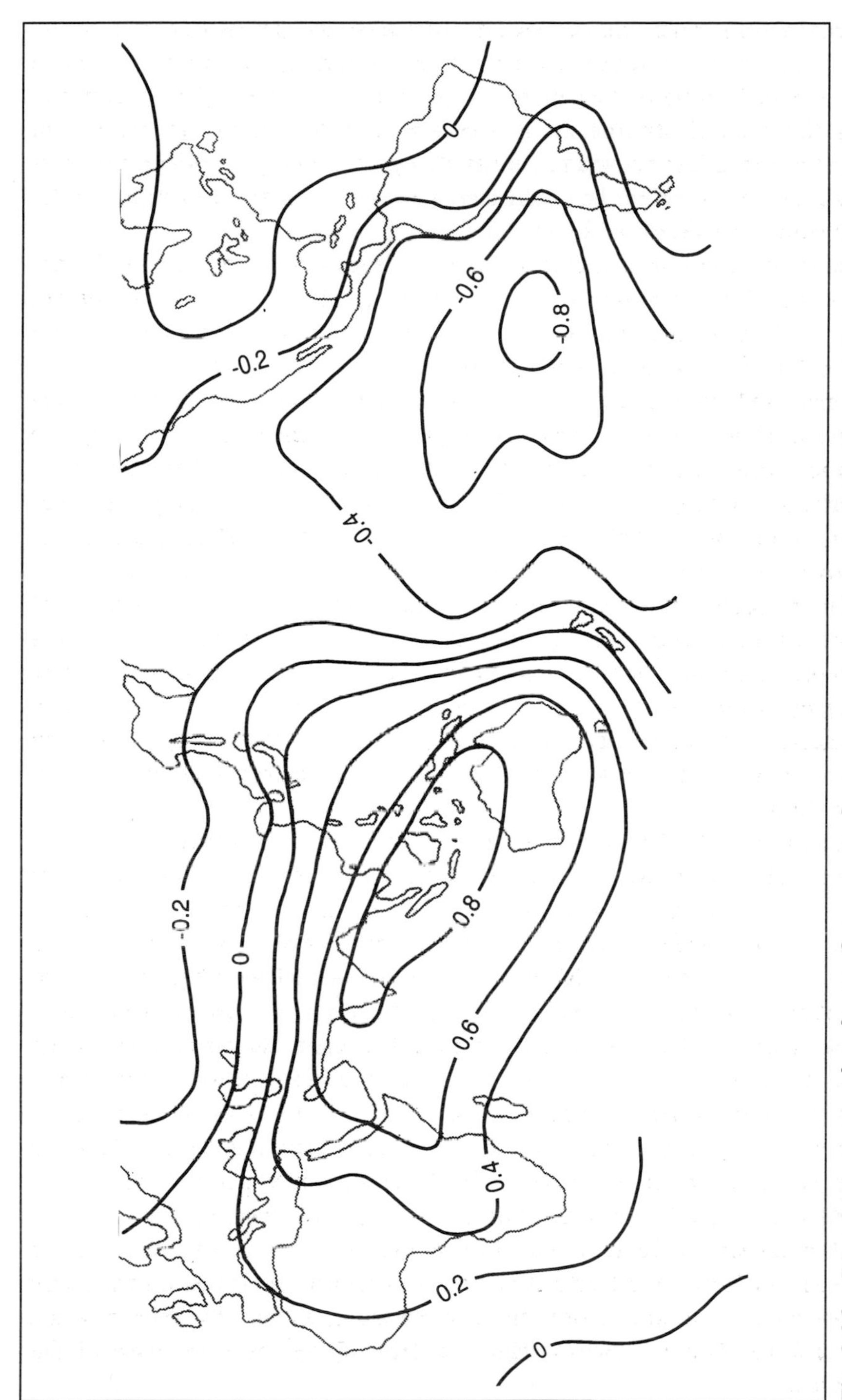

Fig. 2.16 Pressure systems within the Pacific Ocean. High pressure in the south-eastern Pacific is negatively correlated with low pressure over the islands of south-east Asia. The pressure difference between Tahiti (HI) and Darwin, Northern Australia (LP) is known as the Southern Oscillation Index. After Philander (1990)

Global warming, sea level rise and the coastal future

As we have seen at the beginning of this chapter, sea level has experienced important changes over the last 15 000 years (as well as over much longer time spans) as a consequence mainly of the changing cover of ice at the end of the last glaciation. There has been considerable recent concern that human activities, coupled with natural fluctuations, may be creating an additional 'forcing function' likely to produce a rising global sea level. Since the early 1980s there has been a plethora of predictions over the likely magnitude and rate of such sea level changes, but the general consensus now seems to be that the rise will be much less dramatic than first suggested. Thus, the biogeomorphological response of individual coasts will be of great importance in conditioning their future, and the magnitude of the hazard for local populations. We look below at some of the major issues about future sea level rise; Warrick and Farmer (1990) and the Report of the Intergovernmental Panel on Climate Change (IPCC) (Warrick and Oerlemans, 1990) and update (Wigley and Raper, 1992) provide more detailed information.

In many ways the term 'sea level' is a misnomer, because over the globe the height of the sea varies considerably, as the surface relief of the oceans has an amplitude of up to about 180 m. This form of sea surface topography is called the geoid. Causes of the strange shape of the geoid include changes in lithospheric loading, variations in regional water balances (glaciation, evaporation etc.) and tectonic changes in ocean basin shape. The geoid means that sea level is relatively high near Papua New Guinea and in the North Atlantic, and relatively low around India. The geoid changes over time, thus meaning that any changes in ocean water volume may produce complex spatial changes in sea level.

For practical purposes we talk about a 'mean sea level' which refers to the long term average of high and low tide levels at a particular place. This average is usually taken over a period of at least 19 years. So what causes mean sea level to change? We can recognize a range of important factors affecting sea level over different time-scales, i.e. short term (10–10^3 years), medium term (10^3–10^6 years) and long term (10^6–10^9 years) following the classification of Fairbridge (1983). Essentially, as we saw earlier in this chapter, changes in mean sea level are caused either by movements of the land (isostasy), or changes in the volume or mass of water (eustasy), or a combination of the two, set in the context of a slowly changing geoid. In order to investigate the nature and causes of sea level change, it is necessary to have a good series of measurements. Most of our current data come from tidal gauge records, which as Pirazzoli (1993) indicates, are affected by local, as well as regional and global, conditions. This makes such records difficult to compare in a meaningful way.

Furthermore, the distribution of tidal gauges is very patchy, and particularly biased towards areas with a recent history of deglaciation. The completion of the Global Sea Level Observing System (GLOSS), new altimeter satellite data, as well as Very Long Baseline Interferometry (e.g. Diamente *et al.*, 1987) will, however, improve the reliability and coverage of our sea level data sets in future.

Carter (1988) lists the major causes of short term sea level change as global warming, glacier retreat, changes in ocean circulation and subsidence. Over the past 1000 or so years, from a range of evidence it appears that sea level rose between 0.1 and 0.2 mm a^{-1} because of continuing (if declining) isostatic crustal adjustment following deglaciation, not because of added water mass in the oceans (Warrick, 1993). During the last 100 years, thirteen different studies have estimated the mean sea level rise at between 0.5 and 3.0 mm a^{-1}, with most estimates concentrated between 1.0 and 2.0 mm a^{-1} (Warrick and Oerlemans, 1990). However, according to Warrick (1993) it is very difficult to disentangle a 'mean sea level rise' given the range of influencing factors such as vertical land movements, atmospheric and oceanographic effects (ENSO etc.) as well as anthropogenic impacts (e.g. changing sediment loadings as a result of deforestation). The recent apparent order of magnitude increase in sea level rise is probably related to climatic factors, especially ocean thermal expansion and the melting of land ice. These climatic influences are, of course, the major focus of concern for future predictions of sea level change.

Even if no new water is added to the oceans, volume increases can occur as density changes (known as the steric effect). At its present salinity of *c.* 35 per cent, seawater has a maximum density at 0°C. If temperature increases, density will decrease and seawater will expand. Thus if a 500 m deep layer of ocean water were to be warmed throughout by 1°C, sea level would rise by between *c.* 5.7 and 14.6 cm, assuming typical ocean temperatures of 5°C (temperate latitudes) and 25°C (tropical latitudes), respectively. Models which assume a rise in global temperature of 0.5°C between 1880 and 1990 suggest a sea level rise of 2.7–5.6 cm; an expected rise in temperature of a further 2.5°C by 2100 translates into a total thermal expansion of 30 cm for the period 1880–2100 (Wigley and Raper, 1993).

Global warming will also change sea level through changes in the mass balance of small glaciers and large ice sheets. Detailed analysis of historical records has shown that over the last 100 years there has been a general worldwide retreat in alpine glaciers (Grove, 1988) which has probably contributed 1.5–7.4 cm to the sea level rise in the period 1880–1990 (Meier, 1984; Wigley and Raper, 1993). However, 99 per cent of the earth's land ice is held in the ice sheets of Greenland and Antarctica and their present behaviour is difficult to establish and their future behaviour difficult to predict. Ice loss processes (ablation) and ice gain processes (accumulation) are both temperature dependent, but in different ways over different temperature ranges (Fig. 2.17). Thus for relatively warm climate ice sheets, like Greenland, global warming is likely to lead to increases in both accumulation and ablation but with ablation being dominant, so producing a decrease in mass balance and sea level rise (region b, Fig. 2.17). However, under a very cold climatic regime, as in Antarctica, the dominant term is increased accumulation with rising temperature, promoting an increase in mass balance and hence sea level fall (region a, Fig. 2.17). Calibrating these relationships against the present behaviour of both ice sheets is, however, extremely difficult and it is not certain whether or not these ice sheets are in balance at the present time. Both may still be reacting at depth to the last glacial–interglacial transition (Oerlemans, 1993). Additional problems are present in west Antarctica where ice streams from the central plateau feed into marginal ice shelves. The ice shelves are grounded on pinning points which hold the shelves in position and prevent their breakup. The Domesday scenario for west Antarctica is that melting at the bottom of the ice shelves might cause them to thin and lift from the pinning points, leading to the disintegration of the shelves and an increased supply of ice from the plateau. Current modelling exercises suggest that the ice shelf thinning rates would need to be very high

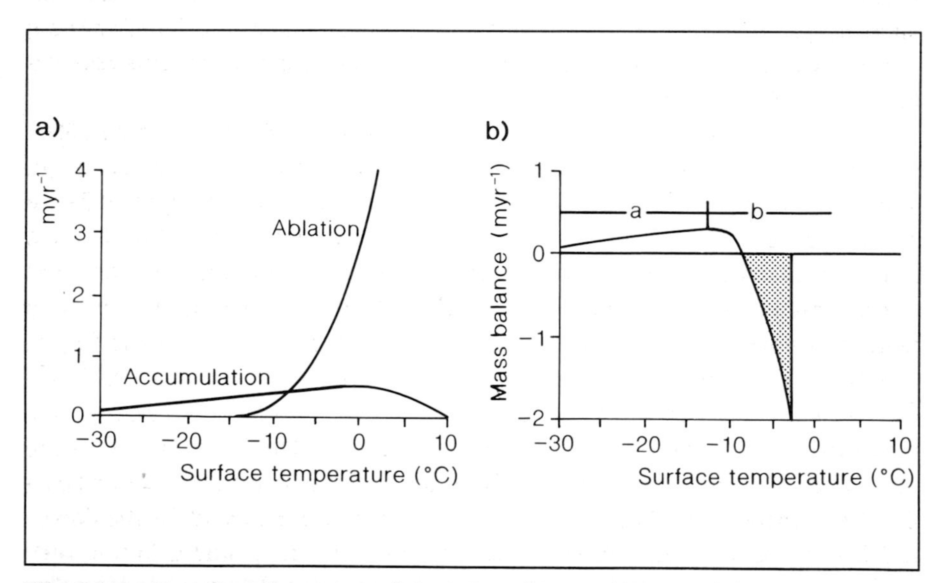

Fig. 2.17 Complicated response of glaciers to climatic change. a) Dependence of ablation (evaporation and runoff) and accumulation on annual surface temperature. b) Dependence of net annual balance on temperature. After Warrick and Oerlemans (1990)

(10–100 times present values) for this sequence of events to occur, but the uncertainties involved in such calculations are very large indeed (Oerlemans, 1993).

Future changes in sea level are not only likely to be the result of natural processes. Tide gauge records show human activities are increasingly aiding such vertical land movements through encouraging subsidence, by removing oil and other subsurface fluids from sensitive areas. For example, human-induced subsidence has accounted for *c.* 50 per cent of the sea level rise between 1872 and 1985 in Venice, and locally in Bangkok, subsidence caused by groundwater pumping has reached rates of up to 13 cm a^{-1}. Human activities can both increase or decrease the amount of freshwater reaching the sea by increasing runoff (through deforestation), and by damming rivers and creating reservoirs. Newman and Fairbridge (1986) estimated that between 1932 and 1982 damming and addition of water to aquifers removed *c.* 0.75 mm a^{-1} of potential sea level rise. On the other hand, Sahagian *et al.* (1994) suggest that groundwater extraction and diversion of surface waters for irrigation leads, eventually, to the evaporation of most of the mined water which then ultimately returns to the oceans, leading to a rise in sea level. Land use changes are also seen to be important, reducing the water-holding capacity of soils through, amongst other things, deforestation and wetland drainage, and adding to the water returned to the oceans. In all, Sahagian *et al.* calculate that an average rise in sea level since 1960 of 0.54 mm a^{-1} can be attributed to these sources, which represents at least 30 per cent of the observed sea level rise, if a figure of 1.75 mm a^{-1} is used.

Increasingly, attempts have been made to synthesize all the potential different short term influences on sea level and develop predictions based on an increasing anthropogenic input of carbon dioxide into the atmosphere. There are a number of problems involved with such an exercise, including the vast range of influencing factors which can interact with one another, and the uncertainty of the future emissions of carbon dioxide and its interrelationship with the global climatic system. The most recent estimates (Wigley and Raper, 1993) suggest a 'best guess' sea level rise of 46 cm for the period 1990–2100 (with low and high estimates of 3 and 124 cm, respectively). Such an estimate indicates a rate of sea level rise of 4–5 mm a^{-1}, a considerable increase on the 1.0–2.0 mm a^{-1} rate of rise from the historical tide gauge record described earlier. Nevertheless, the IPCC's 1992 estimate continues the trend for the downwards revision of the magnitude of sea level rise (Fig. 2.18) and is in line with studies of tide gauge records which have failed as yet to detect an acceleration term in the record of sea level rise over the last 100 years (e.g. Woodworth, 1990). Thus analyses of coastal problems that assume widespread *in situ* drowning and passive responses of coastal landforms over the next century

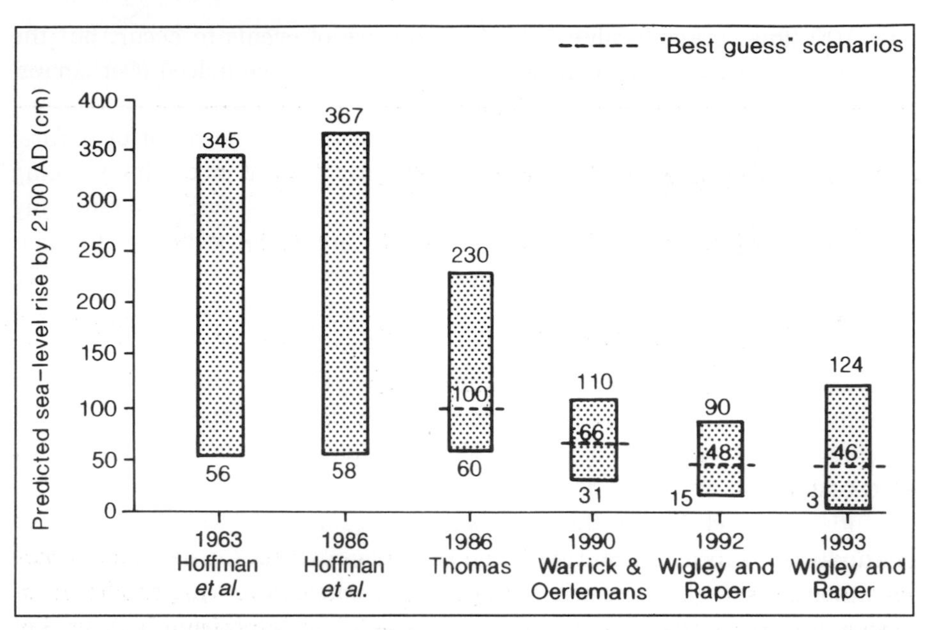

Fig. 2.18 Revisions of anticipated sea level rise by 2100. After French *et al.* (1994a)

need to be replaced with approaches that consider the responsiveness of coastal systems to more subtle future environmental change. Global warming will lead to changes in the distribution of some biological communities, changes in groundwater at the coast and many changes to terrestrial land use and ecosystems, which will have consequent effects on the coastal zone. It is this nexus of changes, combined with a whole host of smaller scale disturbances (both predominantly physical and human in causation), which provides the real challenge for the coastal future.

Chapter Three

SANDY COASTLINES: BEACHES AND DUNES

Introduction

Beaches and dunes are key sites for recreation and tourism, as well as important coastal sediment stores which change in response to varying conditions of erosion and deposition. Over recent years surveys have revealed a high level of erosion on many of the world's sandy coasts (Plate 3.1), coupled often with pollution which may have serious ecological effects. Erosion and pollution

Plate 3.1 Eroding sand dunes, Ainsdale, Lancashire coast, England

problems have come together in highly populated, and often vulnerable, sandy coasts where development and natural hazards have combined to produce a stressed coast (e.g. some of the eastern USA coastal barrier islands).

Beach and dune distribution

Beaches are found in all latitudes where there are suitable sediments, favourable combinations of basement relief (for the accumulation of sediments) and sea level changes. Beaches are the most widely distributed of any coastal sedimentary environment (Davis, 1985). Dunes also have a near-global distribution (see Fig. 3.1) and their presence is encouraged by active sand supply, strong onshore winds, low precipitation and humidity, shallow beaches and large tidal ranges. Coastal dunes have been stated to be largely absent from humid tropical shores (e.g. Davis, 1985), but they are relatively common in many areas such as North Queensland, Australia (Pye, 1983). They are also found in high latitude areas, such as Iceland. Beaches are generally orientated towards major wave and wind approach directions, with coastal orientation also having an influence. Notable barrier beaches stretch almost continuously for hundreds of kilometres along the eastern USA coast, and calculations show that they cover 23 per cent of the North American shoreline (Davis, 1985, p. 380). In Nigeria, there is a *c.* 200 km barrier lagoon coastline stretching eastwards from the Benin border (Ibe, 1988). Coastal dune systems show a similar pattern to beaches, being found extensively on the Atlantic coasts of north-west Europe, south-west Africa and the Pacific coasts of North America and south-east Australia. The best developed dune systems are located on downwind margins of ocean basins, or in areas near massive glacial deposits. Thus, for example, dunes are found along the entire west coast of North America, from Alaska to southern California. They reach their greatest development along the shorelines of the states of Oregon and southern Washington; the oblique dune ridges of the 72 km long Coos Bay dune sheet reach crest heights of 50 m and ridge lengths of 1200 m (Cooper, 1958). Both beaches and dunes also occur on lake shores.

Beach and dune sediments

Beach and beach ridge sediments vary in composition from fine sands (<1 mm diameter) to cobbles of 150 mm or more in diameter. Davies (1980) illustrates that as a broad generalization, pebbles are a common constituent of beaches in high latitudes (40° N and S) and are rarely found in tropical beaches. This geographical variation depends upon variations in the energy of waves (which controls their ability to quarry rocks) and the extent to which there is a ready

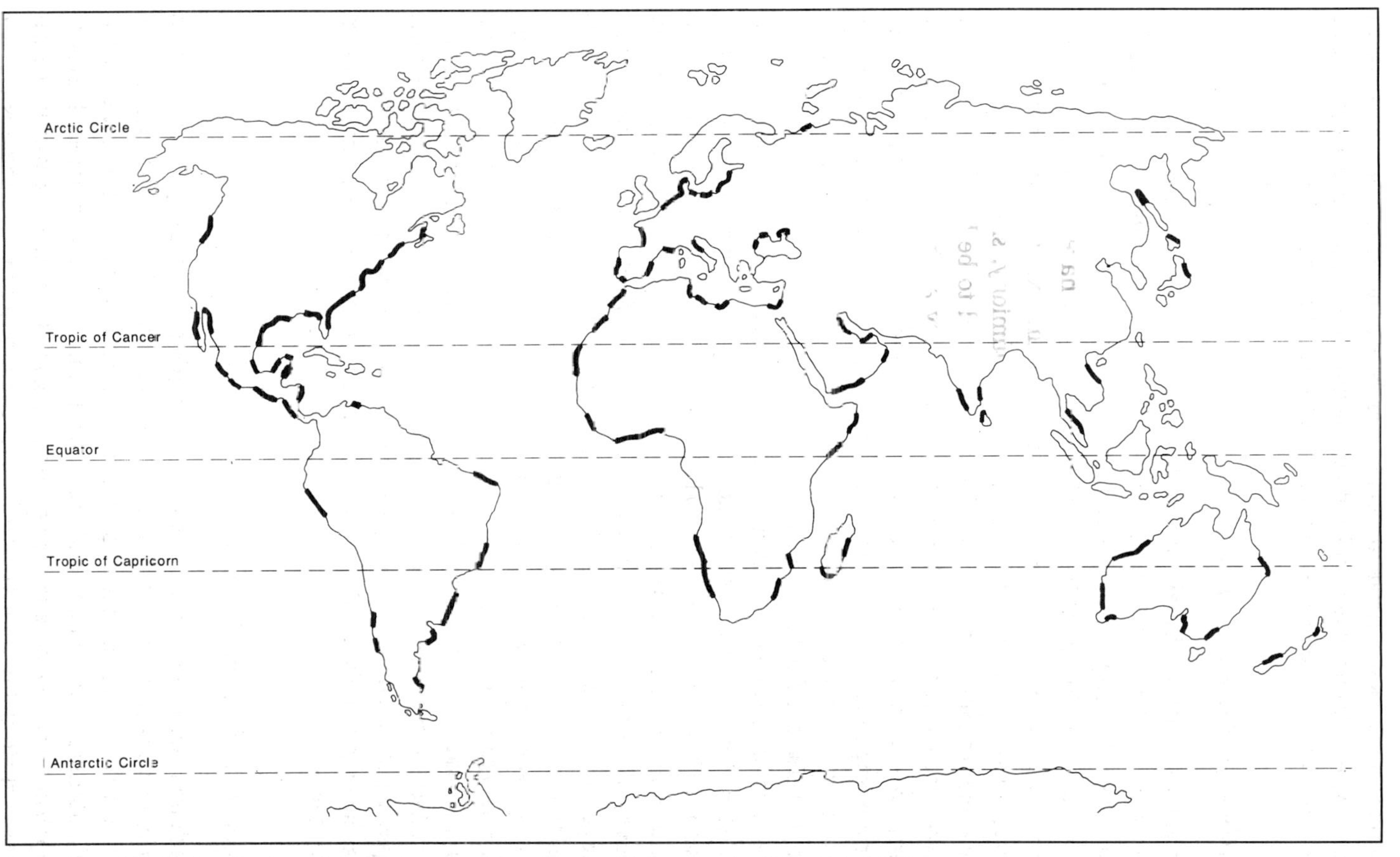

Fig. 3.1 World distribution of major dune and beach areas. After van der Maarel (1993)

supply of pebble-sized sediment (related to lithology and processes such as glaciation). Many high latitude pebble beaches are known, with McKay and Terich (1992) describing three gravel barriers (where gravel is defined as sediment > 2 mm in diameter) along the Olympic coast of Washington state, USA, similar to others reported along the Irish, English and Nova Scotian coasts. Important gravel and shingle beaches occur in England, such as the well studied Chesil Beach, Orfordness (whose conservation value has been assessed by Green and McGregor, 1990), Hurst Castle spit and the Dungeness foreland.

In general there is a strong relationship between beach particle size and beach slope: gravel beaches are steeper than those composed of sand (Shepard, 1963). However, for any given grain size protected beaches are steeper than more exposed ones (classically demonstrated by Bascom (1951) at Half Moon Bay, California) and eroding beaches are reduced to lower slopes (see Chapter 2). More complex relationships between grain size, sorting and beach slope angle characterize another category of beaches, mixed sand and gravel structures, which are characteristic of parts of southern Ireland (Orford and Carter, 1982) and many of the eastern coasts of both the North and South Islands of New Zealand (Kirk, 1980).

Beaches and dunes also vary enormously in their mineral composition. Potter (1984) indicates how the mineralogy of beach sands in South America is controlled by plate tectonics, with active plate margin associations along the Pacific and Argentine coasts rich in volcanics and plagioclase, and a passive margin association from Argentina to Trinidad characterized by high quartz content. At a rather smaller scale, variations in coastal and drainage basin lithology are important controls on sediment mineralogy, as are offshore factors. Thus, many beaches and dunes are rich in calcium carbonate from broken shell and coral fragments. In general, beach material appears rich in siliceous minerals where there are high inputs of sediment from rivers, and high in carbonate minerals where offshore sources dominate. Araya-Vergara (1985) has analysed the mineral composition of beaches in central Chile, for example, and found that the sand is primarily of fluvial origin here, mixed by strong wave energy, with a possible minor component from the sea floor. Darby (1990) shows how the elemental composition of ilmenite indicates the source of sand on the continental shelf of the USA. Over 50 per cent of samples studied could be traced back to the Hudson River and adjacent areas. The sediment was probably moved in huge quantities during glacial meltwater phases. Many beaches around tropical islands with associated coral reefs, such as Aldabra Atoll in the Indian Ocean, have beaches with very high calcium carbonate contents. In Britain, beach and dune sands show great variability in carbonate content depending on local setting. Beaches on exposed parts of western Scotland and Cornwall have high calcium carbonate contents generally speaking, whereas beaches on the sheltered North Wales coast are dominated by siliceous sediments (*see* Goudie, 1990, for a full explanation).

Beaches

Beach types

Beach planforms at the coastal zone scale are determined by the nature of regional coastal wave–sediment cells, which are themselves dependent upon geological controls on coastal configuration, wave approach direction(s) and the nature of coastal sediment supply and transport. Several basic types of coastal alignment are recognized:

- under conditions of limited or terminated sediment supply, beaches become swash-aligned. The lack of oblique wave attack removes the potential for longshore movement of sediment (see Chapter 2).
- where sediment is available in sufficient quantity to meet the demand for longshore transport, beaches may prograde in a line parallel to the line of maximum drift, at *c.* 40–50° to the direction of wave approach. Such beaches are termed drift-aligned. Examples include the Somers-Sandy Point coastline in Westernpoint Bay, Victoria, Australia (Bird 1985b) and many central Chilean beaches, which are aligned in relation to a northward drift (Araya-Vergara, 1985).
- cuspate forelands and angled spits. These incorporate both drift- and swash-aligned beaches. Recurved spits are common in macrotidal, storm wave environments with drift-aligned spits and barriers. Notable examples are the Dungeness and Orfordness forelands in England and the Culbin Sands in Scotland.
- zeta-form, log spiral beaches. These are asymmetric curved bays from headland to headland and are basically swash-aligned features responding to the refraction and diffraction of obliquely directed waves. Morphology and grain characteristics both change along such beaches and there may also be changes in beach state, from dissipative profiles at proximal locations to reflective forms at distal ends. Such beaches are found along many of the world's swell-dominated coasts, such as in the UK where swell-wave energy penetrates the North and Irish Seas. Other good examples are many tourist beaches along the east coast of Peninsular Malaysia in Pahang and Johor (Wong, 1990).

Wider context

These contemporary settings need to be placed within the wider context of the balance between aeolian and nearshore processes, given tectonic and historical controls. Thus, for example, on the south-east coast of Australia, present sea level was reached 6000 years ago. Here, under a strong wave environment, littoral sands have been driven onshore to form a series of beach ridges and intervening lagoons. In comparison, the combination of high sediment supply from the mobilization of glacial margin sediments, a wide continental shelf, a declining rate of sea level rise and a generally lower wave energy climate than that experienced in Australia, has resulted in the formation and survival of barrier islands backed by lagoons on the east and Gulf coast shores of the USA. Along the coast of Louisiana, USA, sand supply is limited to the release of sand from old deltaic distributaries and beach ridge deposits, e.g. the Chandeleur Islands of the Mississippi delta (Otvos, 1979). These inputs, activated by shoreline retreat on high rates of sea level rise associated with the subsidence of fossil deltaic complexes, do not offset sediment losses which take place offshore and through tidal passes. Sand dune deposition is, therefore, taking place by recycling a limited, and probably reducing, sand volume. In this setting, dune volumes are small and dune lives short, for growth is constrained not only by sand supply but also by periodic storm erosion. Raccoon Island and Whiskey Island, two of the four barrier islands which comprise the Isles Dernieres, another sediment-starved barrier system in the Mississippi delta region, were expected to disappear below sea level in 2001 and 2016, respectively. Following severe sand-stripping by Hurricane Andrew in 1992, however, both islands are now projected to be drowned before the end of the century (Bush, 1992).

Beach ecology

Despite their initial barren appearance, many beaches support a diverse flora and fauna. As with morphodynamics, the major controls on beach ecology are wave action and tidal characteristics. Beaches do not support living aquatic macrophytes, although dead, washed-up seaweed from the offshore zone may be found. Beach flora is composed of benthic microalgae and phytoplankton, with diatoms often the dominant form in both categories. Beach fauna consists of a range of invertebrate macrofauna and interstitial forms, zooplankton, fishes and birds. They are essentially marine systems and the resident fauna is mainly of marine origin. A valuable introduction to beach ecology is provided by Brown and McLachlan (1990). Ecologists view beaches in a rather different light to geomorphologists, with the important controlling factors being exposure rating and filtration of seawater in the intertidal and subtidal zones, as well as beach morphodynamics.

The benthic microflora of beaches includes bacteria, cyanobacteria, autotrophic flagellates and diatoms, living attached to sand grains, or in pores, or on the sediments. On many exposed beaches rich accumulations of diatoms occur in the surf zone, forming 'blooms' which look like oil slicks. Invertebrates, such as flat worms, nematode worms, rotifera, polychaete worms, lugworms (e.g. *Arenicola marina*), oligochaetes, sipunculids, whelks, bivalves, copepods, amphipods, isopods, swimming prawns, callianassa (burrowing prawns), ocypodid crabs (e.g. ghost crabs), collembola (or 'springtails'), wrack flies and beetles, are all common on beaches and many are tube-dwellers. As storm waves can excavate beach habitats extremely rapidly, the few large species present are fast and efficient burrowers, like the razor clam, *Siliqua patula*. Burrowing is an important adaptation to sandy beach life, and many animals indulge in surfing, such as the whelk *Bullia* which uses the surf zone to skim across the beach. Feeding strategies encompass: filtering particles from suspension in water just above the sediment surface, grazing on particles at the sediment surface, or ingesting particles deposited on or in the sediments (Mann, 1982). Filter feeders tend to dominate in sandy sediments because high energy levels keep organic detritus in suspension (see below). In many localities these strategies must follow the rise and fall of the tide.

Beach morphodynamics seem to exert a major control over community characteristics as intertidal faunal species richness is greatest on dissipative beaches and lowest on reflective ones. Also, dissipative beaches have a large surf zone and circulation cells. Diatoms here provide much primary productivity that is retained in the system, which then does not have to rely on marine inputs. Biomass increases logarithmically with modal breaker height. Spatially, the distribution of macrobenthos animals shows patchiness (often related to small scale beach topography such as cusps) and cross-shore zonations. Several zonation schemes have been proposed, similar to those suggested for rocky shores, and Brown and McLachlan (1990) have produced a generalized scheme (Fig. 3.2). There is an important division at the water table, with air breathing creatures above, and water breathers below. Such zonation may be very variable over time, for example in monsoon climates, as beach water table levels vary.

Several ecosystem investigations have been made on beaches focusing on carbon and nitrogen budgets and food webs. There are four main components of food webs here, i.e. primary producers, macrofauna, interstitial biota and water column microbes. As an example, there have been several studies of the Alexandria beach and dune system in Alogoa Bay, South Africa. Extensive longshore sand transport in the surf zone reaches 5–9 $\times$ 10^5 m^3 a^{-1} on a high energy intermediate beach characterized by longshore bar–trough configuration. Longshore and rip currents produce a surf circulatory system. The surf

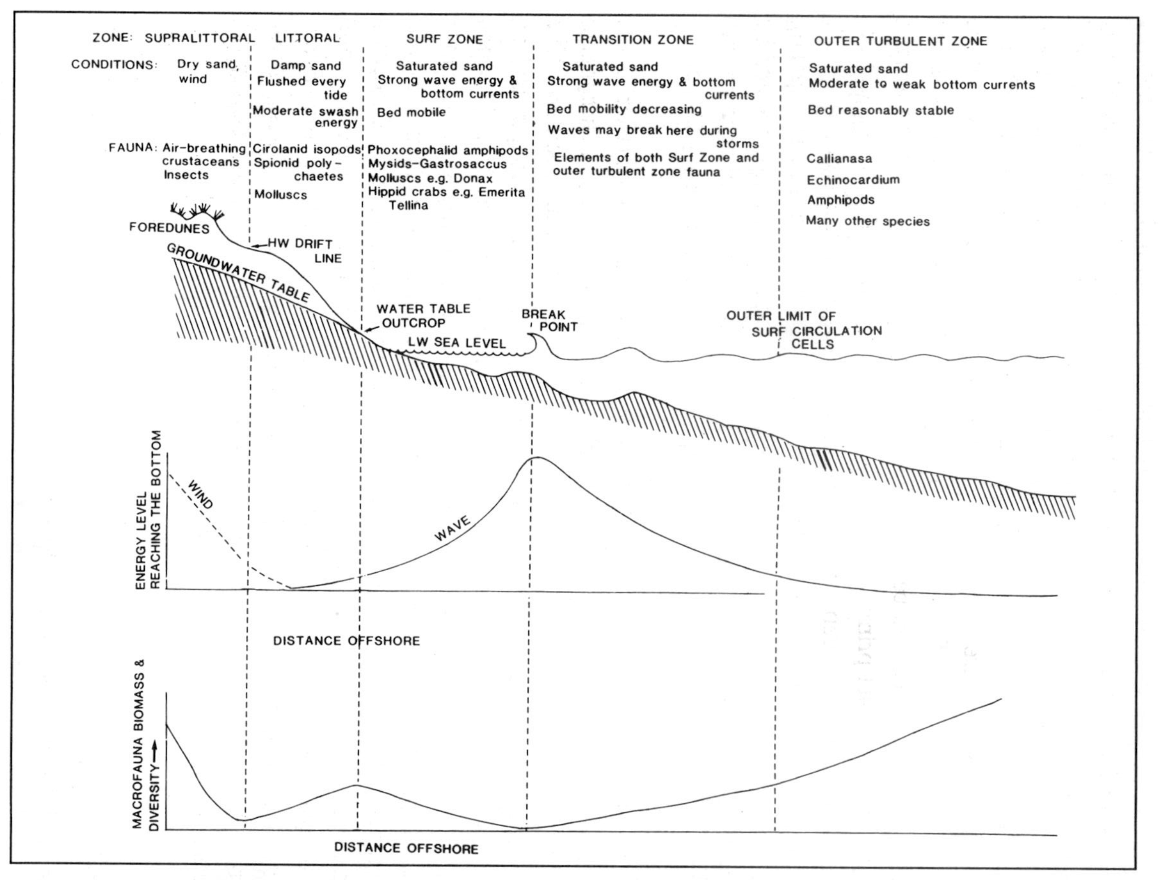

Fig. 3.2 Ecological zonation across a sandy beach. After Brown and McLachlan (1990)

zone here is dominated by diatoms, particularly *Anaulus australis*. Three food chains have been recognized by McLachlan (1990): the macroscopic, the interstitial and the microbial loops (composed of microorganisms in the water). The interstitial food chain includes bacteria, protozoans and meiofauna in the pore spaces of beach sand. It is fuelled by organic matter pumped into the sand by waves and tides and acts as a filter, purifying the surf. The microbial loop includes bacteria, flagellates, ciliates and other microplankton in the water and has only recently been studied, but appears to consume much of the primary productivity here. Brown and McLachlan (1990) have related the relative importance of the three food chains to morphodynamic state, following the models of Wright and Short (1984) and others, as shown in Fig. 2.11. Little trophic interaction occurs between these food chains, which recycle nitrogen and phosphorus. Net primary productivity is limited here not by nutrients but by wave energy. Nitrogen cycling studies from eastern Cape beaches, South Africa, show that the interstitial system is more important to nutrient cycling than the macrofauna system. In some beaches nitrogen inputs are boosted by stranded kelp in the surf zone, and there is often a net surplus of nitrogen in the surf zone. Accreting beaches may be sinks for nitrogen, but generally the excess is exported to sea (Brown and McLachlan, 1990).

Gravel beaches possess a rather different flora, including many higher plants on the upper parts above high water mark. In Britain, for example, Dungeness (2035 ha) contains a complete succession from bare shingle to woodland. The storm ridge at the seaward side has almost no vegetation, but moving inland on to less mobile areas, *Beta vulgaris* ssp. *maritima*, *Crambe maritima*, *Glaucium flavum* and *Rumex crispus* colonize. *Festuca rubra* characterizes more stable areas. Vegetation is very patchy on gravel forelands and requires the accumulation of organic matter for moisture and nutrients. Along the gravel ridges a community of lichens (*Cladonia impexa*), mosses (*Dicranum scoparium*) and higher plants (*Festuca tenuifolia* and *Silene vulgaris* ssp. *maritima*) become dominant, with prostrate shrubs (e.g. *Ilex aquilifolium*) developing in places (Boorman, 1993). In general, diversity of gravel beach and foreland vegetation increases with increasing stability of the shingle substrate.

Dunes

Coastal dune origins and types

Two main models of coastal sand dune formation have been proposed, i.e.

- dune formation associated with rising sea level conditions (transgression) — as sediments are pushed onshore from continental shelves. This seems to have occurred for most of the north-west European dunes at the end of the Pleistocene/early Holocene. The earliest of the dunes in the UK today date to about 5000–6000 BP, although there is also evidence of dune building activity in the late Devensian (Tooley, 1990).

- dune formation in association with falling sea level as exposed offshore sandy accumulations become prone to wind deflation.

Christianson and Bowman (1981) find evidence for both models in different areas of north-west Jutland, Denmark. Thus, in the Klim area monitoring from 1968 onwards shows rising sea level and dune advances of 80 m, whereas in the Hanstholm–Hjardemål area a major phase of dune building activity occurred between 1550 and 1750 AD, when sea level lowered. However, in this second case human activity may well have encouraged the dune development. Dune morphodynamics are still incompletely understood, and thus a classification based on process–form linkages is still incomplete, although Short and Hesp (1982) have related beach morphodynamics to dune forms. Other classifications make distinctions between vegetated and unvegetated dunes, and primary and secondary forms. Short and Hesp (1982) find stable foredunes associated with low wave energy situations, and blowouts to parabolic dunes in moderately high wave energy situations (*see* Fig. 3.3). Finally, transverse dune sheets are found in high wave energy conditions.

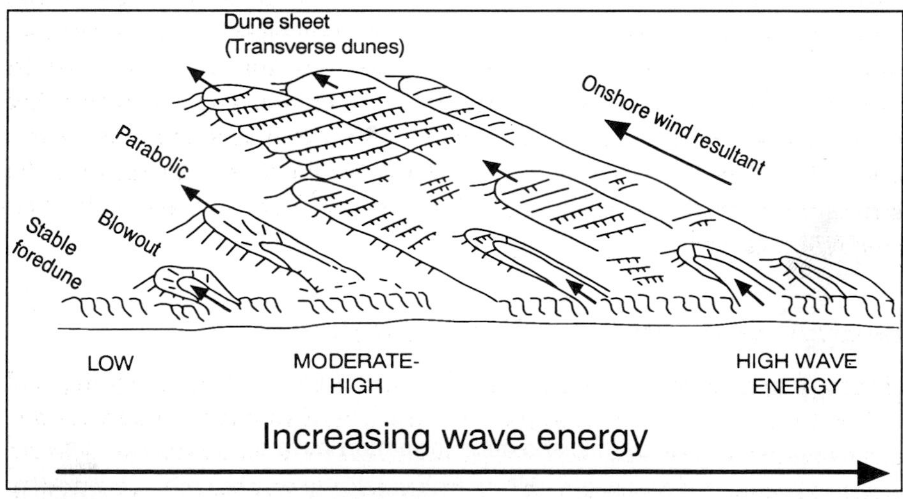

Fig. 3.3 Dune types of the south-east Australian coast related to wave energy characteristics. Modified from Short (1988)

Foredunes

Foredunes develop at the back of beaches in the litter zone. The tidal litter provides important nutrients, moisture and shelter for growing plants. Embryonic foredunes lead to bigger dune accumulations and have been classified by Hesp (1988) according to the percentage vegetation cover (producing 'morpho-ecological' stages). Dune erosion occurs during storms at the scarp face. Foredunes show a wide range of depositional and erosional morphologies.

Transgressive dunes

Behind the foredunes a range of mobile dunes, partly or wholly unvegetated, form and these are known as transgressive dunes. Sediment becomes transported away from the coastal zone in blowout episodes (on vegetated dunes) or as mobile, unvegetated forms move inland.

Blowouts

Any partly vegetated coastal dune may be prone to the development of blowouts. These are erosional hollows or depressions within a dune complex. Once initiated, they develop through an interplay of deflation and collapse, sometimes forming complex erosional dunescapes (Carter *et al.*, 1990). Observations of the development of two blowouts at Island Beach State Park, New Jersey, showed a clear morphological zonation of the throat (dominated by sediment transport), lateral walls (characterized by large-scale erosion), and a floor and blowout rim (sites for deposition). Gares (1992) concluded his study of this site by acknowledging that blowouts are an integral part of coastal foredunes.

It is useful to make a distinction between coastal dunes and dunes at the coast. True coastal dunes, the foredunes, are the only ones that are necessarily related to coastal zone processes. The other forms are less tied to the coast. In some cases, dunes are capable of climbing up steep drift slopes. It seems that many coastal dunes occupy a transitional state between stable and mobile forms (Carter *et al.*, 1990). Human activity has often been blamed for the increasing mobility of formerly fixed dune systems, but the situation is often much more complex.

Dune morphodynamics — the role of vegetation

Although aeolian sand transport was first quantitatively described by Bagnold in 1941 the prediction of sand transport from beaches has been only moderately successful over short monitoring periods. This is partly due to moisture controls introduced by sand supply from periodically wet beaches (as found by Sarre, 1989a on Braunton Burrows, south-west England), but even along the dry coast of Israel where such effects are minimized, actual sand transport was

Plate 3.2 Vegetation clearance and sand removal for development, Seven Mile Beach, Grand Cayman Island, West Indies. Such activities aid the erosion of the backshore during storms and release sediments from the beach ridge sediment store

found to be only 40 per cent of theoretical transport (Goldsmith *et al.*, 1990). The main reason, therefore, that coastal dunes differ fundamentally from desert dunes is because of the key role of vegetation. Vegetation acts as a soft baffle, increasing the thickness of the calm air layer and absorbing the energy of saltating sand grains, thus increasing sand deposition and encouraging the dune to grow vertically in place rather than through slipface deposition and migration (Goldsmith, 1989; Sarre, 1989b). Whilst the role of a dense vegetation canopy is well established (and can be highly efficient as Rosen (1979) calculated that landward sand transport was reduced by 88 per cent upon crossing the vegetated surface of an accreting dune), there are complex relationships between sand transport and the height, aspect area, porosity and spacing of plants in dune wind fields.

The interrelationships between sand transport/aerodynamics, vegetation and dune form are particularly well seen in the morphology of incipient foredunes. Shadow dunes result from the reduction of wind velocities within and downwind of pioneer species which establish on the backshore in storm or high tide debris lines (Hesp, 1989). Thus dune initiation and initial growth pattern are clearly influenced by patterns of seedling recruitment and survival and their interaction with beach geometry and swash/backwash processes near the high tide limit. At low plant densities individual roughness elements may act independently of one another: accelerated flows around well-spaced individual vegetation clumps may mean that a sparsely vegetated area may deflate faster than an unvegetated surface (Willets, 1989). However, as plant density increases, the degree of near-surface flow penetration decreases, drag increases and the rate of downwind deceleration increases (Hesp, 1983); the morphological expression of these processes is for dune height to increase and dune length to decrease as vegetation density increases. However, as the canopy is pliable and thus of variable roughness, at the early stages of dune growth these relationships are highly dynamic. As wind strength and sand transport vary so dunes flatten or grow vertically as the vegetation canopy adjusts to the prevailing wind field. As well as variations in density, variations in plant height play an important role in determining dune growth and accretion. Experiments have shown that flow penetration is less in tall as opposed to short vegetation and this results in increased sand deposition over shorter distances. These differences help explain why hummocky ridges develop in the presence of *Ammophila* whereas *Spinifex* dunes are typically low sand platforms (Hesp, 1989). Inter-species tolerance to sand burial also becomes important and determines whether or not the roughness element is further renewed on dune growth. However, once a dune becomes established, subsequent lateral plant spread by both shoot and rhizome colonization and sand build-up leads to the development of hummocky dune ridges.

However, although dune vegetation to some extent creates its own environment, dune vegetation and form is also itself affected by environmental controls. Thus dune microtopography influences the distribution of soil nutrients — particularly phosphorus and potassium — which in turn affects plant rooting patterns. Furthermore, dune surfaces are subject to well-documented progressive changes in organic content, calcium carbonate content and pH with age. These processes may have geomorphological implications. At the same time, however, dunes may be subject to sudden changes such as when wet dune slacks are filled by blown sand or when dune blowouts develop. Animals may also play a role in dune morphodynamics, either indirectly through grazing, or directly through trampling and burrowing. On the de Blink dunes in the Netherlands rabbits excavate burrows and caves, with up to 1 kg of sand moved per burrow (Rutin, 1992). Although no estimate has been made of their overall contribution to the sediment budget, they have a clear local influence on dune topography, producing distinctive stepped dune profiles.

Dune ecology

All pioneering coastal dune plants are characterized by very high salt tolerance, tolerance of burial by sand, and extensive vertical and horizontal root systems.

Across a relatively stable dunefield, abiotic stress levels decline landward, whereas biotic stresses (such as competition) increase. Blowouts may act to bring the abiotic stresses common on foredunes further back into the dunefield (Hesp, 1991) with concomitant effects on vegetation patterning.

The vegetation of dunes, because of its key role in dune sedimentation processes, has been well studied. In the USA, the pioneer colonizers of coastal foredunes are *Ammophila arenaria* (marram grass), *Salsola kali* (saltwort), *Ammophila brevigulata* (American beach grass) and *Uniola paniciliata* (sea oats) as discussed by Goldsmith (1985). The most commonly used species for artificial dune planting in the USA are *Ammophila brevigulata* along the mid and upper Atlantic coast, *Ammophila arenaria* along the Pacific north-west and Californian coasts, sea oats along the south Atlantic and Gulf coasts and the panic grasses *Panicum amarum* and *Panicum amarulum* along the Atlantic and Gulf coasts. On natural dunes on North Carolina barrier islands, sea oats are the primary dune builder, and *Spartina patens* becomes important on older dune terraces (Godfrey and Godfrey, 1973).

In Britain, marram grass, sand couch grass (*Agropyron junciforme*), saltwort and sea rocket (*Cakile maritima*) are all important pioneer species. In dune slacks *Salix repens* (weeping willow) develops on calcareous soils, and *Potentilla anserina* (silverwood) and *Carex nigra* (common sedge) under acidic conditions. In the UK, forest plantations on dunes are often composed of exotic pines, such as *Pinus nigra* spp. *Laricio*, and *P. contorta* (Doody, 1989). In the southern hemisphere many South African foredunes are colonized by *Scaevola thunbergii* (McLachlan, 1990), and in Australia *Cakile maritima*, *C. edentula* and *Spinifex serviceus*, *S. hirsutus* and *S. longifolius* are all common foredune species (Hesp, 1989).

We know a growing amount about dune fauna and ecosystem processes. Arthropods and vertebrates (especially insects, birds and mammals) are major components of dune fauna, with crustaceans important on the seaward side. Insects are usually dominant, especially Hymenoptera, Diptera and Coleoptera. There is also a rich interstitial fauna, which has only recently received scientific attention (McLachlan, 1991). Animal zonation tends to follow that of plants, with increasing diversity to the landward side consequent upon decreasing stress. At Wilderness on the South African coast, for example, fourteen arthropod orders were found in the dunes as well as two rodent species. Sand movement was found to be a major control of the spatial patterning of both vegetation and the rodents (Masson, 1990). McLachlan has studied part of the dunefield at Alexandria, South Africa. On the 20 per cent of the dune area which is vegetated, he found that species diversity increases from three species on the beach foredune to more than forty species 2–3 km inland. This increase in diversity is accompanied by an increase in cover, height and woodiness.

Crustaceans dominate the foredune fauna (80 per cent of the biomass) whereas insects dominate further inland. Unique species in this area include sand frogs, Damara terns and gerbils. Three food chains, involving macroscopic grazers, macroscopic detritivores and interstitial fauna, respectively, are recognized in the dunes (McLachlan, 1988, 1990). On many dunefields grazing animals have a large impact on vegetation composition.

Beach–dune interactions

Beach–dune interactions are just one part of a web of interrelations between landward and offshore areas, and between various places along the coast. The hierarchical nature of such interactions means that change to one small part of the system may have consequences for a much larger area. Such consequences have often been unforeseen and dangerous, as we shall see in the following sections, and may also take a long time to become apparent. In these circumstances, simple cause–effect relationships may not be present. We can recognize four types of interaction between beach and dune, all related to energy movements, as shown in Fig. 3.4, i.e. movements of sand (beach to dune and dune to beach), salt spray (beach to dune), groundwater (dune to beach) and organic material (beach to dune and dune to beach). These movements are associated with erosion and geomorphological changes, as well as nutrient inputs and outputs. Of these, the movements of organisms and sand are the best studied.

Sand movements

Landward aeolian sand transport is greatest on wide, low angle dissipative beaches and least on narrower reflective beaches. The level of flow disturbance also plays a part: coarse-grained sediments, often organized into local cusp topography, on reflective beaches promote flow disturbance, reduced wind velocity gradients and lower volumes of sand transport at given wind speeds (Short and Hesp, 1982). Foredunes of reflective beaches therefore tend to be small, of limited lateral extent and relatively immobile, and show rapid recovery (<1 month) after erosion events. Sand dunes associated with dissipative beaches are potentially of much greater volume and dynamism. Foredunes may exhibit a range of forms from stable, well vegetated, laterally continuous dunes to highly mobile, poorly vegetated forms. Unstable foredunes may lead to the erosion of more landward-lying dunes and perhaps even the ultimate formation of wide, transgressive sand sheets.

At regional and local scales, actual transport volumes are influenced by tidal range, which determines the limits to tidal wetting and the time available for

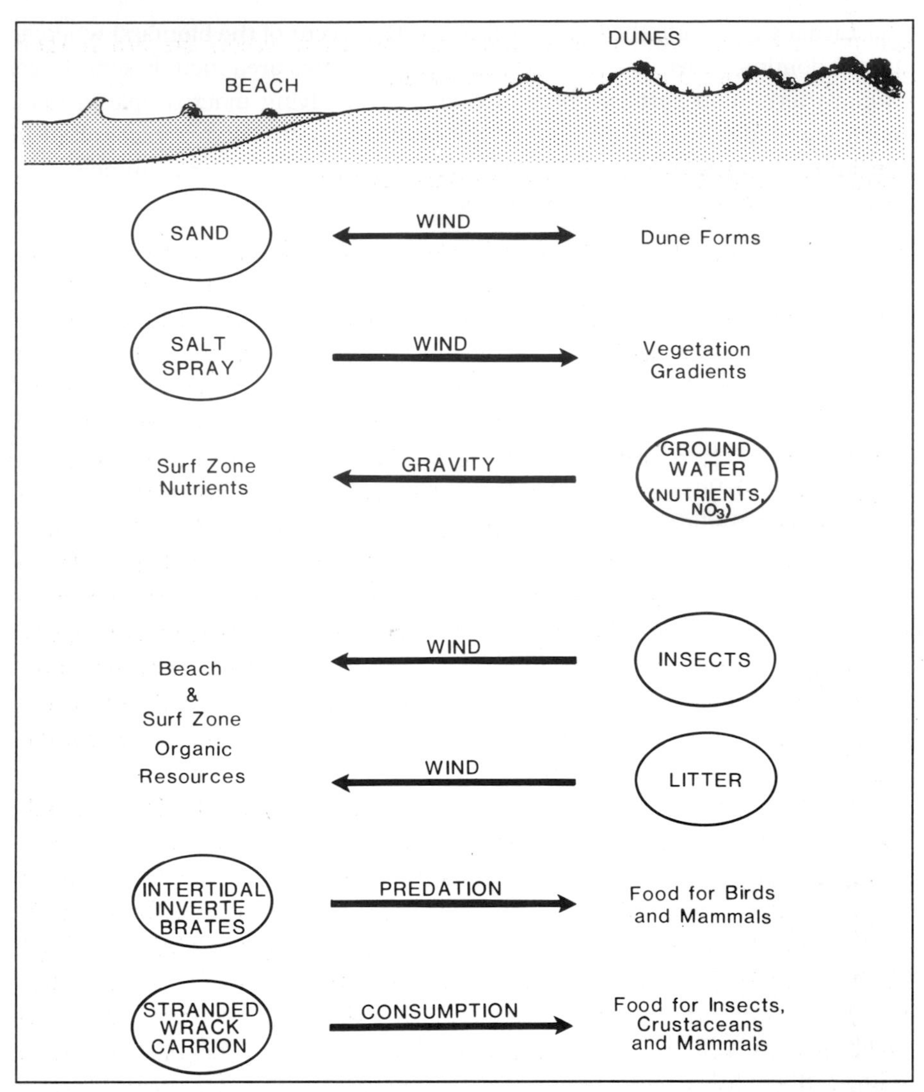

Fig. 3.4 Interchanges of material between beach and dune. After McLachlan (1991)

sand transport from the intertidal zone, and to beach topography. Intermediate beach states (Chapter 2), with water filled or damp troughs between sand bar source areas, introduce obstacles to saltating sand grains. Occasionally, beach surfaces become armoured, either by a protective layer of shells or pebbles, or by the formation of a salt crust. Thus, Ritchie and Penland (1990) note that oyster shells can form pavements limiting sand movement within the South Louisiana dune-beach coastline. Pye (1980) recorded the presence of ephemeral beach 'salcrete' in Queensland, Australia, when the beach sand became consolidated temporarily by seawater.

What problems affect sandy coastlines?

More than 70 per cent of the world's sandy coastlines appear to have been undergoing net erosion over the past few decades (Bird, 1985a); areas of progradation are generally restricted to areas uplifted as a result of tectonic or glacio-isostatic processes or sites close to abundant sediment supply at river mouths and from offshore sand shoals. The reasons for coastline retreat can be classified into those that result from long term changes in the relationships between sea level rise and sediment supply, the effects of human pressures on the coastal zone and its hinterland, and the possible first signs of some consequences of man-induced climatic change, which may intensify in the future. Here, we concentrate upon anthropogenic stresses in the coastal zone. These may be direct or indirect, and include: dredging operations, extraction of minerals, sands and gravels, building of dams to reduce fluvial sediment inputs, reduction in sediment supply from naturally eroding cliffs because of basal protection, and interference with longshore sediment transport from the construction of piers, jetties and breakwaters associated with port and coastal recreational activities (Plate 3.2).

In the USA, Morton (1979) reports how dams are reducing the flow and sediment load of Texas rivers to the Gulf of Mexico. Thus, the Brazos River now delivers *c.* 860 000 m^3 of sediment per year to the beaches. This is about 30 per cent of the volume moved annually before dams were built on the river. In Italy, out of a 7500 km coastline, beaches stretch for some 3250 km, and about 35 per cent of these are eroding because of extraction of beach sediment, damming rivers, inappropriate coastal defences, subsidence and perhaps changing climate (Caputo *et al.*, 1991). In Japan, most rivers are dammed and there is much subsidence along the coast caused by overpumping from coastal aquifers (Walker and Mossa, 1986). Along the west coast of Aruba, Netherlands Antilles, erosion of beaches is causing problems for tourism. The effects of the removal of sediment supply are particularly well seen in deltaic environments where high sediment inputs are required to offset the natural subsidence and sea level rise associated with downwarping of the delta itself. Since the final closure of the Aswan High Dam in 1964, there has been pronounced erosion along parts of the Nile Delta front. Between 1970 and 1987 retreat rates at Rosetta Point, for example, were 80–120 m a^{-1} (Sestini, 1992).

Sand mining has been seen to cause problems of beach erosion (Hilton, 1989), and can be accidental, as in California where it has occurred in association with the removal of kelp (Hotten, 1988). In Canada, beaches on Prince Edward Island have been extensively mined for aggregate, but in other parts of the country (such as Newfoundland–Labrador) such mining is controlled by law (Earney, 1990). Along the southern Namibian coast, there has been extensive mining of beaches for diamonds, which has involved constructing huge seawalls 350 m seaward of the high water mark to enable mining to take place on the landward side at up to 12 m below sea level (Earney, 1990). The seawall must be maintained continuously because of wave action in this high energy environment. Beach rock may also be quarried, affecting coastal morphodynamics, as shown along the Kuwait City shoreline where beachrock was mined to provide armouring for a highway built on landfill (Kana and Al-Sarawi, 1988).

These examples show that human activities often deplete sediment supplies to coasts or interrupt the longshore transfer of sediments. Such sediment deficiencies often result in coastal erosion. The response is often the construction of further structures: to lessen incoming wave energy (e.g. breakwater installation), to prevent further erosion (e.g. seawall construction), or to retain remaining sediments (e.g. groyne building). Unfortunately, these structures in turn interfere with both shore-normal and alongshore processes and may generate further coastal problems, initiating a vicious circle of yet more coastal protection and locking coastal managers into a never-ending programme of artificial shoreline maintenance and control.

Shore-normal processes and the role of shoreline structures

Engineers build seawalls to protect the land, not the beach. Yet such structures may set in train processes that can lead to the loss of the protective structure itself. Chapter 2 showed how beach profiles range from dissipative to reflective states. Dissipative profiles are characteristic of storm wave, spilling breaker conditions: the beach profile widens under such attack to reduce the wave energy per unit area of beach. Seawalls prevent the development of this extended profile, but cause high energy levels to be concentrated on a fixed beach width. Increased backwash volumes comb material from the beach. In addition, higher beach water tables saturate beach sands. As a result, reduced frictional strength between sand grains because of high pore water pressures may promote landslide-type failure of beach sands (Silvester, 1974). Thus seawalls often trigger the narrowing and lowering of the beaches that front them; the wall is threatened not by direct wave attack, but by the prospect of undermining as beach scour proceeds. On the Bayou Laforche headland, Louisiana, USA, a 1.8 km long seawall was built in 1986 from cement-filled bags to help stop erosion along a vulnerable low-lying beach and dune coast affected by periodic hurricanes. As with many such schemes, a combination of seawall (hard engineering) and beach nourishment (soft engineering) was utilized. Subsequent monitoring has shown that after two hurricanes in 1988 high rates of coastal erosion occurred, higher than on most nearby natural beach–dune systems (Nakashima and Mossa, 1991). Also in the USA, beach erosion at North Padre Island, Texas was almost nil between 1937 and 1969, but after the construction of a seawall the situation changed. Beach slope stayed relatively constant, but the beach retreated by *c.* 25 m and its elevation was lowered by *c.* 0.5 m. Between 1975 and 1987 erosion rates of *c.* 2 m per year have been measured here (Morton, 1988). Carter (1988) illustrates the problem well for Porthcawl, South Wales: the 1906 seawall had to be replaced with a more seaward structure in 1934 and this second structure is now also in danger (Fig. 3.5). Typical responses to beach lowering are either to protect the wall base with ugly 'rip-rap' boulders or concrete tetrapods (Plate 3.3) or to create an artificial beach in front of the structure (see later in this chapter).

Alongshore processes and the role of shoreline structures

Seawalls

Erosion also often occurs downdrift of many seawalls, because the reflected wave energy then interacts with incoming wave energy. This produces a high-energy environment, with no sediment to transport thus encouraging erosion downdrift. The temptation, therefore, is to extend the wall downdrift to prevent the new erosion problem — a solution that is likely to simply shift the problem

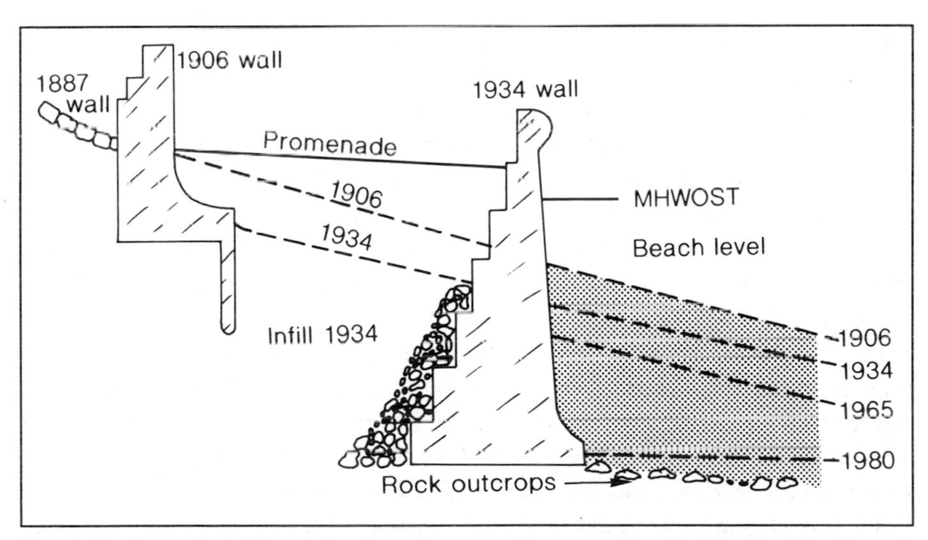

Fig. 3.5 Lowering of beach levels and progressive seawall replacements at Porthcawl, South Wales. After Carter (1988)

Plate 3.3 Tetrapods awaiting deployment, Lamberts Bay, Republic of South Africa (photograph: H.A. Viles)

even further downdrift. In this way it is not difficult to see how seawall structures have rapidly expanded along many coastlines. Walker and Mossa (1986), for example, have calculated that there are *c.* 8000 km of seawalls and dykes in Japan, which occupy approximately 25 per cent of the total Japanese coastline. Seawalls modify longshore processes but do not prevent them. A particularly prevalent coastal problem, however, is the disruption of longshore drift by three types of structure: jetties, breakwaters and groynes (or groins).

Jetties

The effect of jetties, structures which extend beyond the breaker zone, on alongshore sediment transport is a combination of updrift accumulation and downdrift erosion (Fig. 3.6). A classic example of this coastal problem is provided by the history of coastal sedimentation at Santa Barbara harbour, southern California, which is subject to the west to east drift of fluvially derived sediment. The harbour breakwater at Santa Barbara was connected to the shore in 1930 and immediately the updrift area to the west of the jetty began to fill with sediment. By 1937 the embayment was full and sands were beginning to move along the breakwater and fill the harbour itself. At the same time, the beaches to the east of the harbour were not being replenished with sediment and shoreline erosion in excess

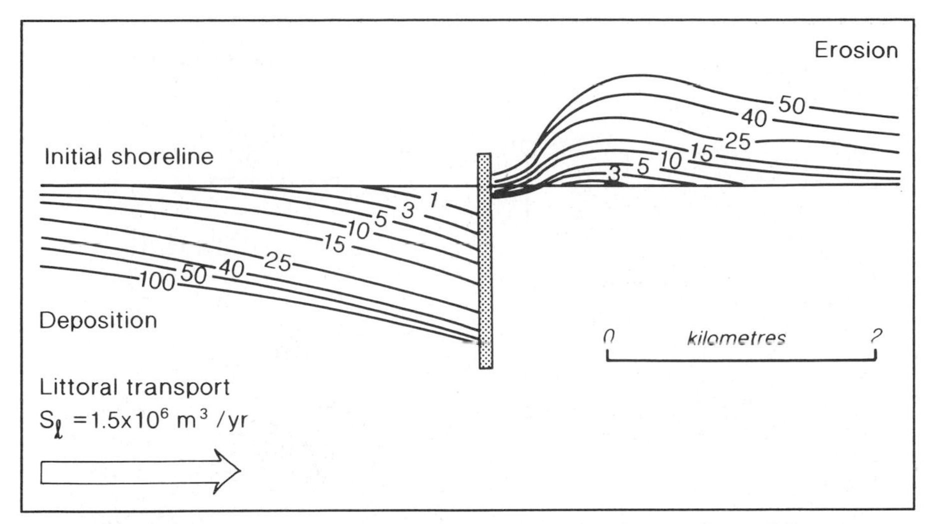

Fig. 3.6 The blockage of littoral sand transport by a breakwater from a laboratory model, showing shoreline changes in years. After Komar (1976)

of 100 000 m^3 a^{-1} was taking place. The initial attempt to solve this problem, in 1935, was to dump 154 000 m^3 of sand at −7 m water depth 300 m offshore; the dump simply stayed there. In 1938, 440 000 m^3 of sand was successfully pumped from the harbour shoal to the eroding beach and this operation continued until 1952 when the pump was over-loaded by sand movement during south-easterly storms. Now, the harbour shoal is itself protected by a breakwater and periodic dredging is used to transfer 270 000 m^3 a^{-1} of sand to the downdrift beaches (Wiegel, 1964). This solution, of mimicking natural processes by artificial means, is also characteristic of fixed inlets on barrier islands. Here by-pass pumps are used to transfer sediments across tidal passes and maintain alongshore transport. Thus, the Fort Worth inlet in Florida currently moves 100 000 m^3 a^{-1} of sand, about 60 per cent of the estimated littoral drift (Fig. 3.7).

Breakwaters

To alleviate these problems, why not build a breakwater parallel to the shore to protect the shore, but not interrupt the longshore drift? The effect of a breakwater is, however, to reduce wave energy in its lee; as longshore current velocities are driven by energy from the breaker zone (Chapter 2) sediment deposition takes place shoreward of the breakwater. Thus breakwater construction in Santa Monica, California was followed by dredging to maintain access to the protected pier (Fig. 3.8). At Preston Beach, near Weymouth, England, a long strip of heavy rocks was deposited in the late nineteenth century to prevent erosion. However, it appears that these caused more damage and accel-

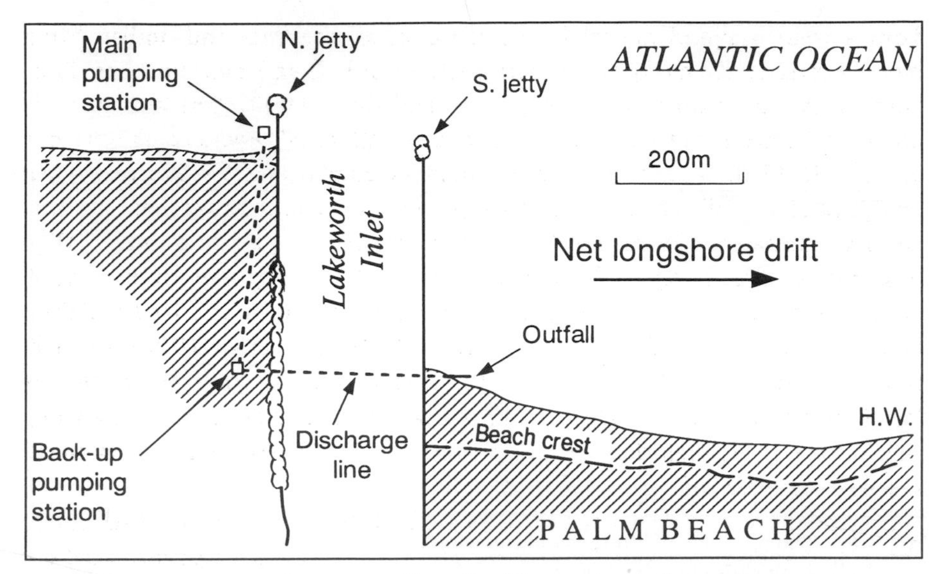

Fig. 3.7 A variant on the jetty problem. Sand by-passing jetties used to maintain inlets, Florida

erated erosion of sand by the production of eddies. They were later blasted and removed at great expense (Owens and Case, 1908).

Groynes

These are shore-normal walls, extending across most or all of the beach profile, and built in regularly spaced lines along a coastal stretch. Groynes are made

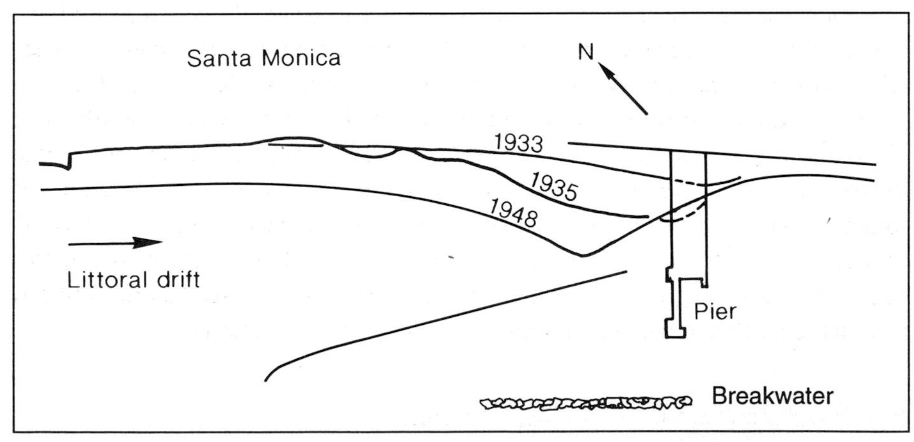

Fig. 3.8 Sedimentation problems at Santa Monica, California after breakwater construction in 1934. After Komar (1976)

from a wide range of materials, including stone, concrete and timber. They have been used for many years, with early records from Dover harbour, England in the seventeenth century (Owens and Case, 1908). The aim of such groyne schemes is to trap beach material in small 'bays' produced by adjacent groynes. In 1850, a Captain Vetch proposed the emplacement of fifteen groynes on Spurn Head, UK which had become prone to erosion after removal of shingle. The estimated cost was £400 (Royal Commission on Coast Erosion, 1907). It is unclear what became of this scheme, but in 1864–6 groynes were built at Spurn Head from Dantzing timber piles connected by planks, and by 1877 they had collected *c.* 1 250 000 tons of shingle (Owens and Case, 1908). On Sylt Island, on the north German coast, the first wooden groynes were built in 1865, concrete ones in 1913 and 1924, and iron ones after 1927 (Kelletat, 1992). Their use has declined here since 1972 when beach nourishment schemes were started.

Unlike jetties or breakwaters, groynes are built to trap a portion of the longshore drift and thus restore beach volume. Like a jetty, they trap sediment on their updrift sides and lose material on their downdrift sides. Over time, however, the trapped sand or gravel evolves to an orientation normal to the incoming waves, so forming a miniature swash-aligned coast between adjacent groynes. In such circumstances longshore transport is zero and beaches downdrift of such a system of groynes become starved of new sediment inputs. As with seawalls, the management response is often to erect a further groyne system downdrift to conserve the remaining beach material. A further difficulty is that groynes may promote sediment loss offshore. If storm waves arrive from the opposite quadrant to the persistent swell then a rip current may be set up along the downdrift side of the groyne. Angled groynes, in an *en echelon* design, may solve such problems (Fig. 3.9). Groynes can, however, be a successful form of coastal management as once they are full longshore drift becomes re-established, either by flanking or overpassing the landward margin or by endpassing the seaward end. Bray *et al.* (1992) have found that groynes are essential and effective parts of coastal protection schemes in southern England on shingle rather than sand beaches. In many countries, groynes are now an extremely common part of the coastline with Japan, for example, possessing some 10 000 groynes and jetties spread over some 32 000 km of coastline (Walker and Mossa, 1986). Frequently, groynes are used in conjunction with seawalls and offshore breakwaters in an integrated attempt to manage coastal erosion.

Beach nourishment — the soft option?

As an alternative, or additional, strategy artificial beach replenishment using sand from elsewhere is a 'more natural' method of limiting beach erosion which

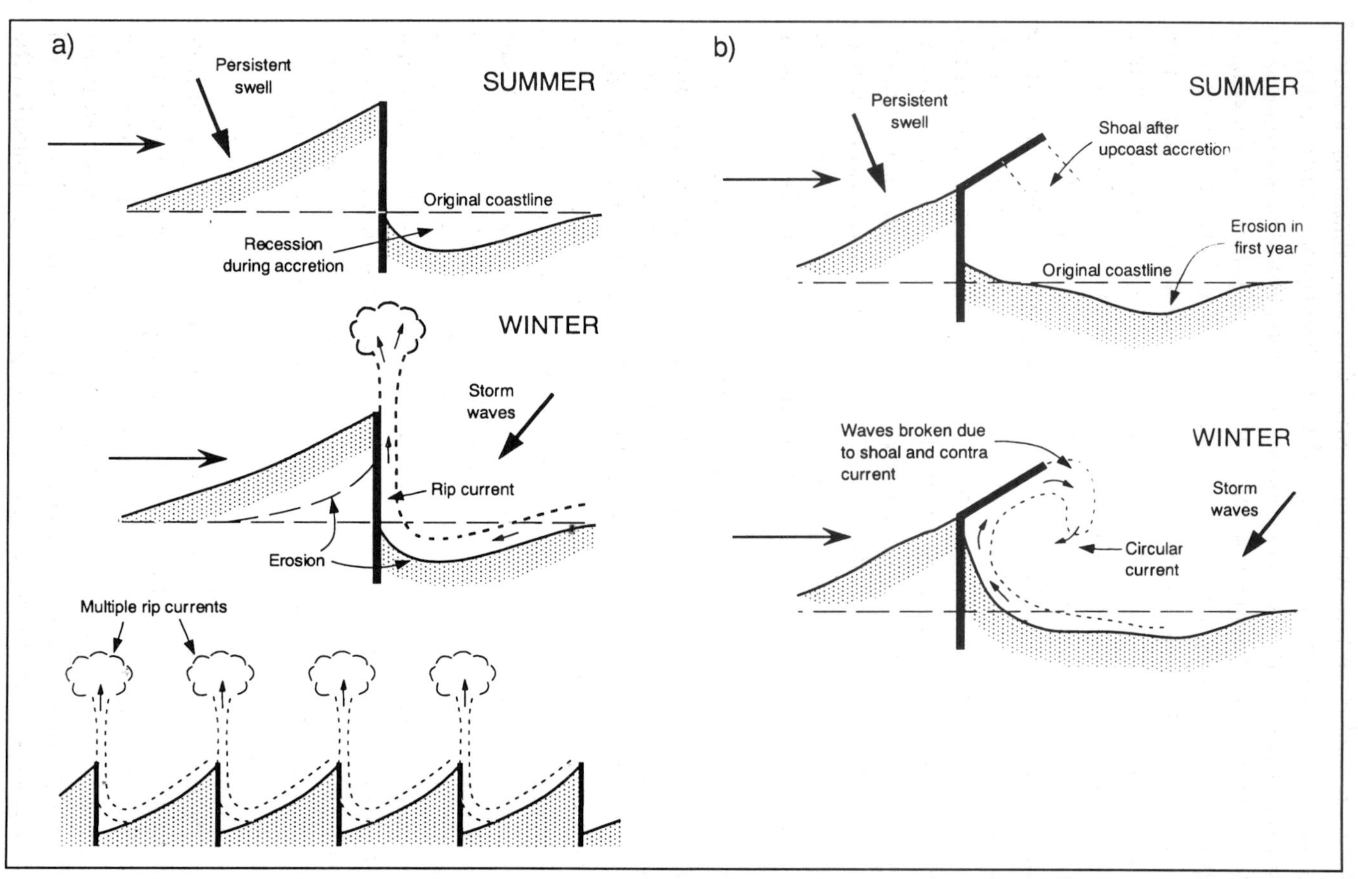

Fig. 3.9 a) Sediment losses from a groyne field subject to bi-directional wave attack. b) A possible solution. After Silvester (1974)

has received much recent scientific attention. In theory, using beach fill both protects the backshore and 'saves the beach' from despoilation by a suite of groynes and walls. In southern England major beach nourishment schemes have been carried out in Poole Bay, Christchurch Bay, Bracklesham Bay and on Hayling Island (Bray *et al.*, 1992). Beach nourishment has also been attempted on gravel coasts, as at Dungeness where *c.* 30 000 m^3 of shingle is moved every year to protect the nuclear power station there. Drawbacks of this method are that it is usually expensive, may only work on a short time-scale, and involves the removal of sand from some other deposit. At Cape Hatteras, North Carolina, USA nearly half a million dollars was spent in 1972 on beach nourishment at Buxton. After 10 months all the new sand had disappeared! A major constraint for many beach nourishment schemes is a suitable, and economically viable, source of sand. Along the Louisiana coastline, for example, serious erosion has prompted plans for beach nourishment in collaboration with barrier island restoration. A search for suitable sand has been carried out on the continental shelf off Louisiana, as nourishment will only be cost-effective if a nearby source is available (Suter *et al.*, 1989). Offshore dredging itself may cause severe coastal erosion problems, as was found at Hallsands on the Devon coast of southern England in the nineteenth century.

Whether beach nourishment schemes succeed or not depends on the natural sediment budget conditions of the area. In the USA, for example, beach nourishment schemes generally last less than 2 years in New Jersey and up to 7 years in Florida south of Cape Canaveral (Pilkey, 1991). An interesting study of two adjacent beach fill schemes on the Algarve coast of southern Portugal clearly illustrates the spatial variability of response. At Praia de Rocha, following fills in 1970 (*c.* 900 000 m^3 of sand used) and 1983 (*c.* 150 000 m^3 of sand used), sediment has been lost very slowly, at less than 2–5 per cent per year. At Praia dos Três Castelos, to the west, a fill of 200 000 m^3 in 1983 was rapidly eroded, losing 15–20 per cent per year, and returning to pre-fill conditions after 5 years. Psuty and Moreira (1992), who used a range of photographs and postcards to elucidate the beach changes over time, suggest that the success of the Praia de Rocha scheme was due to the sheltered nature of the beach there.

Building new beaches

There are several examples of the production of new beaches by deliberate beach fill schemes, and by the dumping of quarry and mine waste. Two million tonnes of colliery waste have been dumped annually along the north-east coast in England, mainly in the Durham area. Bourman (1990) discusses an interesting example in southern Australia where from the 1940s until 1982 *c.* 1.5 million tonnes of limestone gravels from quarrying nearby were dumped in the aptly named Rapid Bay to form an artificial gravel beach. Long term monitoring has shown that the beach prograded from 1949 to 1975, and retreated between 1975 and 1984, with minimal erosion since then. The transported sediment has buried nearby rocky reefs and damaged aquatic habitats to the north-east. Although longshore drift has been relatively slow, where the sediment is polluted ecological damage may well start to occur. In 1977, development and improvement of the Kuwait City waterfront began, including the construction of a number of artificial beaches, such as one on the rocky intertidal terrace of the central headland (Kana and Al-Sarawi, 1988).

Beaches and future sea level rise

The concept of an equilibrium beach profile, and its relations and adjustments to mean sea level, has been used by coastal engineers in recent years (Pilkey *et al.*, 1993). The most widely applied semi-quantitative description of the response of the equilibrium profile to sea level rise has been that of Bruun (1962, 1983). The so-called Bruun rule starts from the premise that an equilibrium shore profile is maintained with an optimum depth of water in the nearshore zone. As sea level rises, deposition must take place on the sea floor to re-establish the water depth, with the sediment coming from the subaerial beach itself. The beach therefore undergoes beach face retreat. This can be represented mathematically as follows:

$$W = \frac{XS}{Y}$$

where:

W = width of beach erosion (in m)
X = horizontal length from shore to limited depth of sediment transport (closure depth) (in m)
S = sea level rise (in m)
Y = vertical length of profile (in m)

Shoreline retreat depends upon the average slope of the shore profile. Thus, for the eastern seaboard of the USA, a 1 m rise would generally cause beaches to retreat 50–100 m from Maine to Maryland, 200 m along the coasts of North and South Carolina, and 100–1000 m along the Florida coastline (Titus *et al.*, 1991). A test of the model on climatically induced fluctuations in Lake Michigan by Hands (1980, 1983) expanded the Bruun rule into three dimensions, and has been the only satisfactory case which clearly shows it in action. He discovered significant lags between the changes in the forcing function and the beach profile response. Bruun's model has been criticized as X is difficult to define in many cases and the offshore topography may be more complex. More fundamentally the assumption of an 'equilibrium profile' has been questioned. It is unclear over what time-scale equilibrium is assumed, and the model does not consider the vital processes of longshore drift and onshore sediment movements which complicate sediment budgets along many coasts (making them three-dimensional, not two-dimensional). The model also allows for progressive adjustments in profile only; there is no consideration of stepwise movements or thresholds.

Dubois (1992) has attempted to modify the Bruun rule to take account of a bar and trough form of offshore profile (rather than the simple concave shore profile recognized by Bruun). Dean and Maurmeyer (1983) expanded the Bruun rule to model the landward migration and upward growth of an entire barrier island system. Dubois (1992) has adapted the Bruun rule rather differently to barrier island systems, suggesting that the shoreface will transgress but that the ramp (where Bruun predicted deposition to occur) will be abandoned. This work implies that a fundamental premise of Bruun's rule may be false, i.e. rising sea level may not be linked to the transfer of sediment from shoreface to the offshore zone (ramp). An alternative approach uses dynamic equilibrium principles to create a numerical model of onshore–offshore transport (Kriebel and Dean, 1985). An intensive study of these and other models by SCOR Working Group 89 (1991) showed that 'the status of models for the beach response to elevated water levels is far from satisfactory, and predictions of the associated shoreline recession rates yield uncertain results'.

Bray *et al.* (1992) have used a three-dimensional form of the Bruun rule (after Dean, 1991) including a full sediment budget for the area to predict the effects of future sea level rise on coasts in south and south-west England. On the South Haven dune complex in Dorset, for example, three coastal segments showed different predicted responses. The two open coast sites on Studland Bay responded similarly, except that northward littoral drift would cause erosion in the south and accretion in the north. The impact of sea level rise would thus be overshadowed by longshore drift effects in the northern sector. At Shell Bay, recession would probably be controlled by sediment loss to the Poole Harbour entrance channel, thus future sea level rise will only contribute weakly to erosion here. The South Haven example backs up the conclusions reached by many workers that sea level rise is only one, and often not a major, factor controlling coastal erosion. As Bryant (1985) showed at Stanwell Park, New South Wales, Australia, high rainfall events are a major control on erosion rates as they increase the height of the water table, promoting liquefaction and erosion of sand. Thus, future climatic changes associated with global warming may also contribute to the changing nature of coastal erosion, with an increase in precipitation and storminess.

Of course, the presence of a multiplicity of coastal protection structures will complicate the response of sandy coasts to sea level rise in the future. Where seawalls are present, for example, the retreat strategy is unavailable as sediment movement onshore is prevented. Few predictions have been made, but the general feeling is that protected beaches are likely to be especially vulnerable to future sea level rise. On protected coasts, scour in front of seawalls would increase, allied with an increase in the frequency of overtopping, and an increase in the probability of structural damage. Groynes, which do not protect against onshore–offshore sediment movements, would not help if Bruun's rule applies. Where littoral longshore drift occurs, groynes may help to protect the upper part of the coastal profile, but will presumably themselves become eroded. Offshore breakwaters should help protect the beaches, but not the offshore zone on their seaward side. They may also be subjected to scour and damage on the seaward side (Bray *et al.*, 1992). Beach nourishment would have to be continued at an increasing rate to prevent erosion, thus becoming an even more expensive option in the future (see later in this chapter).

Pollution of beaches

In August 1993, Michael Saltmarsh, owner of a beach at Croyde Bay, Devon, England, took the South West Water Company to court for polluting his beach with sewage, condoms and other waste from an outflow updrift. During April 1994 he won his case, with concomitant compensation. This illustrates an increasing problem of beach pollution. As beaches are major foci for recreation, pollution is a serious issue, which can also affect the natural beach ecology. Beach pollution arrives via a range of pathways, i.e. rivers and groundwater, landfill and dumping, wind, onshore currents, or alongshore currents. Sources of pollution include dumping from ships at sea, or from oil drilling activities, sewage pipelines and outflows, industrial and agricultural effluent into rivers and the atmosphere. Some pollution events are bizarre — in July 1992 a beach on the island of Jersey was polluted by effluent from 3500 t of fermented potatoes dumped in an adjacent field.

Beaches around the world have become polluted, and the European Union, for example, has instigated a stringent programme of beach pollution monitoring in Europe. In the UK, out of 404 registered bathing beaches 76 per cent passed the European Union standards for sewage-related bacteria in 1991. However, beaches such as those at Blackpool and Southport have frequently had bacterial and viral levels above European Union recommended levels. Various surveys carried out by volunteers have revealed a chronic pollution problem on British beaches. Beachwatch '93 collected 33 t of rubbish from 121 British beaches. Elsewhere the problems are also serious. On the Israeli coast, a survey of six beaches found that the average monthly count was 36.77 pieces of litter per 5 m wide beach transect (Golik and Gertner, 1992). Over 70 per cent of the rubbish was plastic. In contrast, a study of twenty-nine beaches in Jamaica found problems with persistent tar accumulations, especially on the east coast and near Kingston harbour (Jones and Bacon, 1990). In Egypt, in a recent study, seven out of twelve bathing beaches were found to greatly exceed international standards of microbiological contamination. The other five beaches were still heavily contaminated (Sestini, 1992).

Although all coastal types are prone to pollution, beaches (especially sheltered ones) may be particularly vulnerable as sediments act as a pollution trap. Thus, sandy beaches may absorb and store pollution which can cause ecological damage long after pollution inputs have ceased. Oil pollution, for example, affects the meiofauna of beaches for up to a year and may clog up interstitial pores and reduce oxygen availability (Strain, 1986; Owens *et al.*, 1987; Bodin, 1988). Dispersants used to clean oil from beaches may also affect beach ecology (Brown and McLachlan, 1990). Some species are very sensitive to pollution, others such as the polychaete worm *Capitella*, thrive on it. In the Inner Thermaikos Gulf, Greece, near the city of Thessaloniki, beaches have become heavily polluted by domestic and industrial sewage and agricultural fertilizers, producing an adapted benthic flora (Georgos and Perissorakis, 1992).

Dune management

Dunes are used for a wide range of human activities, some of which conflict with each other, and several of which impinge on natural processes (Table 3.1). On the other hand, natural dune movements threaten some coastal activities, are perceived as a 'problem' and conflict with the idea of fixed coastal dunes as a form of coastal defence. Because of their dynamic nature and their interdependence with beaches and nearshore sedimentation, their management cannot be considered in isolation. Overmanagement has become a problem in some areas, as dunes have become fixed and vegetated, preventing sediment movements. Important human activities affecting coastal dunes include the alteration of cliff and beach processes and sediment budgets by seawalls, mining etc., damage to vegetation cover by trampling, direct removal of dune sand, introduction of alien species, alterations to groundwater and pollution from oil spills and seeps.

Dune stabilization, planting and destabilization

For many years the use of various plants, especially marram grass, has been recommended as facilitating dune stabilization. Carey and Oliver (1918) repro-

Table 3.1 Dune uses and their impacts on morphodynamics

Use	*Impact*
Extraction:	
sand mining	Erosion
water extraction	Lowers water table, may increase deflation and change ecology in slacks
Conservation/protection:	
coastal protection	Stabilizes dunes; disturbs sediment movements
nature reserves	May restrict some species
Recreation:	
biking, riding and walking	Damages dune vegetation and may encourage blowouts
golf courses	Create dense grass sward; may need protection from erosion
Agriculture:	
cultivation	Fertilizer use affects nutrient balance
stock grazing	Damages vegetation
rabbit grazing	Damages vegetation
afforestation	Changes species mix; felling may increase vulnerability to erosion
Development:	
military uses	Bombs and tracked vehicles may accelerate erosion
transport, housing and pipelines	May need careful management to avoid local disturbances

Source: compiled from Ranwell and Boar (1986)

duce a chart showing the experimental planting scheme followed by Sören Biörn in 1795, and Pye (1990) records the encouragement of marram planting in the eighteenth century in the Formby area, England. A large range of techniques are now available for those wishing to manage dunes (Table 3.2). It is becoming clear that not all soft engineering schemes involving biophysical methods are as 'environmentally friendly' as perhaps at first they appeared, and some may have serious geomorphological consequences. Most dune initiation or stabilization programmes involve the use of brushwood, or paling, fences to trap sand, planting of vegetation (especially *Ammophilia arenaria* or marram grass) to aid dune development (Plate 3.4), and the cordoning off of the 'seeded' areas allied with the use of wooden walkways to prevent damage by trampling. Many coastal dunefields in Europe have been 'restored' in this way. Often the fencing is set too low in the profile, leading to summer accretion followed by erosion in the winter. Fences should be set near the first eroding dune, not down on the beach.

The success or failure of such schemes is often difficult to assess and long

Table 3.2 Coastal dune management techniques

Aims	*Techniques*
Stabilization	Seawalls
	Beach nourishment
	Shore protection with geofabrics
	Groynes
	Gabions
	Fencing of brushwood or palings
	Thatching with brushwood
	Contouring
	Binding with chemical binders
	Mulching
Vegetation development	Planting of strandline annuals, dunebuilding grasses, perennial dune turf grasses and woody perennials
Vegetation management	Fertilizer application
	Controlled stock and rabbit grazing
	Scrub control
Destabilization	Controlled trampling
	Controlled disasters — artificial sediment disturbance

Source: compiled from Ranwell and Boar (1986) and Doody (1993)

Plate 3.4 Watering of planted dunes at Houts Bay, Republic of South Africa (photograph: H.A. Viles)

term monitoring is necessary. Many succeed in establishing a dune ridge, but one which does not behave naturally, thus upsetting the larger sediment budget. The causes of dune erosion or instability need investigating to start with, as well as the overall relationship between dune–beach and offshore sediment supply. Dunes damaged by localized, specific human impacts will be easier (and more appropriate) to restore than those characterized by long term, natural instability and change. In Spain, for example, dune regeneration has been successfully achieved (at least so far) at Devesa del Saler, Valencia. Here, dunes as part of a barrier island were destroyed by a tourist development in 1970–3. Saler Beach is a low energy, dissipative beach with a tidal range of less than 50 cm. Over a 16 year period there has been a net accumulation of sand and the area is now a natural park (Sanjaume and Pardo, 1991). The Braunton Burrows dunefield, North Devon, England required extensive restoration after the Second World War during which it had been used for military training involving tracked vehicles and explosives (Kidson and Carr, 1960).

A range of plants and fence types can be used, mimicking as closely as possible the natural flora. Planting has often been quite unnaturally done in the past, failing to copy successional pathways, using regimented rows of too dense planting. Such 'overplanting' can lead to a dramatic change in the morphodynamics of such dunes which then become prone to intermittent slope failure. As found by Carter and Stone's (1989) study of dunes on the Magilligan foreland, Northern Ireland, dense vegetation affects the style of failure, but does not reduce the overall rate of dune erosion by wave attack. Under natural circumstances, coastal dunefields reflect the interplay of depositional and erosional processes mediated by patchy vegetation cover. Now many dune management plans involve a 'disturbance regime' to prevent overstabilization occurring. Thus, at Braunton Burrows post-war conservation and restoration schemes have led to the loss of species-rich grassland, which is being replaced by invasive scrub. Now grazing is being reintroduced, as well as some disturbance of the sediment regime (Doody, 1993). Recent observations of natural stabilization of blowouts by algal crusts (composed of cyanobacteria and green algae) show another potential management strategy (Pluis and de Winder, 1990). If blowouts are to be encouraged, perhaps such natural stabilizers can also be encouraged to mediate especially small scale blowout activity.

Maintaining natural dune dynamics: the seaward fringe

Godfrey and Godfrey (1973) compared the biogeomorphology of a natural barrier system (Cape Lookout) and a developed barrier (Cape Hatteras), both on the Outer Banks, North Carolina, USA. Sea level here has been rising by about 30 cm per century over the past 5000 years, with an 8 cm rise between 1960 and 1970. Hurricanes and north-easterly winter storms both affect these shores. The Cape Lookout shore has an overwash barrier with dunes and marshes behind (Fig. 3.10a). The opening and closing of temporary inlets is also another facet of the flexible response of a natural barrier. At Cape Hatteras, by contrast, a continuous artificial dune line has been planted which acts as a dyke, preventing overwash (Fig. 3.10b). This affects the succession in the plant communities behind, which develop associations less tolerant of flooding. Inlets are fixed. The management here means that the islands are eroding at the front, but

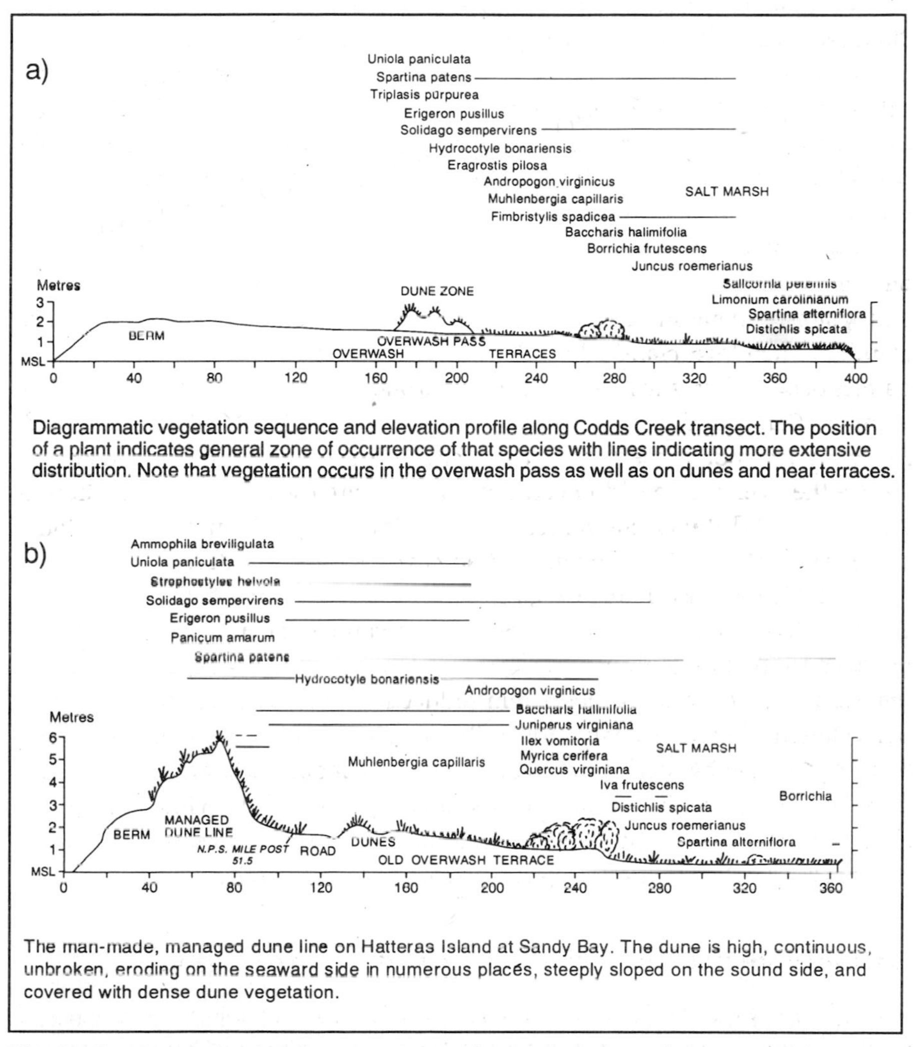

Diagrammatic vegetation sequence and elevation profile along Codds Creek transect. The position of a plant indicates general zone of occurrence of that species with lines indicating more extensive distribution. Note that vegetation occurs in the overwash pass as well as on dunes and near terraces.

The man-made, managed dune line on Hatteras Island at Sandy Bay. The dune is high, continuous, unbroken, eroding on the seaward side in numerous places, steeply sloped on the sound side, and covered with dense dune vegetation.

Fig. 3.10 Comparison of the vegetation and geomorphology of a) a natural (Cape Lookout) and b) a managed (Cape Hatteras) barrier island. After Godfrey and Godfrey (1973)

not retreating as a unit — as would be the case for natural barrier beach systems. Such managed islands respond less well to storms, being eroded more seriously and taking longer to recover (Fig. 3.11; Dolan and Godfrey, 1973) as is well illustrated by responses of natural and fixed dune systems to Hurricane Ginger. Thus, both the sensitivity and resilience (Chapter 1) have been affected.

Dunes and future sea level rise

As with the linked beach environments, natural and managed dunes may respond very differently to sea level rise. Also, natural dunes will show a suite

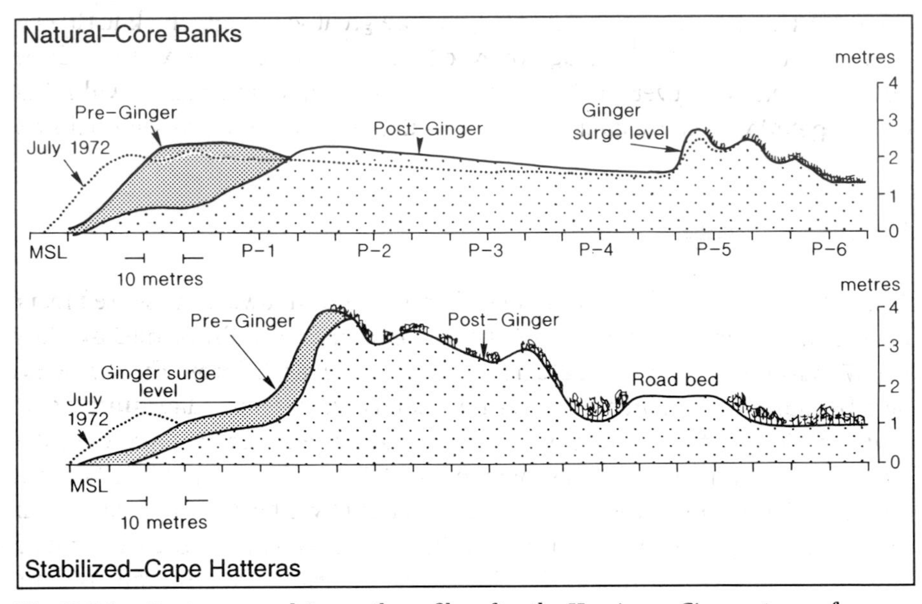

Fig. 3.11 Pre-, post- and 9-month profiles after the Hurricane Ginger storm of September 1971. Note post-storm washover deposit and dune front recovery on natural coastline. After Dolan and Godfrey (1973)

of responses depending upon the controls on their sediment budget. As discussed later in this chapter, artificial dune ridges on barrier islands can prevent most overwash and may be very difficult to 'repair' once attacked by high storm water levels. In comparison, natural, patchily vegetated coastal dunes usually recover quickly and where unimpeded can migrate landwards in response to rising sea levels. One of the major problems for future dune management is that a very large percentage of many dunefields has already been reclaimed (i.e. transferred to other uses) and therefore the sediment is lost, and the dunes cannot transgress over this area. In north-west England, for example, between Liverpool and

Southport one of the largest dune systems in England has largely been eroded by the building of railway, roads, industry, housing, an airport and golf courses, leaving only approximately 10 per cent of the original area (Doody, 1989).

Various future scenarios for Dutch coastal dunes, which protect around 80 per cent of the country's coast, have been proposed. Dune migration inland might occur, with a resultant loss of land. On the other hand, such rejuvenation of the dunes could be seen as a good thing, following years of intensive management and stabilization. Such movements might be accompanied by drift of sand and salt spray inland, thereby affecting inland vegetation. Indirect effects would include changes to the dune groundwater regime, with concomitant impacts on slack vegetation (van der Meulen, 1990). The impacts have not been worked out with any precision, given the great variability in climatic predictions, and lack of knowledge of exactly how climatic forcing affects dune environments. However, regional changes to precipitation regimes will influence vegetation thereby indirectly conditioning dune response to sea level rise.

Barrier islands

The integrated nature of nearshore–beach–dune–landward zone sediment movements is clearly illustrated by the case of barrier islands. Defined by Hoyt (1967) as 'elongate islands, parallel to shore and separated from the mainland by a bay, lagoon or marsh area', barrier islands are coastal sedimentary accumulations characterized by the ability for water and sediment to go over the crest as washover. Usually, as shown in Fig. 3.12, they consist of beaches and dunes penetrated by occasional inlets and backed by a landward wetland area and/or lagoon (Plate 3.5). Found worldwide, they are common along the coasts of south and west Australia, the Netherlands, north Germany, the Atlantic coast of South America, parts of the Atlantic coast of Africa, and particularly in North America where they occupy a large part of the coast from Alaska and Cape Cod down to Mexico.

Barrier island morphologies

Two main types of barrier island are found, classified according to the importance of wave and tidal processes. Where wave action dominates, as on the south Texas coast in the USA, longshore transport dominates leading to long, continuous islands with few inlets (Fig. 3.12a). Perdido Key, Florida, is a good example of a wave-dominated barrier shoreline with a diurnal microtidal (0.4 m mean annual tidal range) environment (Jagger *et al.*, 1991). Tidally transported sediment accumulates in back-barrier flood deltas; ebb-tide deltas

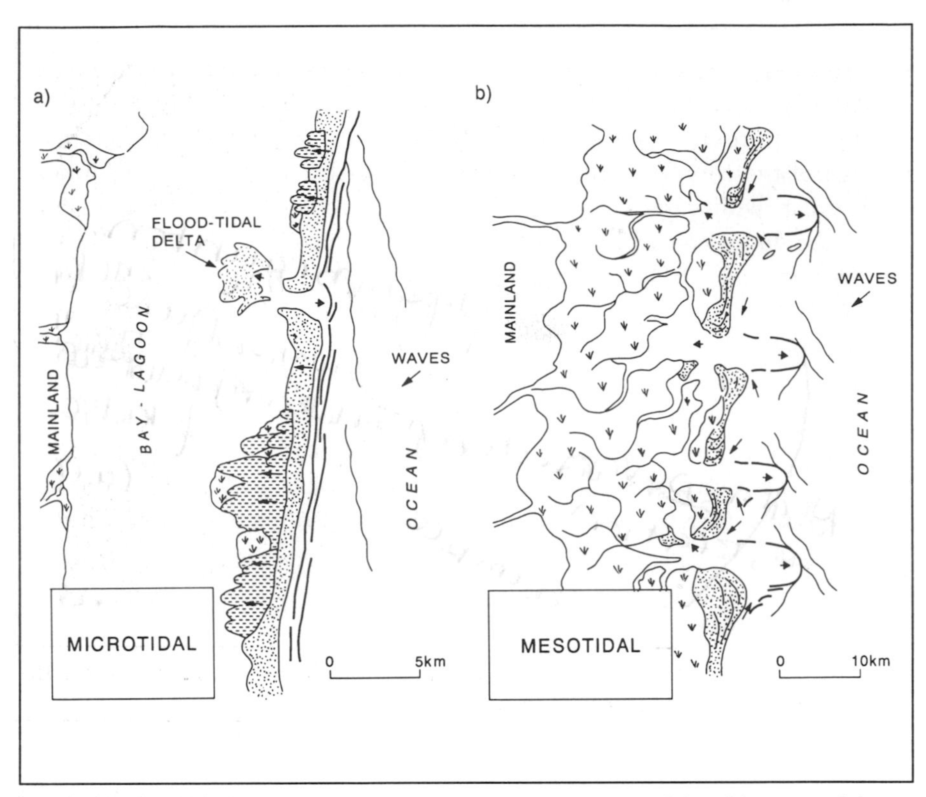

Fig. 3.12 The geomorphology of barrier islands in a) microtidal and b) mesotidal settings. After Hayes (1979)

are rare with low sediment entrainment and subsequent rapid dispersal by incoming waves. The dominant onshore process on wave-dominated barriers is overwash, which overtops the barrier during storm conditions and on particularly high tides. The subsequent deposition of sediment produces a washover accumulation on the lagoon side, whose exact form depends on the interaction between overwash volume and barrier height (Carter, 1988). Such overwash deposits provide substrates for marsh accumulation and also, in the longer term, supply a surface over which the barrier can migrate under conditions of rising sea level.

Where tides dominate the barrier environment (e.g. along the south Georgia coast of the USA, and the Friesian Islands of the Netherlands) barrier islands are relatively short and wide, with a typical 'drumstick' shape (Fig. 3.12b). It is the combination of tidal jets within inlets and the deposition of flood- and ebb-tide deltas, wave refraction of the incident wave-field over tidal shoals, and down-barrier longshore sediment transport which produces this shape of island. The

Plate 3.5 Barrier island morphology and habitats (beach, sand dune and salt marsh) at Scolt Head Island, Norfolk coast, England

inlets are only stable if sediment transport inwards is balanced by downdrift sediment movements. If this sediment transfer is impeded, then updrift accumulation will take place and the inlet will migrate downdrift; conversely, if the transfer is highly efficient and sediment removal exceeds supply, then the updrift side of the inlet will erode and the tidal pass will migrate updrift. In many places, such as on the eastern USA coast, barrier islands show a long term trend of migration inland as a response to rising sea levels, waning sediment supply and changing storm and wave activity. Barrier islands move inland through a combination of overwash and inlet formation (Dolan *et al.*, 1982). On average, for example, barrier islands along the mid-Atlantic coast of the USA have eroded and retreated at a rate of 1.5 m per year, with the lowest rates in the southern part of the coastline. The movement of barrier islands comprises two major processes; progressive shoreface retreat, and during occasional high rates of sea level rise, 'surf-zone skipping', or stepwise retreat (Sanders and Kumar, 1975).

Detailed studies have revealed the nature of barrier island geomorphic change. On Padre Island, Texas, for example, a cycle of erosive activity begins when a hurricane breaches a foredune ridge. This breach gradually 'heals' except at the downwind end where an aeolian chute develops, forming an aeolian fan and dunefield at the back of the island. This sand transport system may last for up to 30 years, before it returns to pre-hurricane conditions (Mathewson and Cole, 1982). A similar cycle has been identified on the southern Louisiana coast, as shown in Fig. 3.13, where a major hurricane occurs every 10–12 years (Ritchie and Penland, 1990).

Integrated barrier island management

As we have seen, barrier islands are dynamic features, which in many parts of the world are migrating onshore as a result of relative sea level rise. They have also been settled and managed. Thus, a survey of 282 barrier islands along the Atlantic and Gulf coasts of the USA found *c.* 70 to be urbanized, 80 were state and local reserves or recreation areas, 15 of the largest were owned by the US federal government for nature reserves of National Seashores, and the other *c.* 120 were privately owned and largely undeveloped (Dolan *et al.*, 1982). These figures mask a very rapid increase in the area covered by urban development, which has increased by over 50 per cent since 1945.

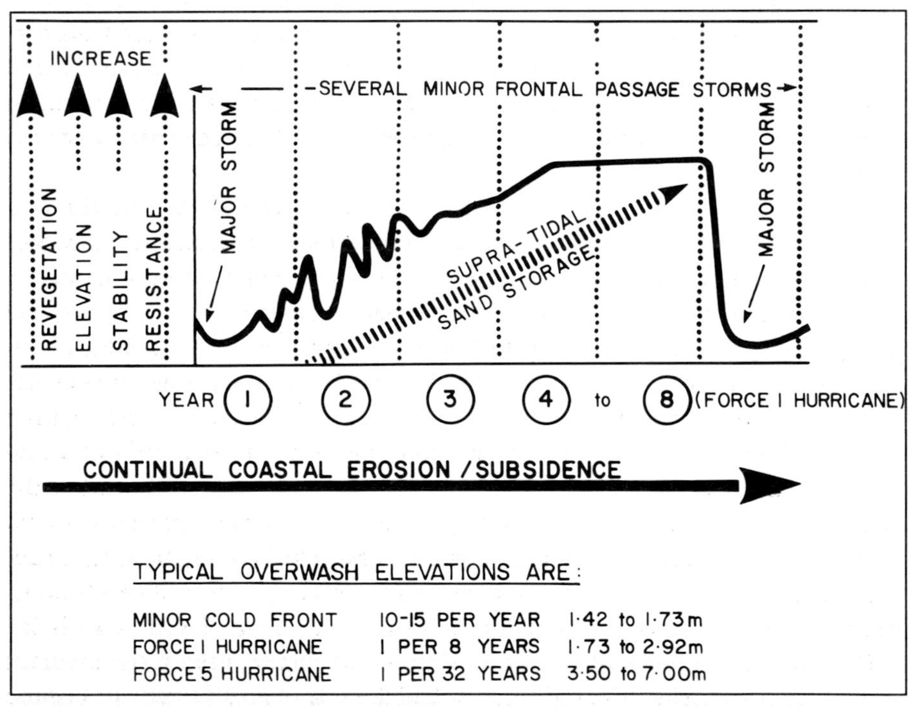

Fig. 3.13 Cyclical changes in south Louisiana barrier coast dunes related to overwash and hurricanes. After Ritchie and Penland (1990)

The regular, ordered urban developments on barriers aim for a permanence which seems at odds with the dynamic nature of these environments. Most early attempts at barrier island management tried to impose such regularity and orderliness on the barrier environment using planted dune lines and/or engineering structures to attempt to 'fix' the island in place. As we have seen, such schemes have often made the protected areas more vulnerable to high magnitude storm events. So how can we improve our understanding of the morphodynamics of barrier islands and feed such knowledge into better management plans?

Clearly, barrier island morphodynamics are related to sediment movements at a range of scales, all of which need to be understood before any manipulation can take place. In the past, mistakes have been made by trying to control small-scale sediment transfers (e.g. over foredunes to the landward side) without considering the medium-scale and large-scale contexts. Thus, the morphodynamics of the eastern US barrier islands are controlled at the larger scale by long term sediment availability and sea level trends. Within this setting, Dolan *et al.* (1982) have identified systematic spatial variations in the distribution of erosion and damage from major storms along the coast, producing zones of consistently high erosion. Within one barrier island system, we can see that sediment movements are integrated with, for example, aeolian sand transport along beaches leading to dune accumulations (as found at Westhampton Beach, New York by Nordstrom *et al.*, 1986). Storm events which appear to lead to 'catastrophic' erosion must be seen as part of this sediment redistribution system.

Considerable research effort has recently been put into elucidating the way barrier islands work, and the exact effects that various management practices have. For example, at Westhampton Beach three areas have been compared; a developed area with groynes; an adjacent developed area without groynes; and an undeveloped sector downdrift to the west (Nordstrom *et al.*, 1986). Beach erosion here has increased rapidly over the past 50 years, possibly in response to the creation and stabilization of two inlets. Groynes, beach fill, dune fill, sand fences and vegetation planting schemes have all been implemented to try and solve the problems. The coast becomes increasingly starved of sediment to the west, because of sediment trapping by groynes. Aeolian sand movements tend to work in the opposite direction from west to east. The investigations here suggest that, for optimal success, sand fences should be built before beach nourishment, and that the beach and dune system should be managed as a whole.

On a rather smaller scale, work on Fire Island, to the west of the previous study, has investigated the influence of buildings and sand fences on onshore sand movements (Table 3.3). The results show that elevated houses at low to medium density trap relatively little sand (and therefore do not interfere much

Table 3.3 Calculated rates of onshore sand transport by wind and quantities of sand trapped at obstructions across a 1 km length of dune crest under different conditions of development

Development status	*Rate*[a]	*Amount*[b]
Without sand fences		
Natural area (100% vacant)	195.6	0.0
Elevated houses, 25% density	168.6	27.0
Elevated houses, 43% density	149.2	46.4
Houses on ground, 25% density	146.7	48.9
Houses on ground, 43% density	111.5	84.1
With sand fences along 100% of crest		
Fences only	111.5	84.1
Fences with elevated houses, 25% density	96.1	99.5
Fences with elevated houses, 43% density	84.9	110.7
Fences, houses on ground, 25% density	83.6	112.0
Fences, houses on ground, 43% density	63.6	132.0

[a]Rate of transport across dune crest, $km^3\ a^{-1}$.
[b]Quantity of sand trapped at man-made obstructions, $km^3\ a^{-1}$
Source: Nordstrom and McCluskey (1985)

with natural flows), whereas houses on the ground and sand fences together trap a large amount of onshore-moving sand (Nordstrom and McCluskey, 1985). Such interference should be minimized if barriers are to be allowed to function more naturally.

These two studies reveal an inherent tension in much modern barrier island management, i.e. between trying not to interfere with natural processes, whilst also trying to encourage natural processes to provide added coastal protection. Clearly, however successful such schemes are on the short to medium timescales, they must bear in mind the overall long term constraint of onshore migration, which is most likely to continue in the future.

Barrier islands and the next century: options in the face of sea level rise

What management options are available if sea level rise does take place as expected over the next 100 years? Over the last decade, the United States Environmental Protection Agency (EPA) has considered a range of options for the management of developed barrier islands. On the assumption that property owners will pay for the bulk of protection costs and will decide which course of action to take within national and state legal frameworks, the EPA have identified four possible courses of action: a no protection strategy, engineered retreat, raising barrier islands *in situ* and encirclement by seawalls (Titus, 1990; Titus *et al.*, 1991; Fig. 3.14).

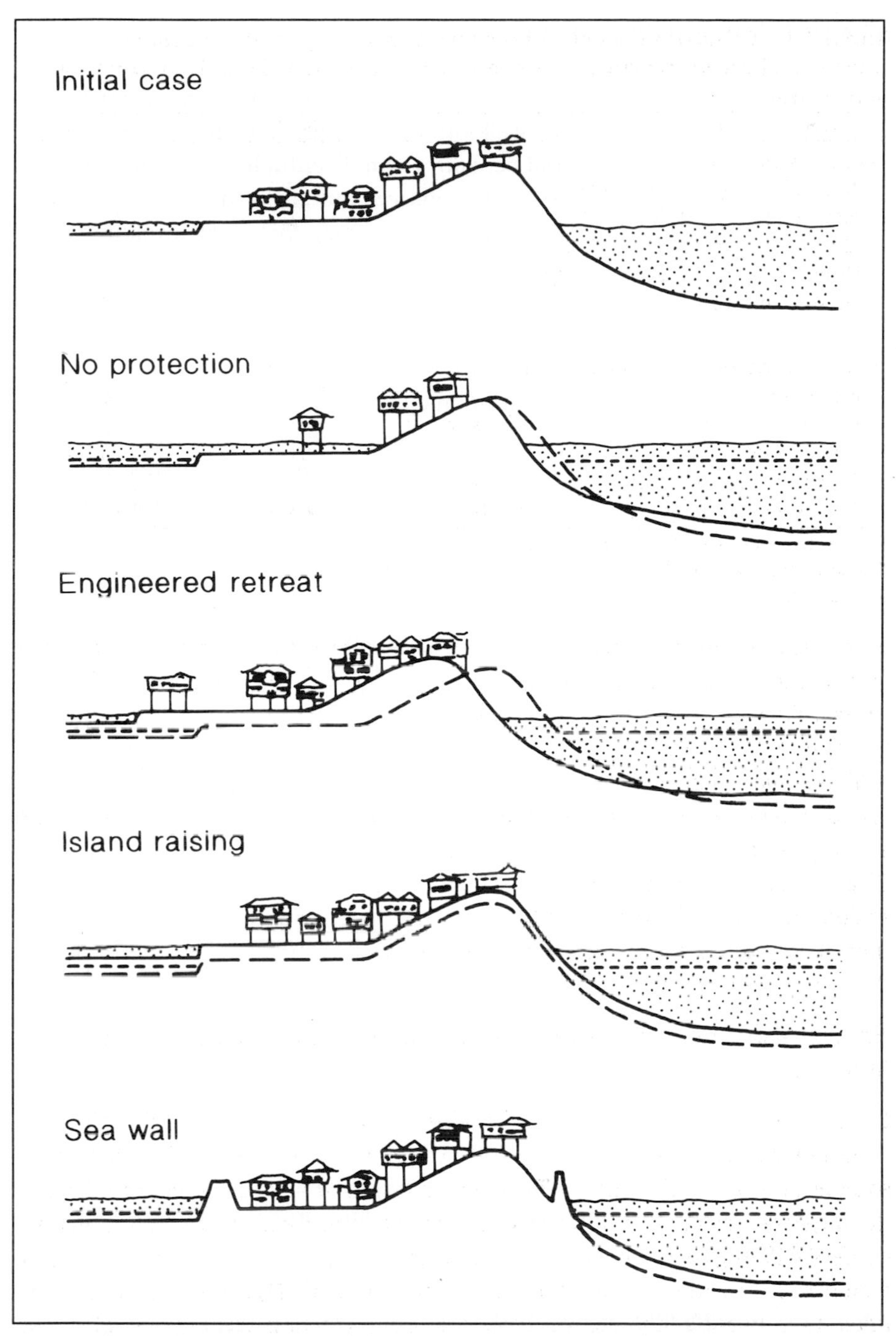

Fig. 3.14 Responses to sea level rise for developed barrier islands. After Titus (1990)

Option 1: Letting nature take its course

Observations and theoretical considerations strongly suggest that sea level rise will be accompanied by the loss of the oceanside beach on barrier islands, with oceanfront properties, built on well-founded pilings, remaining *in situ* as the beach disappears in front of and beneath them. Eventually these buildings will become uninhabitable and have to be removed but given the value of property with a beach frontage most owners will try to maintain their dwellings for as long as possible. This situation is undesirable: a beach is a critical community asset which should not be sacrificed for individual property maintenance and remaining structures may interfere with sediment transfer processes and thus the health of backbeach dune systems (see above). Thus in some US states, Texas and Maine for example, legislation calls for the removal of buildings once they lie seaward of the vegetation line. A better alternative, however, is not to allow the shoreline to reach this stage. This objective can be pursued by one of three approaches: the delimitation of setback zones for new construction, post-flooding bans on property reconstruction, and denial of flood insurance for rebuilt or new structures at the coastal margin.

Set-back zones require a compromise between expected sea level rise and associated shoreline retreat on the one hand and coastal property values and rental income on the other. It is difficult to change inadequate set-backs once development has taken place but very large set-backs remove all incentive to development because of the sharp decline in land values with distance from the sea. To date, post-disaster bans on reconstruction have met with considerable public opposition and the high value of coastal property means that redevelopment may take place even in the absence of insurance cover.

Option 2: Engineering a retreat

Engineering a retreat envisages imitating natural overwash by creating new land, probably by the use of offshore sediment sources, on the bayside of a barrier island as the oceanside erodes. The creation of new land (although probably at the expense of bayfront wetlands) would provide space for the re-organization of barrier island structures, either by the 'leap-frogging' of oceanfront structures to the new bayfront or by the 'shunting' of all properties in a bayward direction. Although this is an imaginative solution to coastal erosion from geomorphological and engineering standpoints, it does raise a series of questions about changing landownership and property rights which may make its adoption difficult.

Option 3: Raising an island in situ

Clearly simply elevating properties *in situ* circumvents the political and legal difficulties of options 1 and 2, which require the mobility of both structures and island inhabitants. Unfortunately, however, these advantages have to be set against the potential costs of such an approach: this option requires much greater volumes of sand than option 2 and is likely to be a much more costly course of action.

Option 4: Encirclement with seawalls

This has been the most common approach to coastal protection for several centuries. It does not require major legal instruments and although some property is lost in the process of seawall construction, physical interference is relatively small. Seawalls lead to problems of beach maintenance and may be aesthetically unattractive but nevertheless this option is often seen as the best compromise from the barrier islander's point of view. However, unlike engineered retreat or *in situ* raising which can be added incrementally as sea level proceeds, initial costs are high, partly as such structures have to be considerably overdesigned to allow for the worst estimate of sea level rise and partly because on long, narrow barrier islands the ratio of perimeter length to area protected is not favourable.

Evaluating the options: the case of Long Beach Island, New Jersey, USA

Long Beach Island, New Jersey, south of New York, is a typical recreational barrier island of the eastern seaboard. The island is 29 km long, 2–4 blocks wide and densely developed with single-storey, individual family houses; few have been elevated above flood levels. Titus (1990) has estimated (i) the volume and costs of sand; (ii) the engineering costs of moving structures and infrastructure and building seawalls and (iii) the value of property that would be lost under a no protection option for a range of sea level rise scenarios to 2100, here reported to an upper figure of 120 cm. Interestingly, most of the sand required is for maintaining the beach, even at high rates of sea level rise (Table 3.4). Total costs by option are given in Table 3.5a. It is clear that once current natural erosion buffers are exceeded (i.e. beyond the first 30 cm of sea level rise) some form of protection is preferable to no protection, even before the preser-

Table 3.4 Sand required for engineered retreat or island raising *in situ*, Long Beach Island

Sea level rise to 2100 (cm)	*Sand volume* ($\times 10^6$ m^3)			
	Engineered retreat	*Island raising* in situ		
		Beach profile	*Bayside*	*Oceanside*
30	4.4	11.9	–	–
60	8.9	23.9	3.5	–
90	13.5	35.9	7.0	–
120	17.9	47.9	10.5	2.4

Source: Titus (1990)

Table 3.5 Cost of sea level rise for four alternative options

(a) Total cost

Sea level rise (cm)	*Levée with beach*	*Raise island*	*Island retreat*	*No protection*
30	52	105	41	55
60	434	285	109	462
90	509	522	178	843
120	584	786	247	1548

(b) Incremental cost

Sea level rise (cm)	*Levée with beach*		*Raise island*	*Migrate*	*No protection*
	Levée	*Sand*			
30	0	52	105	41	55
60	330	52	180	68	407
90	0	75	237	69	381
120	0	75	264	69	705

Note: all costs are in thousands of US dollars
Source: Titus (1990)

vation of rental income that results from options 2–4 is taken into account; this applies also for calculations for the US coastline as a whole (Titus *et al.*, 1991; Table 3.6).

The question, therefore, for the future is what kind of protection should be employed and here an idea of incremental costs is more helpful than cumulative totals (Table 3.5b). These costs show the high capital investment required for seawall (levée) construction; thus only densely developed islands with considerable high-rise developments and high land values (e.g. Miami Beach, Florida; Ocean City, Maryland; Galveston, Texas) will be able to contemplate this course of action. In the short term, existing beach nourishment schemes will probably remain the favoured option. Thereafter, it is clear, as it is already clear on the UK coastline, that engineered retreat becomes an increasingly attractive course of action, particularly for low-rise, relatively low-value dwellings such as on Long Beach Island. Such economic considerations focus the debate on to how migrations on changing landforms might be managed politically and legally (Titus, 1990).

Table 3.6 Costs of coastal defence and no protection strategies for the entire coastline of the coterminous USA for three sea level rise scenarios

	Magnitude of sea level rise to 2100		
	Current sea level trend (12 cm)	*50 cm*	*100 cm*
Cost of coastal defence	4	55–123	143–305
Estimated cost of inundation and erosion	20–51	128–232	270–475

Note: costs are in billions of US dollars (1991 prices)
Source: Titus *et al.* (1991)

Case study 3.1: Erosion problems along the barrier beach coast of Nigeria

The Nigerian coast is dominated by the Niger delta which occupies around 60 per cent of the total 800 km long coastline along the Gulf of Guinea. To the west of this delta is a 200 km long strip of barrier island coastline (Allen, 1965) around Lagos. Altogether, the Nigerian coastal plain reveals a long history of deposition and seaward advance. At depth, quartzose sands are found which represent a marine transgression, and these sediments are overlain by fluvioglacial deposits and littoral sands dating from the late Pleistocene onwards. Thus, the coast appears to have experienced regressive conditions over the entire Holocene.

Along the barrier beach coast there is a low, degraded cliff line some 10 km inland, fronted by a depositional plain covered with parallel trending sandy ridges interspersed with muddy deposits. Some areas of lagoon remain at the landward side of the modern beach ridge, but there has been a general trend towards filling of lagoons over the Holocene. Now, however, the coast is undergoing erosion, perhaps as a response to a recent, small rise in sea level. As Usuro (1985) expresses it: 'Everywhere along the coast sand cliffs characterize the backshore, and beaches are being seriously eroded'.

The coast is a high energy environment, with waves produced by a long fetch across the Atlantic, coupled with predominant south-westerly monsoon winds blowing onshore throughout the year. The beaches are characteristically steep and dominated by plunging breakers. The climate is perennially wet with average annual precipitation of over 3500 mm, and back barrier lagoons are fringed by mangroves such as *Rhizophora racemosa*. The barrier coast sediment regime is complicated because of west to east longshore drift which transports an estimated 500 000–2 000 000 m^3 of sand per year (Oyegun, 1993).

Within this physical setting there has been a long history of settlement, with Yoruba and Edo peoples moving to the area around present-day Lagos in the fifteenth century (Peil, 1991). British occupation and colonial rule began in 1851 and lasted until independence in 1961. Nigeria is now the most populous country in Africa, although estimates of the 1990 populations vary from 88.5 million (from the official census) to 118 million (Taylor, 1993). Lagos itself has a population of over 6 million, and over 60 per cent of all Nigeria's economic activity is concentrated in and around the city. The discovery and exploitation of vast oil reserves in the Niger Delta area brought prosperity to Nigeria in the 1970s and early 1980s, but the economy went into recession in the early 1990s.

The conjunction of a recent, natural trend towards coastal erosion, and a developing economy focused especially upon the port of Lagos has resulted in a serious erosion problem. The port of Lagos is situated where there is a break in the barrier stretching along the coast. As the urban area expanded during the nineteenth and early twentieth centuries around the Lagos lagoon large areas of mangrove swamps were cleared to eradicate disease and also to provide new land. Between 1900 and 1946 nearly 1000 acres of swamp were reclaimed (Mabogunje, 1968). Much of this reclaimed land is prone to subsidence and vulnerable to flooding, especially if the protective barrier beaches are eroded.

Development of Lagos as a major port and capital city necessitated improvement of the harbour which, as part of the barrier coast, had an offshore bar partly obscuring the entrance. Dredging of the harbour began in 1907 and construction started in 1908 of two breakwaters and a training wall to provide a deep and safe entry for ships. These works were completed in 1917. Over 2 million tons of granite were brought some 50 miles from Aro quarries, near Abeokuta, to build the walls. The general west–east longshore drift has been interrupted by the breakwaters (Fig. 3.15), leading to a long term problem of accumulation of sand to the west (on Lighthouse Beach) and erosion to the east (on Victoria Beach). In 1950, erosion led to temporary breakthrough of storm waves into Lagos lagoon round the landward end of the eastern breakwater (Buchanan and Pugh, 1955). Victoria Beach has retreated up to 2 km in places (up to 69 m a^{-1}), and Lighthouse Beach has prograded up to 0.6 km (Ibe, 1988) since the harbour works were started. It has been estimated that around 2.5 km^2 of beach has been lost along this part of the coast since 1900. The beach itself is an important recreation area for the inhabitants of Lagos. Its backshore zone is now built-up, and it also protects the lagoon coast behind from the high energy wave environment.

Buchanan and Pugh suggested in 1955 that:

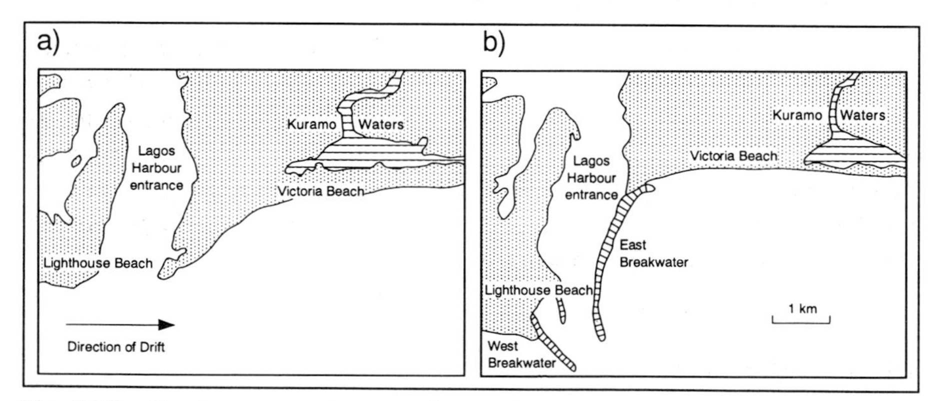

Fig. 3.15 Developments to Lagos Harbour and the erosion of Victoria Beach, Nigeria a) in 1900 and b) in 1988. After Ibe (1988)

> Continued destruction of Victoria Beach will unquestionably lead to a permanent joining up of sea and lagoon east of the present entrance with serious results for the port, and some sort of control is obviously required: short groynes have proved ineffective, and pumping of sand from the west to the east side of the entrance is likely to be the most satisfactory solution. (p. 266)

Since then, several schemes have been implemented which have not proved to be particularly successful. Usuro (1985) reports that beach nourishment began here in 1976 and succeeded in building up the beach in the short term. However, renewed erosion in 1980 led to a dramatic loss of the beach up to the first row of backshore houses. According to Ibe (1988) 2.7 million m^3 of sand were deposited on Victoria Beach between 1980 and 1981 as further nourishment.

The progradation of Lighthouse Beach is also a potential source of problems, as it is likely that sand will soon extend out past the end of the breakwater, creating a new offshore bar system and impeding navigation into the harbour. On the other hand, a reinstated longshore drift would provide much-needed sediment to Victoria Beach. Clearly, long term management of this coastline is needed to ensure that the entrance to the port is kept open, whilst preserving Victoria Beach as much as possible. Any management plans need to be formulated bearing in mind that Nigeria is not a rich country, and further that most past attempts to halt the problem have failed. As Oyegun (1990) puts it:

> Lessons have been learned globally that few engineering responses to shoreline protection have been completely successful, seeing that they are static impositions on an otherwise dynamic environment. What needs to be done ultimately is to design a scheme that provides for a well-nourished beach which itself protects the shoreline.

Long term nourishment coupled with the establishment of structures such as groyne fields to help conserve the new sand will be needed, allied with monitoring to assess the ongoing success of the scheme. Planning strategies, such as establishing a 'set-back' line, will also be necessary to control future development within this eroding coastal zone. There are many Nigerian scientists now working on the sediment regime of the area, and this work will provide an invaluable basis for any future schemes.

Selected references

ALLEN, J.R.L. 1965: Coastal geomorphology of East Nigeria: beach ridge barrier islands and vegetated tidal flats. *Geologie Mijnbouw* 44, 1–21.

BUCHANAN, K.M. and PUGH, J.C. 1955: *Land and people in Nigeria*. London: University of London Press.
IBE, A.C. 1988: Nigeria. In Walker, H.J. (ed.), *Artificial structures and shorelines*. Dordrecht: Kluwer Academic, 287–94.
MABOGUNJE, A.L. 1968: *Urbanization in Nigeria*. London: University of London Press.
OYEGUN, C.U. 1990: The management of coastal zone erosion in Nigeria. *Ocean and Shoreline Management* 14, 215–28.
OYEGUN, C.U. 1993: Land degradation and the coastal environment of Nigeria. *Catena* 20, 215–25.
PEIL, M. 1991: *Lagos*. London: Belhaven Press.
TAYLOR, R.W. (ed.) 1993: *Urban development in Nigeria*. Aldershot: Avebury.
USURO, E.J. 1985: Nigeria. In Bird, E.C.F. and Schwartz, M.L. (eds), *The world's coastline*. New York: Van Nostrand Reinhold, 607–13.

Case study 3.2: Dunes of the Dutch coast

The Dutch dunes are the most extensive coastal dune system in western Europe and are in a relatively natural state, stretching along much of the 400 km Dutch coast. The regime is microtidal to mesotidal. The dunes, except those in the north, are lime-rich. Much of Holland, approximately half in fact, is low-lying, mainly below sea level, and is protected by dykes as well as natural and artificial dunes. Sea level rise here has been 15–30 cm over the last 100 years, with contributions from subsidence of the North Sea basin and human impacts encouraging local subsidence. Since the twelfth century Dutch engineers have been involved in coastal protection and land reclamation. Now over 7000 km^2 of land is reclaimed. Coastal dunes have four main roles in Holland, i.e. coastal protection, a source of public water supply, nature conservation and recreation.

The Dutch coastal dunes have been extensively affected by coastal protection schemes (dune reinforcement and stabilization) and water supply schemes (artificial infiltration using river water). Some areas have also been affected by acid precipitation, such as the Meijendel area, where up to 30 kg N ha^{-1} a^{-1} and 2 kg P ha^{-1} a^{-1} are deposited. Many scientists now feel that the dunes are over-managed and that 'controlled disturbance' is necessary to increase their nature conservation potential.

The Het Zwanenwater area, in the north-west, shows the problems caused by coastal protection works. Here, erosion rates average 1.5 m per year, and over the past 15 years there has been much artificial reconstruction of dune ridges. Such work has led to sand encroachment and damage to wet slack areas. In 1987, 2 million m^3 of dredged sand was placed on the beach as an alternative way of protecting the coast, and the slack area is now managed as a nature reserve (Klomp, 1989). Increasingly, geomorphologists and ecologists have become involved in attempts to reconcile coastal protection and nature conservation imperatives, through more flexible schemes like beach nourishment and the encouragement of dune blowouts (van der Meulen and van der Maarel, 1989).

Attempts are also being made in the Netherlands to ameliorate the impact of dune infiltration works for water supply schemes. Drinking water has been extracted from the groundwater of Dutch coastal dunes for many years, and this has led to the drying out of dune slacks and the creation of brackish groundwater. During the late 1950s infiltration works were built using river water from the Rhine, Meuse and polder canals to recharge the groundwater. This infiltration has produced many changes which have affected the ecology, such as:

- altering natural groundwater level fluctuations (Fig. 3.16)
- changing groundwater flow characteristics
- increasing the nutrient loadings in dune soils
- physical disturbance of dune ecosystems through digging.

In recent years the Dutch government has halted plans to increase the infiltration works and has investigated ways to reduce the ecological impacts (such as increased pre-purification of the infiltrating water). The Stichtung Duinbehoud (The Foundation for Dune Conservation, established in 1977) has been very active in opposing the expansion of infiltration works and in coming up with new management ideas (Salman, 1989).

Finally, the Dutch coastal dunes have also shown the conflicts between nature conservation and recreation pressures. Recreation activities such as hunting have decreased in importance over the years, but many pursuits such as cycling, golf and rambling have the potential to damage dune ecology through trampling. Schemes to encourage access to some parts of the dunes and improve visitor information (the so-called 'nature conservation without barbed wire' schemes) have been ineffectual in protecting vulnerable areas (van der Zande, 1989).

This case study shows how difficult it can be to reconcile different dune functions in one management scheme, and how we need to improve our understanding of natural processes and how they react to disturbance. Problems of over-management and over-stabilization of dunes show that we must not aim to control, rather to work with nature and try a range of solutions over the short and long term to reconcile the requirements of nature and society.

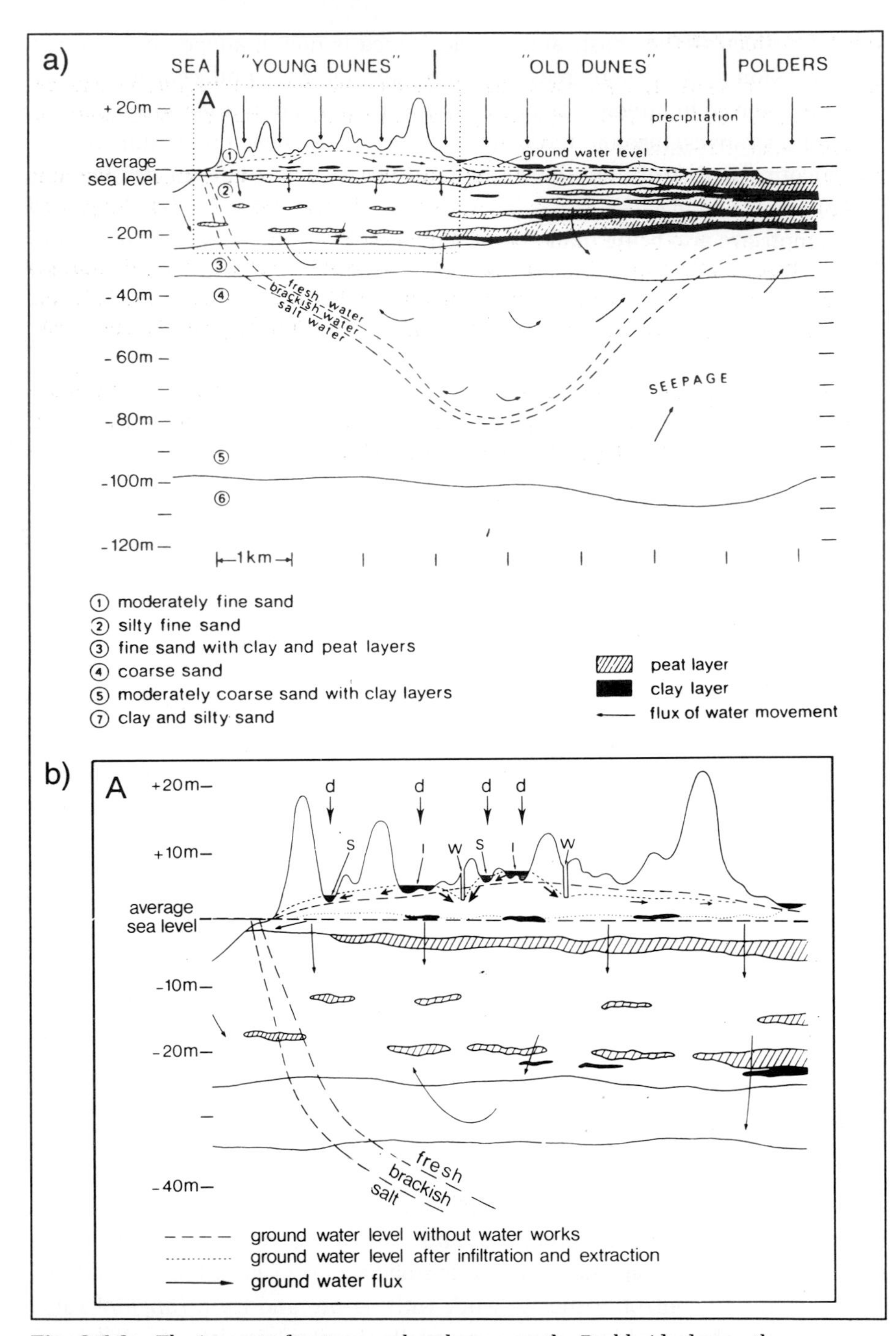

Fig. 3.16 The impact of water supply schemes on the Berkheide dunes, the Netherlands. a) The original water table, b) changes to the 'Young dunes' after infiltration and water extraction. d = depression, s = seepage pool, w = well, i = infiltration pond. Modified from van Dijk (1989)

Selected references

KLOMP, W.H. 1989: Het Zwanenwater: A Dutch dune wetland reserve. In van der Meulen, F., Jungerius, P.D. and Visser, J. (eds), *Perspectives in coastal dune management*. The Hague: SPB Academic, 305–12.

SALMAN, A.H.P.M. 1989: The role of the Foundation for Dune Conservation (Stichtung Duinbehoud). In van der Meulen, F., Jungerius, P.D. and Visser, J. (eds), *Perspectives in coastal dune management*. The Hague: SPB Academic, 239–48.

VAN DER MEULEN, F. and VAN DER MAAREL, E. 1989: Coastal defence alternatives and nature development perspectives. In van der Meulen, F., Jungerius, P.D. and Visser, J. (eds), *Perspectives in coastal dune management*. The Hague: SPB Academic, 183–96.

VAN DER ZANDE, A.N. 1989: Outdoor recreation and dune conservation in the Netherlands. In van der Meulen, F., Jungerius, P.D. and Visser, J. (eds), *Perspectives in coastal dune management*. The Hague: SPB Academic, 207–16.

Chapter Four

ROCKY COASTS: CLIFFS AND PLATFORMS

Introduction

Rocky coasts, which comprise a wide range of forms from steep, plunging granite cliffs, to weakly consolidated coastal slopes formed in Pleistocene deposits, present a number of management challenges. The major issues dealt with in this chapter are geotechnical problems, especially related to cliff recession and failures, pollution, and the response of rocky shores to future climatic changes. Understanding the natural functioning of rocky coast biogeomorphological systems is a vital prerequisite to their successful management, and we review the basic issues below. The determined reader should consult Trenhaile (1987), Carter (1988) and Sunamura (1992) for further details.

Distribution and nature of rocky coasts

Cliffs (defined broadly by Emery and Kuhn, 1982, as 'steep slopes that border ocean coasts') and shore platforms ('intertidal rock surfaces of low slope angle', Goudie, 1994) are usually found intermixed with beaches and other coastal landforms (Plate 4.1). Cliffs occur as headlands between bay beaches, for example, and also as cliff lines at the back of large expanses of beach. Cliffs and rock platforms can, therefore, be regarded as small scale coastal features (third order features of Inman and Nordstrom, 1971). There are some very extensive cliff lines, however, for example the 800 km stretch of 1000 m high cliffs found in arid northern Chile aligned along a fault line (Trenhaile, 1987, p. 178). As they are generally small scale features, it is difficult to establish their global distribution unlike larger features such as deltas and large coastal sand dune plains which are easily identified on air photographs and satellite imagery. Emery and Kuhn (1982) estimate that cliffs (i.e. steep slopes bordering ocean coasts) occur along *c.* 80 per cent of the world's coast, and are found in all latitudes, as

Plate 4.1 Volcanic cliffs, Pitcairn Island, south Pacific Ocean

shown in Fig. 4.1. Shore platforms are similarly common, being found in many parts of the world including Australia, Antarctica, Norway and Singapore. At the national scale the distribution of cliffs and shore platforms may be estimated more easily. For England and Wales, for example, Goudie (1990) shows the distribution of cliffs and platforms and estimates that shore platforms occur on *c.* 20 per cent of the English and Welsh coastline.

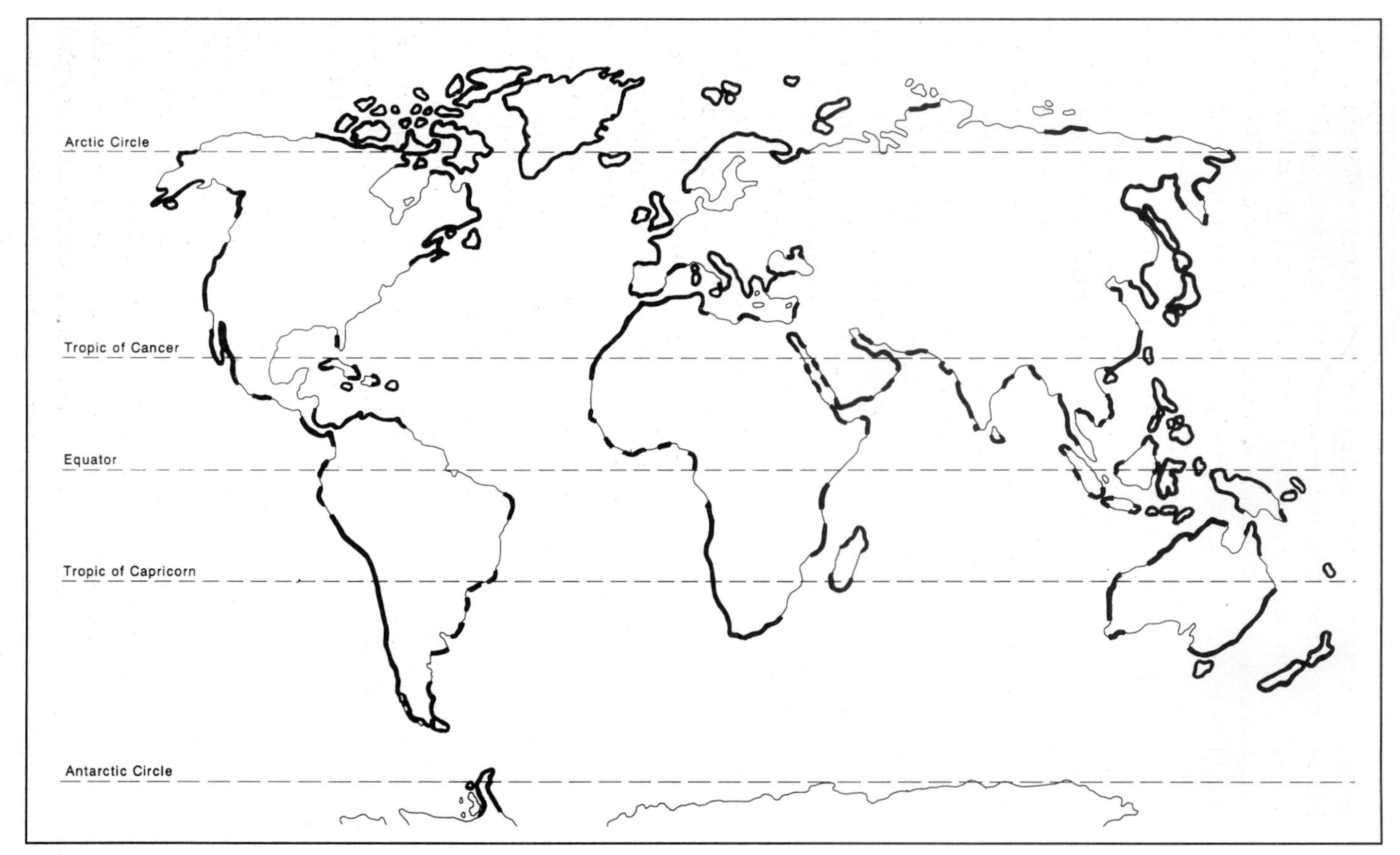

Fig. 4.1 World distribution of rocky coasts. After Emery and Kuhn (1982)

Cliff morphologies vary according to rock or material characteristics, inland topography, tectonics, and present and past erosion regimes (which are in turn controlled by climate, rock type and sea level). Hard rocks with suitable structures are likely to form high, plunging cliffs that are hardly affected by marine erosion. In western Skye, Scotland, for example, cliffs greater than 300 m high have formed in lava flows. Limestone often produces distinctive cliff forms, because of its strength and geochemical properties. At the other end of the scale, cliffs in unconsolidated, or weakly consolidated, material such as clays and glacial deposits are easily and quickly eroded by a combination of marine erosion, sub-aerial weathering and mass movements, often producing complex profiles. Combinations of strata with differing resistance to erosion give characteristic profiles (Fig. 4.2); when combined with variations in strike and dip, complex cliff morphologies result. Furthermore, most cliffs are not cut into simple plateau surfaces, but intersect valley systems produced by fluvial, fluvio-glacial and periglacial processes. Sunamura (1992) has developed a conceptual model of the evolution of rocky coast profiles from five initial shore profiles, for three possible degrees of rock hardness (Fig. 4.3). Assuming microtidal conditions and a stable sea level, a suite of profiles (including cliffs fronted by sloping and almost flat shore platforms, and plunging cliffs) develops according to this model.

It has been argued that marine cliffs are best developed in the high wave energy environments of temperate mid-latitudes. High latitude cliffs often face semi-enclosed seas and wave action is restricted through the year by the role of coastal sea ice. Active glacial and periglacial processes may drape cliff sites in thick mantles of weathered debris. At the other extreme, Tricart (1972) regards true active cliffs as being rare in humid tropical environments, with vegetated cliffs (dominated by terrestrial processes and similar to inland slopes) more characteristic of these zones. In the humid tropics wave energies are usually quite low and shorelines are often protected by offshore reefs. However, several studies dispute this claim. Guilcher (1985) shows that, even in the humid tropics, active cliffs may occur where suitably erodible material is present. He quotes an example of retreating cliffs in Paraiba, north-east Brazil, cut into sandy clays. In Singapore, active cliffs are found on relatively exposed coasts (Swan, 1971).

Tectonic and glacial activity can provide a major influence on cliff development in some parts of the world. Periods of uplift, for example, can produce cliffs with a stepped profile. Faulting often produces spectacular cliff lines, as for example on the Pacific coast of North America. Local subsidence can lead to drowning of the coast and the creation of plunging cliffs. Steep cliffs may form because of glacial erosion and owe little or nothing to marine erosion, as for example many oversteepened fjord-side slopes. In many locations cliff morphology relationships have changed through time and past climatic regimes have

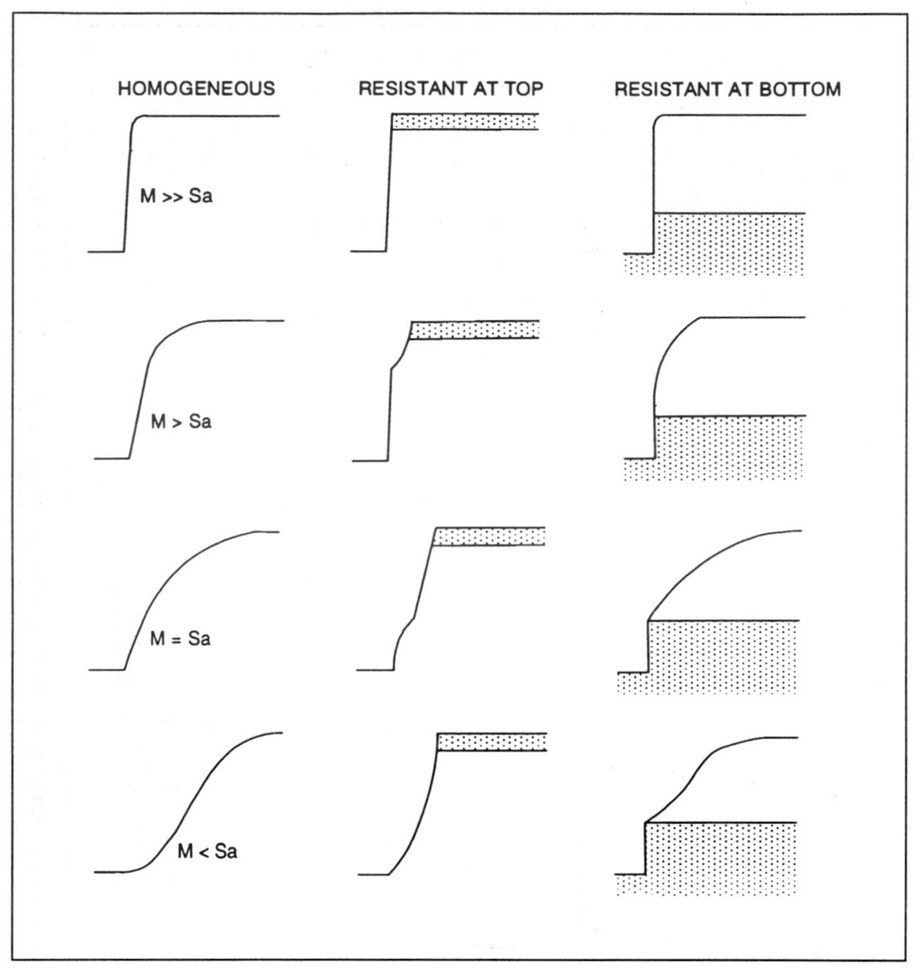

Fig. 4.2 The impact of varying rock strengths on cliff profiles. After Emery and Kuhn (1982)

left a firm imprint on cliff profiles. Dramatic changes in climate and sea level with repeated glacial–interglacial cycles over the last 2–3 million years, with low sea level stands as recently as 18 000 years ago, have produced composite cliff morphologies. During glacial periods, when sea levels were typically over 100 m lower than present sea level and different climatic conditions (usually colder and/or drier) prevailed, slope processes, particularly solifluction processes associated with periglacial climates, were more important than weak marine action. Under interglacial conditions and sea level rise, marine erosion re-asserted its importance against weak subaerial processes on slopes set at low angles by periglacial action. The effect of this change over the last glacial–inter-

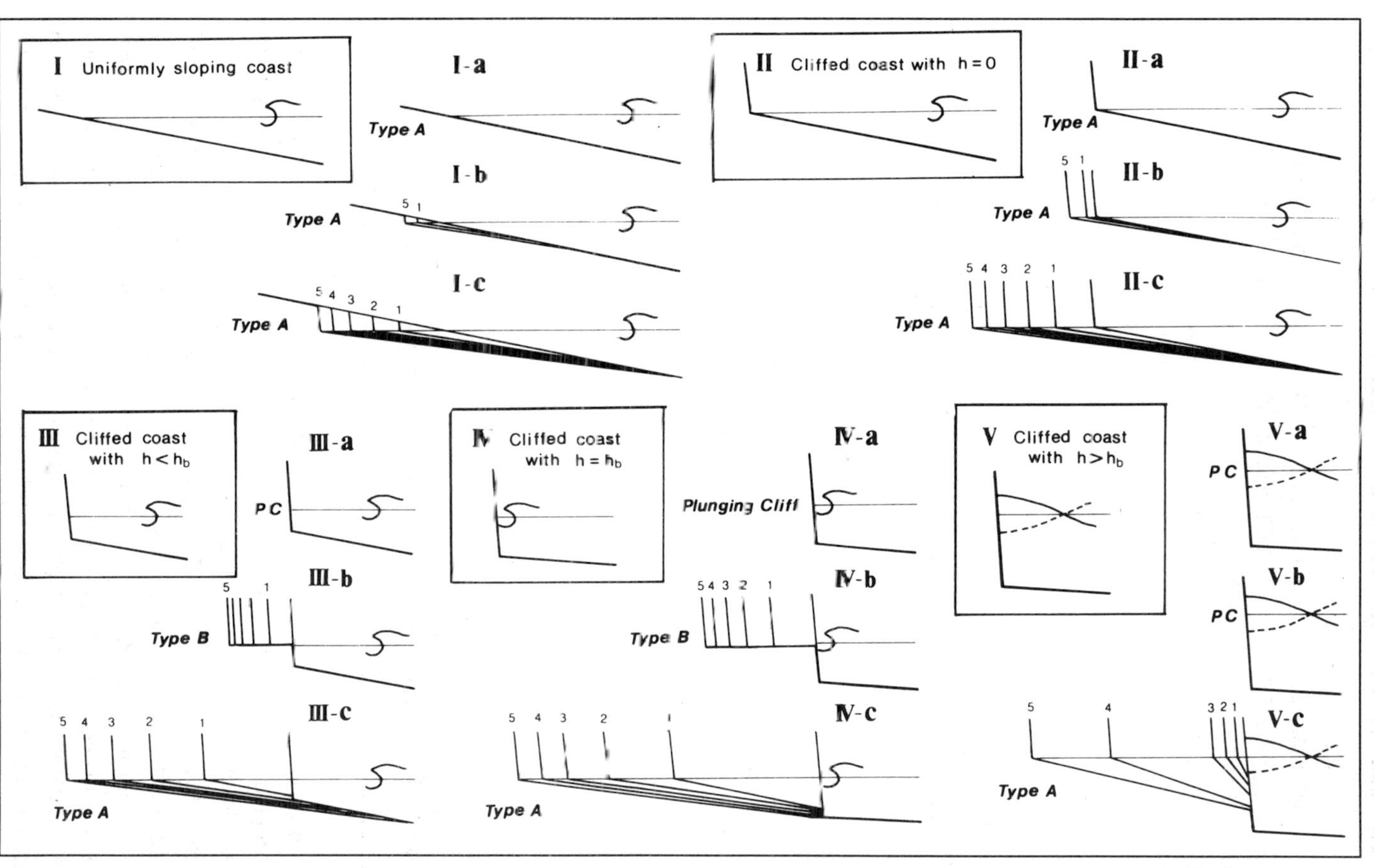

Fig. 4.3 A model of rocky coast evolution based on initial morphology (I–V) and rock hardness (a–c). After Sunamura (1992)

glacial cycle has been to generate typical 'slope-over-wall' cliffs in former ice margin locations (e.g. North Devon, UK; see Pethick, 1984, Fig. 10.4).

This example neatly illustrates the important general principle that cliff morphology is a function of TWO sets of processes, marine erosion AND subaerial erosion and that the form of a particular cliff reflects the balance between these two groups of processes. Taking this line of argument further, Pethick (1984) has suggested that useful explanations of cliff topography and dynamics come from the application of general slope models to coastal cliffs, rather than seeing them as separate phenomena. One relatively successful group of slope models have been 'mass balance' or 'continuity' models. These models consider budgets of sediments passing down the length of the profile, with 'process laws' to describe the rates at which material is transferred into, and then out of, zones on the profile. The behaviour of a series of zones stacked one upon the other then generates a picture of overall slope evolution.

When input is less than output, the profile angle is controlled by the structural characteristics of the materials forming the cliff. Rock fall is a good example of this 'weathering-limited' process as the potential transport is large but the actual rate small, being restricted by the weathering of joint-bounded blocks. Alternatively, where the supply of debris exceeds the capacity for removal then the material builds up into a talus slope whose profile angle relates to the angle of rest of the debris. Simple models (e.g. Fisher, 1866) suggest that as the talus builds up it progressively protects the host rock whose profile develops a parabolic shape, evolving from a vertical cliff to an angle matching the angle of repose of the talus. Both these outcomes are, however, simplistic: on many coastal cliffs both 'weathering-limited' and 'transport-limited' processes operate and interact with one another.

Models which take into account changing relationships between the frictional and cohesive properties of cliff materials, transport mechanisms, slope gradients and the angle at which material will come to rest, show that 'soft' cliffs typically decline in slope angle over time; such profiles fit field data reasonably well (e.g. London Clay cliffs in south-east England). One of the best known examples of cliff evolution is Savigear's (1952) study of a sequence of abandoned cliffs of Old Red Sandstone bedrock between Laugharne and Pendine in south Wales, UK. If one accepts the assumption that the modern spit has developed progressively towards the east, then the most westerly profiles are the oldest, and the most easterly are the youngest, in terms of time since removal of basal marine erosion. Modelling of this sequence (Kirkby, 1984) shows that low rates of subaerial retreat (0.1 mm a^{-1}) give flattened profiles, whereas convexity develops at retreat rates in excess of 0.5 mm a^{-1}.

While these attempts to model changing coastal cliffs are of interest, they can only suggest a smooth transition of cliff morphology through time whereas, in fact, many coastal cliff lines are characterized by high spatial and temporal variability in erosion. Thus while it is possible to show that cliffs in glacial till sediments on the east coast of England have retreated at a rate of *c.* 1 m a^{-1} over the last century (e.g. Cambers, 1976) this figure conceals enormous at-a-site variability. Secondly, most models concentrate upon changing cliff stability/instability measures rather than focusing on the transport processes involved in the movement of materials and the resultant cliff morphologies. Thirdly, model applications have typically concentrated on the evolution of cliffs in the context of the removal of basal cliff erosion and thus the subsequent long-term processes and the long-term loss of strength of materials forming abandoned cliffs. In many cases, however, and often in applied contexts, the need is to predict changes in cliff retreat rate and morphology with the re-application of basal erosion. These questions are better dealt with by the study of particular field case studies later in this chapter.

Hard rock cliffs

Stable cliffs in hard rocks are characterized by small to negligible erosion rates and steep or plunging morphology. Examples include the plunging cliffs in basalt around Banks Peninsula, South Island, New Zealand (described by Cotton, 1951), dolerite cliffs associated with the Whin Sill in Northumberland, northern England, and the granite cliffs in the Seychelles and Singapore. In the latter examples, the presence of deep 'pseudokarstic' fluting on the granite, extending way below present sea level, bears witness to the antiquity of their formation. Often produced by tectonic movements or past processes, hard rock cliffs show little activity in terms of weathering, mass movements, bioerosion and wave erosion today. High cliffs near Cape St Vincent, south Portugal in an exposed situation are almost unnotched (Russell, 1963). Using the comparative data on cliff recession rates collected by Sunamura (1983) we may identify cliffs in granite (commonly eroding at 10^{-3} m a^{-1}), limestone (10^{-3}–10^{-2} m a^{-1}) and flysch/shale (10^{-2} m a^{-1}) as being in our 'hard rock, stable cliff' category. In certain circumstances, however, any of these rock types (especially limestone, flysch and shales) may produce unstable cliffs. We might also add artificial seawalls to this 'hard rock, stable cliff' category — as this is what they are intended to be!

Geomorphological change on hard rock cliffs

Hard rock, stable cliffs are, despite their resilience, affected by both mass movements and basal erosion. Weathering acts to reduce their mechanical strength, paving the way for erosion, and also produces small scale features such as honeycombing and tafoni, which have been ascribed to salt weathering. Salt spray is a potent weapon, travelling inland a great distance on many exposed shores; Mottershead (1989) reports on rapid bedrock lowering on supratidal greenschist in south Devon, England by salt spray weathering. Mustoe (1982) studied the development of honeycomb weathering on coastal arkose sandstone exposures in Puget Sound, USA, and found it developed up to *c.* 3 m above high tide level. High concentrations of soluble salts were found in both rock cavities and the weathered debris, suggesting that salt action may be important. Matsukura and Matsuoka (1991) made similar observations on coastal cliffs in tuffaceous conglomerate on the Boso Peninsula, Japan. Here tafoni and honeycombs, which varied in size from 2–100 cm in diameter and 1–35 cm in depth, were again riddled with salt. However, detailed scanning electron microscopy (SEM) studies by McGreevy (1985) of a honeycombed sandstone cliff from Northern Ireland failed to find any conclusive evidence of salt weathering, although salts were present in pore spaces.

Although deep seated failures are by definition rare on hard rock, rockfalls and topples (*see* Fig. 4.4) may be an important part of profile development. Especially where cliff angles are very high, where the rock is well-jointed, and where cliffs have been undercut at the base, rockfalls occur as, for example, on Lias limestone cliffs in south Wales (Williams and Davies, 1980). Topples are characteristic of rock masses consisting of columns. There are several types of topple, including flexural, block and block flexural types and the secondary types (triggered by other mass movements or the development of tension cracks), i.e. slide head, slide base, slide toe and tension crack types (Trenhaile, 1987). Toppling has been noted from the north Devon cliffs, England, in Carboniferous sandstones and shales.

Several workers have considered the detailed influence of material properties on cliff forms. Compressive strength and jointing frequency and orientation are particularly important factors controlling the nature and frequency of cliff failure and, therefore, profile variations. Allison (1989) investigated cliff changes on the Dorset coast, England in the Portland Limestone, finding that both strength and jointing characteristics were needed to explain the spatial variability of recession rates (Fig. 4.5). In many parts of this coastline recession rates were slow or negligible, although Stair Hole and other areas showed major changes. Three types of mass movement were involved here, i.e. block detachment (where bedding is near-horizontal and cliffs are very stable), wedge failures (dipping beds), and topples (near-vertical beds). Middlemiss (1983), also working in southern England, only this time on chalk cliffs in Kent, investigated the relationship between cliff instability and jointing. These cliffs are retreating at a moderate rate, with jointing the dominant cause of instability, and frost and wave erosion aiding failure.

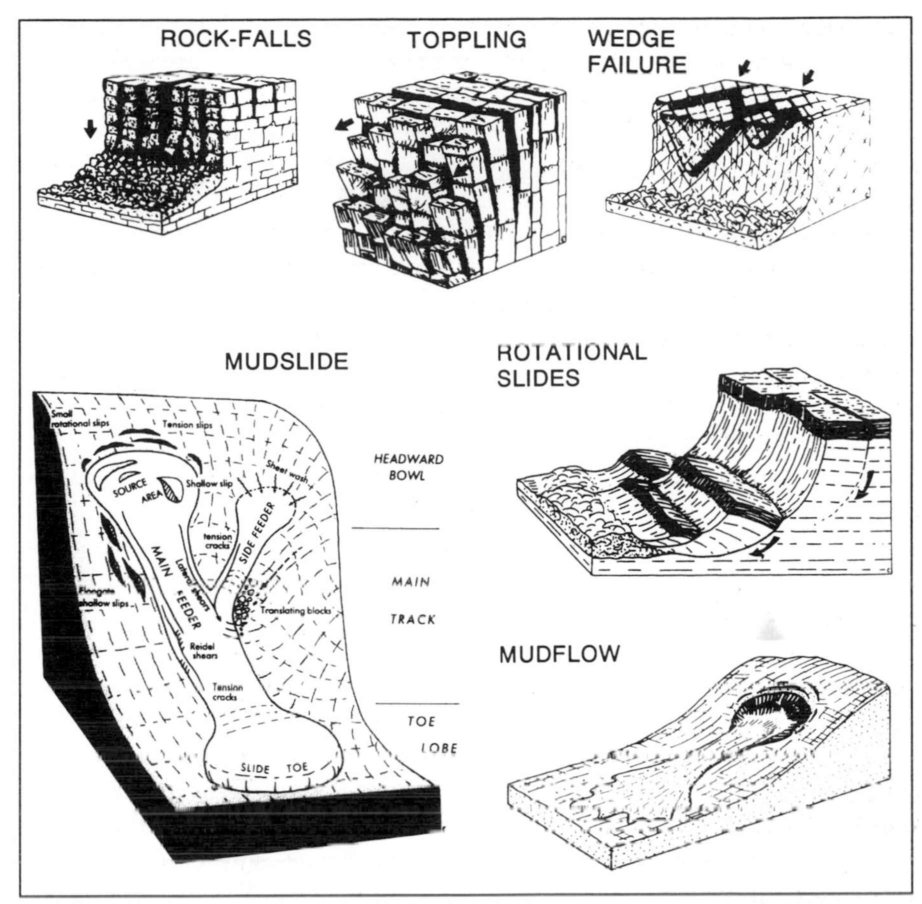

Fig. 4.4 Types of mass movement affecting cliffs. Modified from Allison (1990)

Biological erosion can be a vital feature of hard rock, stable cliffs. Most frequently, biological erosion processes have been observed on limestone cliffs and rock platforms, but other rock types, such as basalt, may be affected. Biological erosion, by boring plants and microorganisms, and boring and abrading animals, occurs dominantly around the intertidal zone. On the other hand, the intertidal zone is also the home of many encrusting organisms (especially along tropical and Mediterranean coasts) which can produce bioconstructional features such as vermetid and algal reefs (see below).

Limestone cliffs are often spectacularly notched at around high water mark; for a long time such features were thought to be the product of dissolution. However, a major problem, particularly on tropical coasts, is that ocean waters appear to be saturated with respect to calcium carbonate. It may be possible for limestone solution to take place at night, or in tidal pools, or peaty mangrove

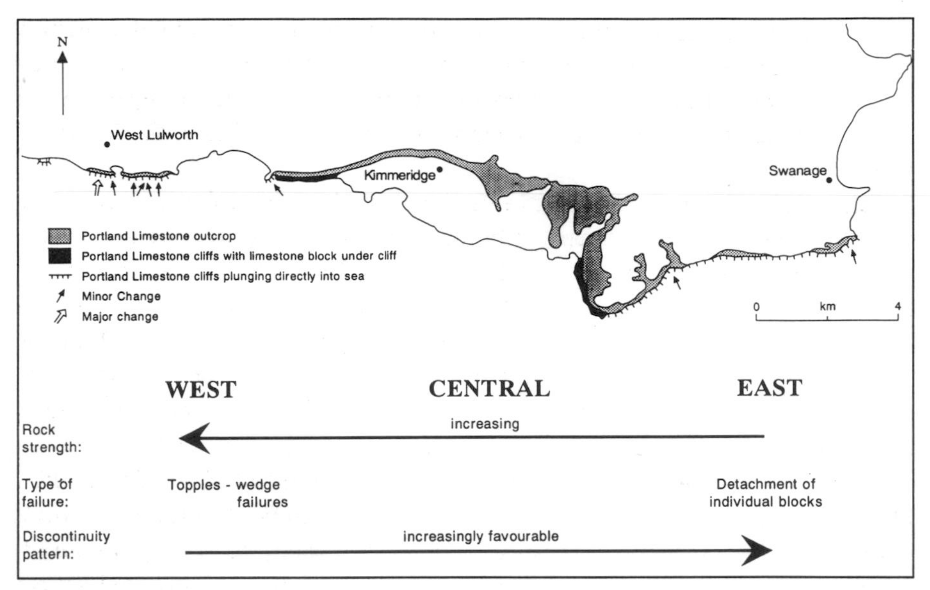

Fig. 4.5 Relating long term (*c.* 100 year) morphological changes on hard rock cliffs, Dorset, England, to fracture patterns and rock strength characteristics. Modified from Allison (1989)

drainage waters (see Trudgill, 1985, for detailed discussion), but other mechanisms of notch formation have also been explored by researchers. Tjia (1985), working in Malaysia, suggested that the smooth surfaces of the abrasional platforms beneath the notches, and the fact that notches are only found on exposed sides imply that abrasion has produced the notch. Bioerosion may also be involved in notch formation (see Spencer, 1988, for a detailed review). The process interrelationships are dealt with later in this chapter. Hodgkin (1970), for example, studied notches in Malaysia (including sites on the wonderfully named Bum Bum Island) and found the rock there to be bored by endolithic algae, and grazed by chitons and browsing molluscs including *Lithophaga lima*. In many cases, bioerosive organisms appear clustered within the area of maximum notch development, suggesting a causal association.

Hard rock cliff ecology

In this section we discuss the ecology of the major part of the cliff profile, excluding the intertidal zone which is dealt with later in this chapter. The main factors controlling the ecology of hard rock, stable cliffs are geology, aspect, salinity, and grazing and competition. In fact, such cliffs, because of their inaccessibility and low economic value, are home to some of the most undisturbed plant and animal communities in the world today. Such inaccessibility, coupled with the relative lack of diversity of communities, has also meant that they have been little studied. There is almost no information, for example, on the ecology of many tropical hard rock cliffs — apart from studies in the intertidal zone.

Salt seems to be a major factor in hard rock, stable cliff ecology. Plants have to be salt-tolerant to live there, but in most cases do not seem to gain any particular advantage from it. Mosses, liverworts and ferns which are in general intolerant of salt, are largely absent from such areas. Certainly, above the fertile intertidal zone, most hard rock cliffs do not seem to be bursting with life — there is normally little or no soil, and conditions on exposed coasts are extreme. Things do grow even on inhospitable surfaces such as these, however. Goldsmith (1975, 1977) has worked extensively on cliff ecology in Britain. At the lowest levels, around sea level, lichens are the dominant life-form. In the middle parts of British cliffs species such as *Armeria maritima* (thrift, or sea pink), *Crithmum maritimum* (rock samphire) and *Festuca rubra* (red fescue) are found. Upper parts are colonized by grasses and inland species. Salinity and aspect seem to be the most important factors controlling the distribution of species here — geology is only important towards the top of the cliffs, and where it provides cracks and ledges lower down. Malloch (1993) describes the vegetation on hard rock cliffs in Britain on a range of rock types, including limestone from Dorset, as shown in Fig. 4.6. In Greece, coastal cliffs on the Sithoniá peninsula in Khalkidhiki contain xerophytic and chasmophytic vegetation, often thorny scrub, whose long roots can penetrate cracks (Lavrentides, 1993). Impressive cliffs, up to 900 m high, occur in calcareous rocks on the southern Turkish coast and are characterized by a low lichen and herb zone, a middle aerosaline zone with xero-halophytic rock-scrub communities, and an upper windswept zone rich in endemics (Lovric and Uslu, 1993).

In the humid tropics many cliffs, including some of the most precipitous, are covered by vegetation except at the very base where salt spray dominates. Tricart (1972), for example, notes the presence of dense shrub, including spiny palmettos, on cliffs at Monrovia and in western Ivory Coast. Where dense vegetation occurs, soil profiles and regolith are produced through weathering. On Aldabra Atoll, low limestone cliffs are covered with endolithic algae and marine organisms at low levels. Algal cover persists right to the cliff top zone, but intense salt spray on exposed coasts prevents all but the hardiest scrub growing on the cliff top. Salt-tolerant bushes (*Pemphis acidula*) take on a krummholz form here because of pruning by salt and wind. Even on sheltered coasts there is a narrow bare zone along the cliff top backed by *Casuarina equisitefolia*, *Cocos nucifera* and other strand-line plants.

Hard rock, stable cliffs are particularly attractive sites for many birds, and in

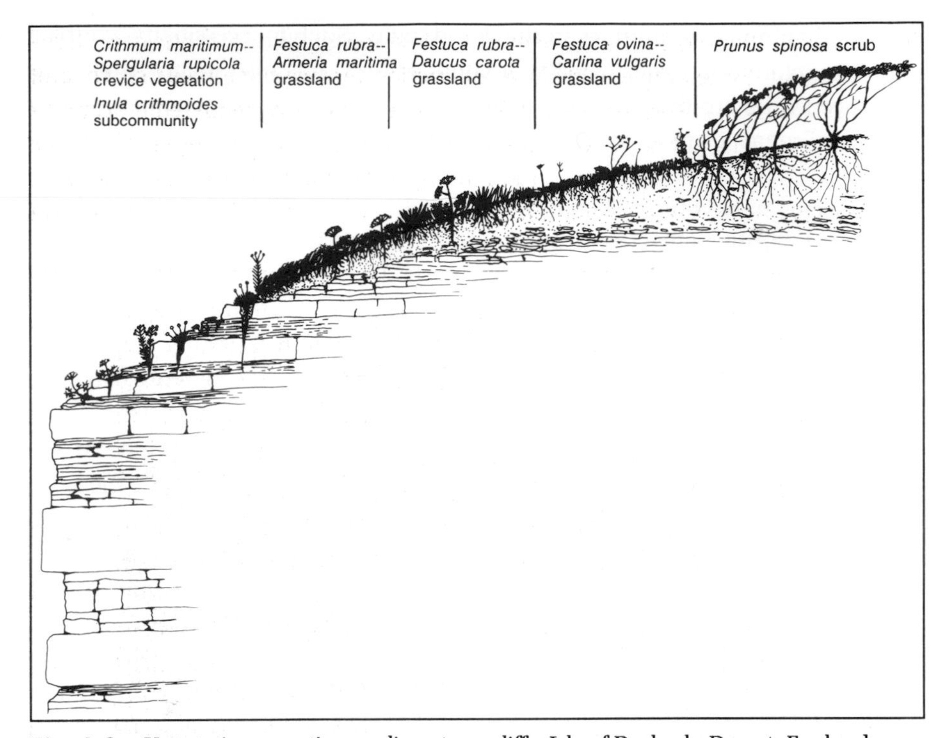

Fig. 4.6 Vegetation zonation on limestone cliffs, Isle of Purbeck, Dorset, England. After Malloch (1993)

turn phosphates and other minerals in guano add to the nutrients available for other species (e.g. lichens). In Britain, Goldsmith (1977) records twenty-four species of sea birds, most of which nest on cliff faces. They are attracted to areas which provide freedom from humans and other predators, favourable currents bringing abundant fish, and suitable ledges for nests. These conditions are found in north and west Scotland and some of the Western Isles. Some land birds also nest on cliff faces and cliff tops.

Cliffs prone to failure

Cliff failure can be a major nuisance and economic problem, although it only rarely leads to loss of life. Two recent examples in England nicely illustrate the impacts. In Scarborough, on the east coast, the Holbeck Hall Hotel collapsed as a result of cliff failure in 1993. During January 1994 cliff failures at Blackgang Chine, on the Isle of Wight, southern England, left five houses perched precariously on the edge. Often, problems are long-standing, deep-seated and very difficult to solve (Plate 4.2). In the Seaton to Downderry area in south-east Cornwall, England, for example, cliffs in a solifluction terrace have failed on several occasions (Sims and Ternan, 1988). In 1974 a cliff failure rendered a property there unsafe, and it had to be demolished. In 1981 cracks appeared in the coastal road along the cliff top, necessitating a large expenditure on a seawall to protect the road from further damage. Carey and Oliver (1918) report on a

Plate 4.2 Cliff erosion near Santa Monica, southern California

cliff failure (locally known as a 'blue slipper') in Essex, England where a field of turnips had slid bodily from the cliff crest to the shore. The turnips were still growing undisturbed.

Cliffs prone to failure are characterized by moderate to high recession rates. They often have complex profiles, produced by a combination of mass movements and marine erosion. Pitts (1986) presents an example from south-east Devon, England of a double undercliff formed by two separate landslides within the 127 m high cliffs. Failure-prone cliffs tend to form in unconsolidated materials (e.g. Pleistocene deposits), clays, sedimentary rocks with a high clay content, e.g. mudstones and shales, porous volcanic rocks, foliated rocks, such as schists, and well-jointed rocks, such as chalks and some weaker limestones. Combinations of permeable and impermeable rocks within a cliff profile lead to especially fragile cliffs, as do massive rocks on top of incompetent materials, and seaward dipping sedimentary rocks in general.

Sunamura (1983) has analysed a huge range of cliff recession data and suggests that rates in failure-prone cliffs are of the following orders of magnitude: 10^{-1}–10^{0} m a^{-1} in chalk and Tertiary sedimentary rocks, 10^{0}–10^{1} m a^{-1} in Quaternary deposits and 10^{1} m a^{-1} in volcanic ejecta. May (1977) partially confirms these estimates in a survey of 160 chalk cliffs facing the English Channel which showed a mean retreat of 0.21 m a^{-1} (method of estimation not given), with a standard deviation of 0.20. However, there are a few problems with comparing such information. First, and perhaps most importantly, cliff recession is by no means a uniform process over time. Long term studies may produce a totally different erosion rate from short term studies at the same site. In the same vein, erosion is highly variable over space and, therefore, results are influenced by the area covered in the study. Second, different methods of obtaining cliff recession estimates may actually be measuring different things. As an example, consider the differences between a comparison of two maps 100 years apart and a 1 year study based on micro-erosion meter measurements (MEM). In the former case, gross change in the cliff geometry is measured; in the latter case, small scale surface recession data are produced and large scale mass movements would not be picked up. Some data should be treated as maximum estimates, and others as minima. Finally, there are very different (and usually unstated) error terms associated with the different techniques. Methods vary from the highly precise (e.g. MEM for which error terms are usually given) to the highly dubious (e.g. anecdotal evidence of the 'well, it was about this big' type).

Geomorphological change — recession and cliff failure

In essence, the geomorphology of active cliffs is controlled by the interplay of subaerial processes (mass movements of various types, and hydrological processes) and wave erosion. As we will discover, such interactions can be very complex and highly variable over space and time.

Mass movements, as shown in Fig. 4.4, include rockfalls, topples, flows and landslides (including mudslides, translational slides and rotational slides). Rockfalls and topples have been discussed above, and it is simply necessary here to point out that they can be equally important on suitable weak cliffs and may trigger off other mass movements further down the cliff profile. Flows involve movement of material with a high liquid content, and may be divided according to the sediment size into debris flows (coarse) and mud flows (fine). They will only occur under exceptional rainfall and groundwater conditions.

Landslides are deep-seated failures which occur when the compressive strength of a rock is exceeded by the load on it. This may occur when material is added (possibly from an upslope mass movement), or the slope angle is steepened through undercutting, or when changes in moisture content of the mass or weathering processes reduce its compressive strength. Translational slips involve movement, often structurally controlled, along a straight plane. Examples are found near Downderry, south-east Cornwall, England (Sims and Ternan, 1988). Rotational slides fail along a deep-seated, concave upwards surface which may be complicated by joints and other discontinuities. Spectacular examples are found along the Devon and Dorset coast in the south-west of England, including Fairy Dell at Stonebarrow Hill (Brunsden and Jones, 1976). A large Late Holocene rotational slip has also been identified just south of Adelaide, Australia, which probably moved at least 300 000 m^3 of material (Bourman and May, 1984). Mudslides are a type of landslide occurring when fine grained sediment slides relatively slowly on a shear surface producing a lobate form. Mudslides are particularly important on the south Devon and Dorset coast, England, forming spectacular features such as Black Ven.

Cliff failures are often triggered by a change in moisture conditions, and/or increased wave action in storms. McIntire and Walker (1964) observed coastal changes in Mauritius after the passage of two cyclones (Alix and Carol) in 1960. Most cliffs showed little change, except along the south coast. Here, lower durable lava beds underlie badly weathered lavas with a high moisture content, issuing out as springs on the cliff face. Storm waves during the cyclones weakened the lower parts of these overlying beds, removed the debris and caused slumping in the upper sections. In extreme cases recession rates of nearly 7 m were noted. Landslides may also be triggered by earthquakes, as in Santa Cruz County, California following the 1989 event (Plant and Griggs, 1990).

Process interrelationships are seen clearly at Stonebarrow Hill in Dorset, England where clays and clay shales are overlain by more competent and well-drained cap rocks. Three morphodynamic zones have been identified (Brunsden and Jones, 1976; Rudkin, 1992). The 40–50 m high sea cliffs are overlooked by a large 'amphitheatre', 1.5 km long, nearly 1 km deep and 85 m high, known as Fairy Dell. Associated with this amphitheatre are three rotational slip blocks: a degraded lower block dating from before the first Ordnance Survey map of 1887 which is thought to have moved 80 m since its formation, a block related to a cliff failure event in 1942 which has since migrated 28 m, and a smaller block activated in 1968 and 1984 (Fig. 4.7). Following rotational sliding, debris from the scar backwall infills behind the tilted block and promotes ponding; this in turn lubricates further movement of the block. As rotation proceeds, so the head of the slope becomes increasingly unsupported and this increases the chance of further failure. This upper platform teminates with the 'undercliff' which leads into the lower platform. Shallow mudslides and mudflows feed from the margin of the rotational slips into feeder tracks and accumulation lobes. The mudslides move along basal and internal shear planes on slopes as low as 3–4° as a result of loading from the upper slopes.

Erosion pin measurements show that there are four scales of movement in the mudslides: random movements, stick-slip movements clustered in time, which typically give 0.08 m of movement from six events in 30 hours as pore water pressures are released; graded movements of *c.* 3.5 m over 17 hours as restraining cohesion is lost; and surge events where the whole mudslide may move 3 m in 20 minutes (Allison and Brunsden, 1990). These movements are seasonal, occurring typically in early and mid-winter when pore water pressures are high and pore salt concentrations reduced (salts contribute to the cohesive component of shear strength) by winter storms. Mudflow debris is ultimately shunted over, and aids the collapse of, the 50 m high sea cliff which fronts the complex. Although mass movements here take place episodically over space and time the whole system is in balance: marine erosion of the sea cliff removes basal debris and then leads to cliff retreat; these processes promote increased rates of debris transport from the undercliff area and then failure of the undercliff itself; this allows the rotational slides to move forward which leads to further failure along the increasingly unsupported wall of the upper amphitheatre.

Such systems are gigantic conveyor belts. At Stonebarrow Hill over the period 1887–1969, it has been estimated that 2200 m^3 m^{-1} width a^{-1} has been translated from the upper landslide scar, compared with an estimate of 1800–2300 m^3 m^{-1} width a^{-1} of material transferred in the basal sea cliff. Controls on, and variations in, rates of basal erosion are crucial to the overall behaviour of such systems. Controlling factors can be grouped into geological structure and stratigraphy, and nearshore processes. Where the strike of the rocks is at right angles to the coast, the key control is the type of material at sea level. Where competent strata are found at sea level, with clays above the zone

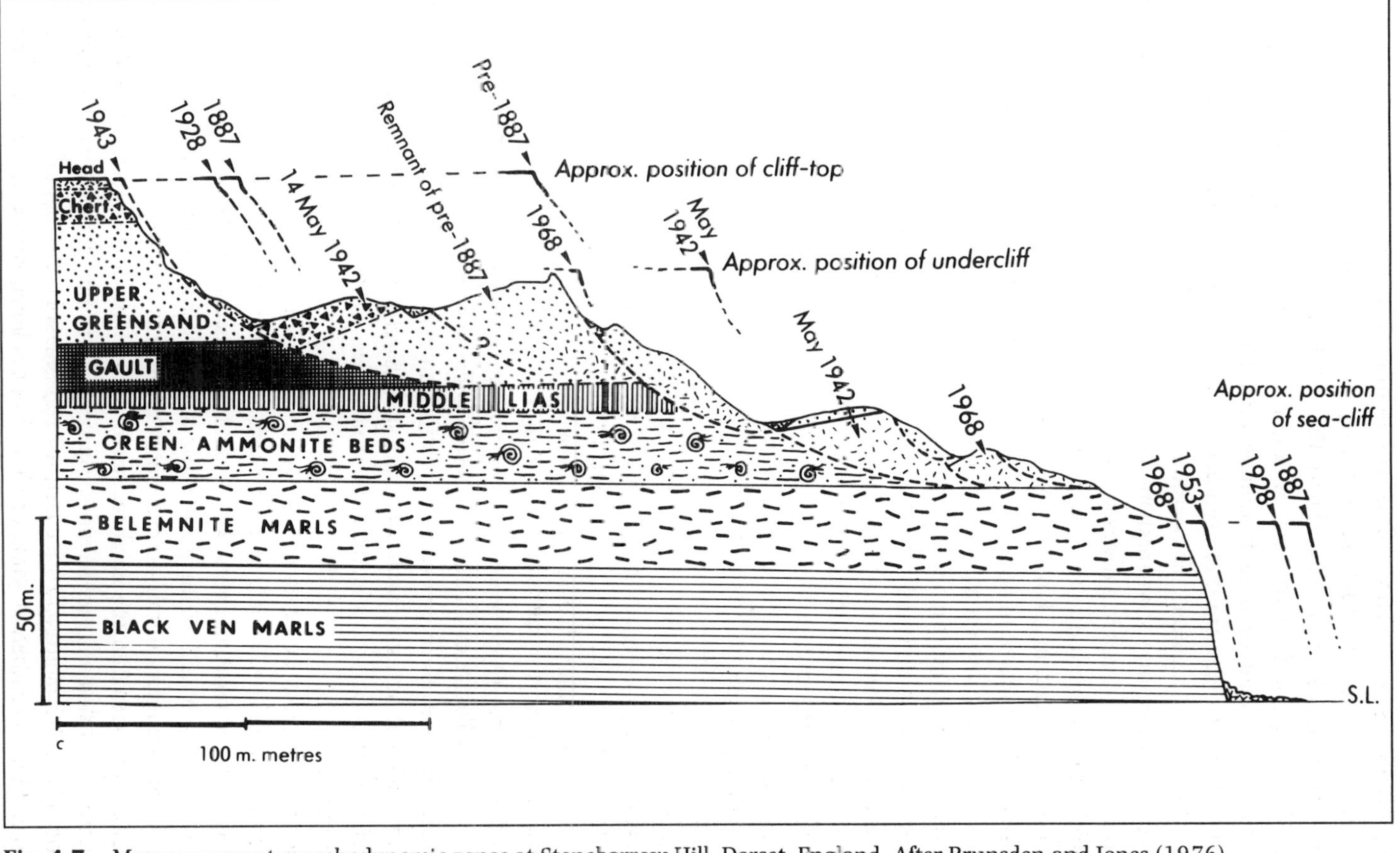

Fig. 4.7 Mass movement morphodynamic zcnes at Stonebarrow Hill, Dorset, England. After Brunsden and Jones (1976)

of wave influence then water is able to drain from the clays and erosion is slow. Where the clay base is 1–3 m above sea level then moderate cliff foot erosion promotes mudslide processes. Where the clay base is below sea level rapid cliff recession takes place and promotes deep rotational sliding (Hutchinson, 1983).

Nearshore sediment transport affects cliff retreat. Thus when beach material becomes depleted in front of a vulnerable cliff, this leads to an acceleration of cliff recession. However, the position is complicated by the fact that the cliffs themselves feed into the nearshore sediment system. Thus in Dorset, cliffs are capped by gravel-bearing deposits which are brought into the cliff system, and ultimately transferred to the beach, by rotational failures. There are long lags within these systems, however: Bray (1992) has calculated that the typical input of gravel has been 4650 $m^3 a^{-1}$ over the period 1901–88, but this will rise to 5600 $m^3 a^{-1}$ when the gravels released by the rapid retreat of the Black Ven complex since 1957 finally reach the beach. These gravels will then feed into the longshore drift within the area (Fig. 4.8). A further complication arises, however, in that mudslide surges onto the beach construct natural groynes which impede movement of beach material, producing cliff undercutting down-drift. Along the Holderness coast in eastern England, variations in cliff retreat (Fig. 4.9) can be explained partly by the great lateral and vertical variability in the glacial tills making up the cliffs, but also partly by variations in basal wave attack. The alongshore migration of features known as 'ords' (Pringle, 1985) is accompanied by the movement of cliff foot channels which expose erodible tills from beneath protective beach deposits.

Temporal variability in basal wave erosion in storms can control subsequent cliff failures, as demonstrated along the Lake Erie coast. Here, surveys every 2 weeks for up to 5 years, showed that bluff foot erosion occurred during north-easterly storms which produce surges of up to 1 m along this low energy coast. A threshold water level (when the beach became submerged) was necessary to cause erosion, which proceeded through quarrying and abrasion (Carter and Guy, 1988). Such temporal variability in cliff retreat may be reflected in long term trends related to variations in storminess. Kuhn and Shepard (1984) in their fascinating study of the coast of San Diego County, California show how between the late 1940s and 1977 mild winters led to relatively slow rates of erosion on unconsolidated alluvium cliffs.

Spatial and temporal variability of cliff failures related to a range of controlling factors are well exemplified by the Oregon coast on the Pacific side of the USA. Here there is a long term tendency towards cliff erosion, which is especially pronounced in the northern sector where low cliffs have developed in a range of uplifted marine terraces. Byrne (1964) has shown that landsliding activity occurs along *c.* 25 per cent of the northern Oregon coast, with rotational slumps and block glides common on sedimentary rock and unconsoli-

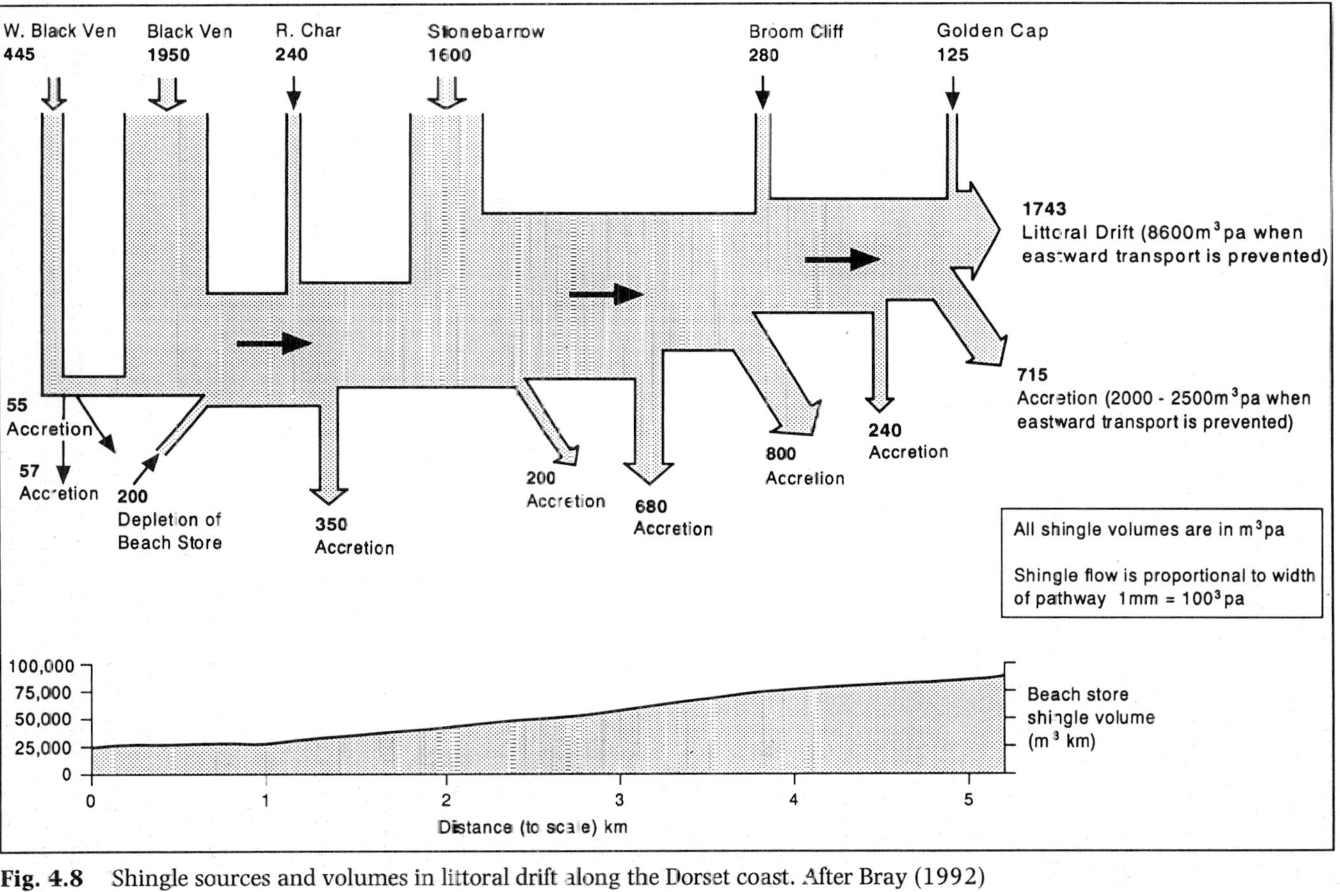

Fig. 4.8 Shingle sources and volumes in littoral drift along the Dorset coast. After Bray (1992)

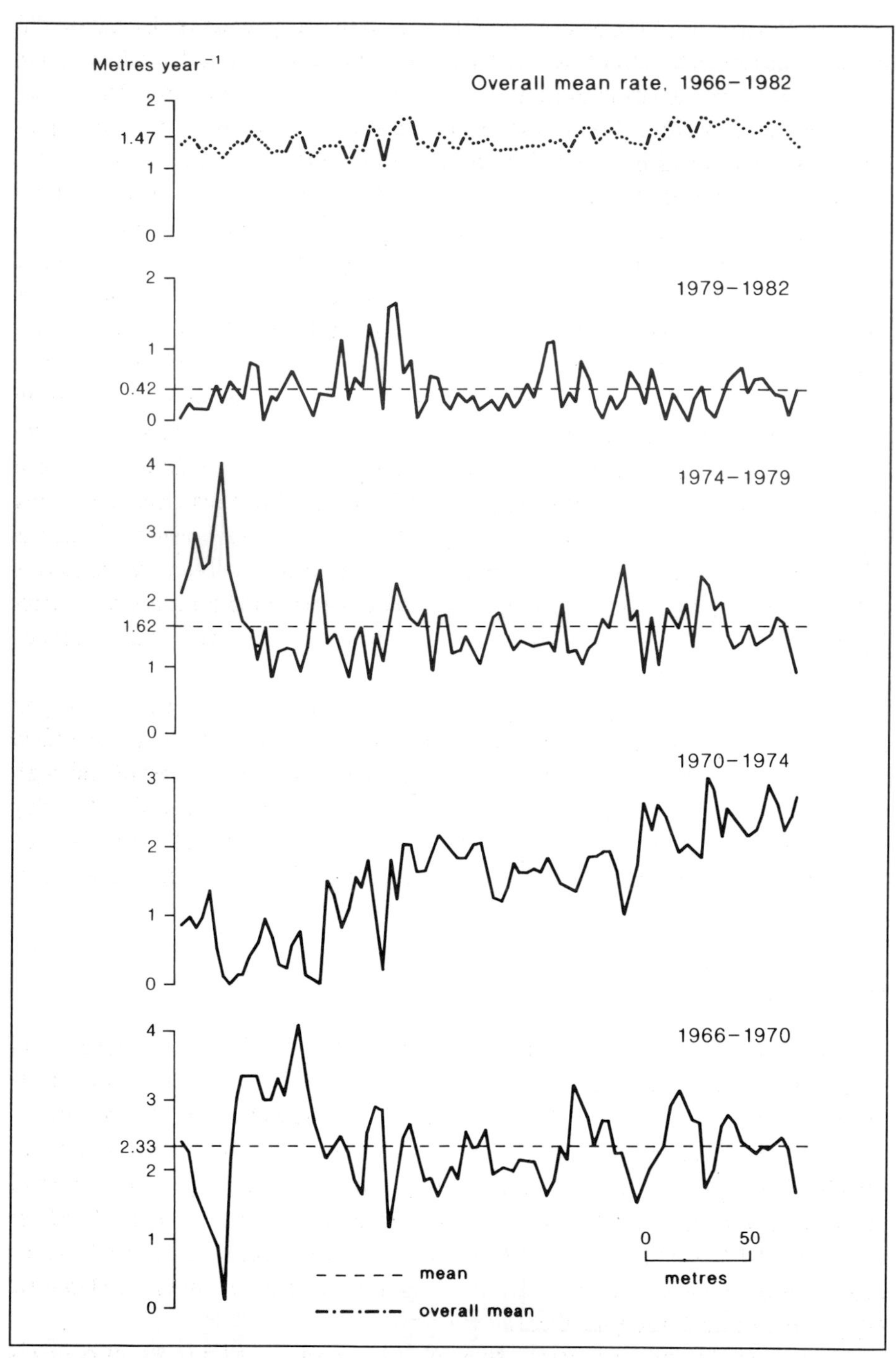

Fig. 4.9 Variations in cliff retreat in space and time, Holderness coast, eastern England. Modified from Richards and Lorriman (1987)

dated sediments, and rockslides and falls common on igneous rocks. These different styles of cliff retreat, as well as more continuous smaller failures, are caused by a variable combination of groundwater seepage and direct rainwash, and basal wave attack. Komar and Shih (1993) explain the highly variable pattern of sea cliff erosion along this coast in terms of a first order factor and a range of second order factors. The overall control is variability in tectonic uplift with the southern half rising faster then eustatic sea level, and the northern half becoming submerged, complicated by periods of abrupt subsidence (Fig. 4.10). In the southern part wave attack is limited and those cliff failures that occur are caused by groundwater seepage and other subaerial processes. On top of the general tectonic imprint, variability in beach conditions and cliff materials adds further explanation. Eight littoral cells have been recognized in the northern sector of the Oregon coast. In some of these the beach has protected the cliffs from major erosion, but the Lincoln City and Beverley Beach cells are in the zone of (albeit small) sea level rise, and have experienced locally high rates of cliff recession and failure. In the Beverley Beach cell the beach is relatively small and provides only limited protection to the Tertiary mudstone cliffs, which are relatively erodible. In the Newport area the influence of geological structure is seen, with steeply seaward-dipping mudstone cliffs particularly prone to failure.

Superimposed on the general spatial pattern of variability along the Oregon coast there are temporal fluctuations. The 1982–3 El Niño event (as described more fully in Chapter 2) affected the Oregon coast with severe storms and high water levels, coupled with enhanced northward sand transport. In the Beverley Beach cell the beach was denuded of sand in the south and cliff erosion was facilitated, whereas the northern end was protected by the accumulation of sand.

Ecology of cliffs prone to failure

The ecology of weak cliffs is controlled largely by the rapidity and nature of cliff recession. Where mass movements occur frequently, vegetation has little chance to gain a foothold. According to May (1977), most plant life on unstable cliffs is derived from cliff-top vegetation, or vegetation on nearby more stable cliff faces. On clay cliffs on the Isle of Wight where cliff top retreat rates of more than 1 m a^{-1} have been recorded, bare areas are colonized by *Tussilago farfara* (coltsfoot) and *Equisetum telmateia* (great horsetail). More stable areas are colonized by a wider range of species including orchids, marsh helleborine (*Epipactis palustra*) and pale flax (*Linum bienne*).

Apart from stability, salt spray, geology and aspect are all important controls of ecology on weak cliffs. As with hard cliff ecology, there seem to have been

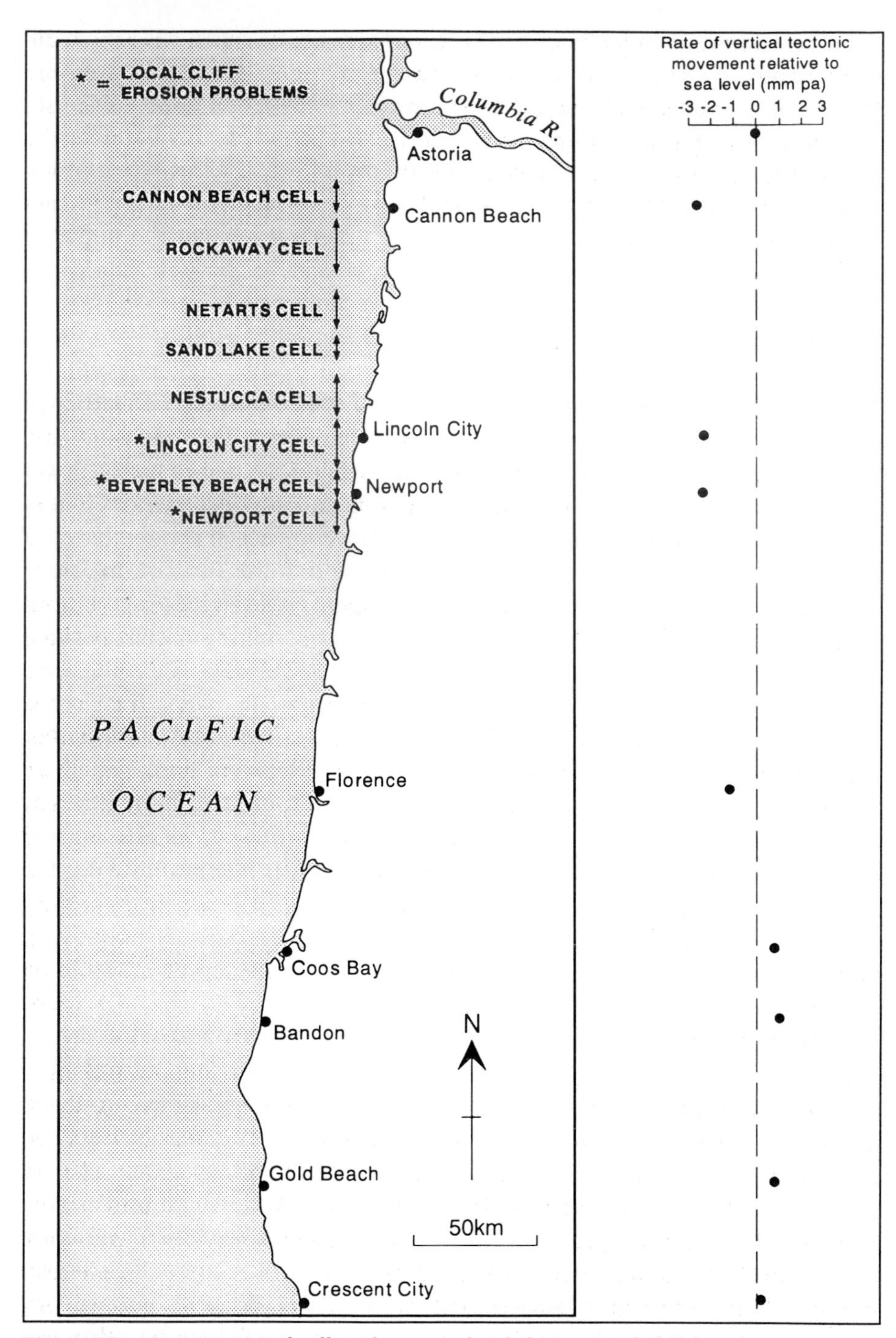

Fig. 4.10 Oregon coastal cells, relative sea level changes and cliff erosion. Modified from Komar and Shih (1993)

few detailed studies. In the case of cliffs prone to instability, there has been some consideration of the possibility of using vegetation to stabilize slopes. Carey and Oliver (1918), for example, note that coltsfoot (*Tussilago farfara*) infests the cliffs at Cromer, UK, but its rhizomes have 'no appreciable effect in stemming the flow of the viscous clay as it makes its way in sluggish cataracts to the foreshore.' (Carey and Oliver, 1918, p. 213). They regard the great horsetail (*Equisetum telmateia*) as being similarly impotent.

Shore platforms

Trenhaile (1987) and Sunamura (1992) make a simple distinction (illustrated in Fig. 4.11) between two main types of shore platform, that is gently sloping platforms (sometimes called ramps) of 1–5° slope, stretching from the cliff/platform junction to low tide level (Type A), and subhorizontal platforms, found in the supra-, inter- or subtidal zone, and generally terminating in a low tide level cliff or ramp (Type B). Sloping platforms are characteristically found in macrotidal environments and subhorizontal ones in meso- and microtidal environments. The nature of the profile appears to be governed by the cross-shore variation in wave force versus rock strength. In simple terms, shore platforms are produced from cliffs by a range of processes which erode cliffs and plane the surface of the platform. There has been much debate, especially in the earlier literature, on how platforms are formed. Clearly, as with most landforms, there are a variety of processes and combinations of processes acting in different areas to produce similar forms, depending on climate, wave environments and rock types (Plate 4.3). Theories of shore platform development all involve the action of one or more of the following processes: chemical weathering, wetting and drying, salt weathering, frost and ice weathering, biological weathering and erosion, wave erosion and mass movements. The theories vary according to which processes dominate and at what level the erosion is focused (e.g. water level, saturation level).

Shore platforms are found on a wide range of rocks, including (according to Trenhaile, 1987, p. 217) weathered granites, some metamorphic rocks, chalk, aeolianites, limestone and particularly alternations of sandstones, siltstones and limestones with mudstones and shales. Most unweathered igneous rocks are too resistant to develop platforms. Swan (1971) found shore platforms in the Singapore Islands developed in granite, quartzites, conglomerates, sandstones, shales, phyllites and tuffs. Mean platform gradient here is nearly 6°. Some relationships have been found between platform width and gradient, and geology and wave environment, but such relationships are complicated by other factors, such as inheritance from the past. Measurements of shore platform erosion

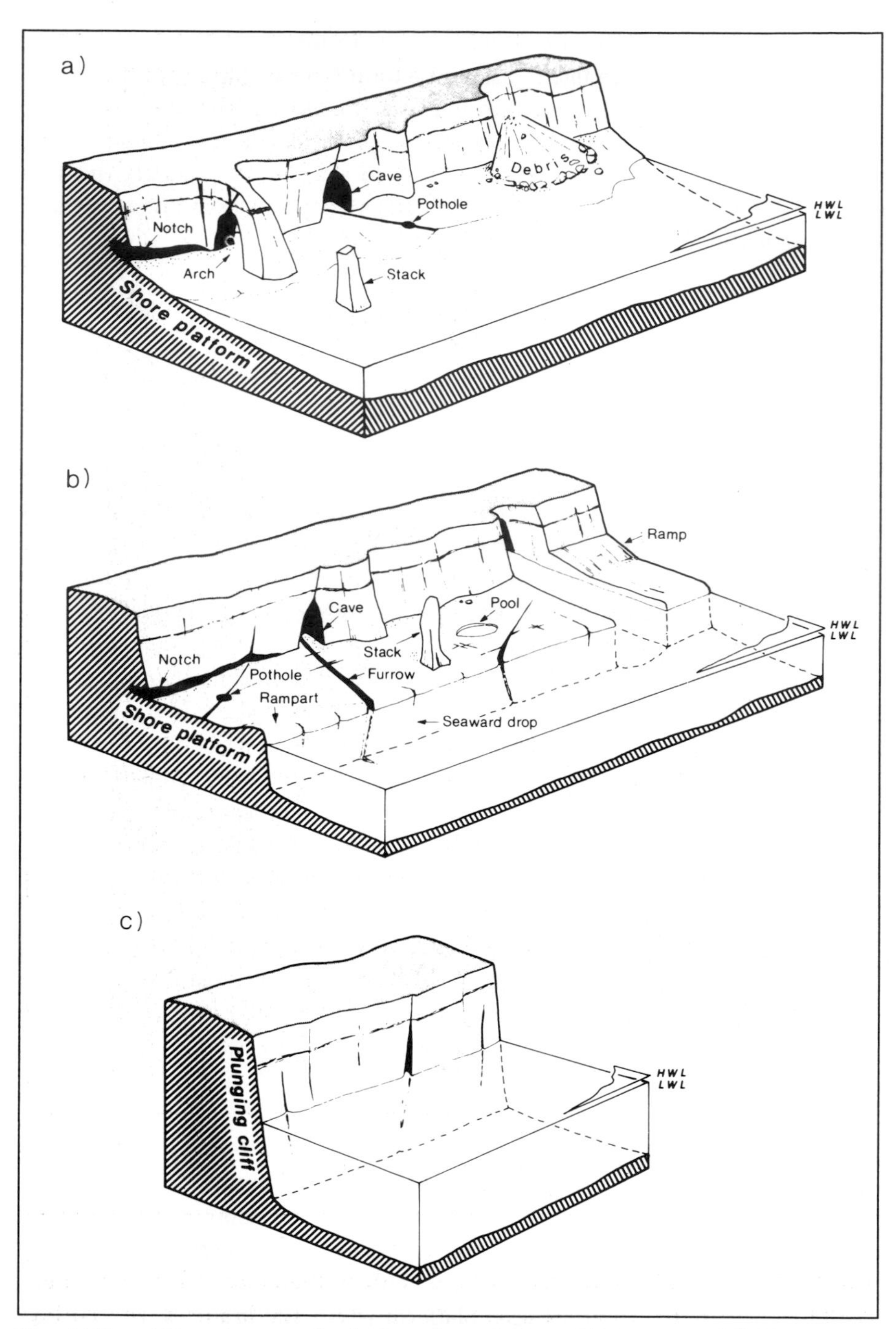

Fig. 4.11 a) Type A, or gently sloping, shore platform, b) Type B, or cliffed/ramped, shore platform, c) plunging cliff profile. After Sunamura (1992)

Plate 4.3 Intertidal platform and notch, Niue Island, south-west Pacific Ocean. The large dimensions of this notch may reflect structural control and/or the incorporation of notches from higher Holocene sea levels

exhibit incompatibility between different studies, as found for cliff erosion rates, depending on the methods used, and whether backwearing or downwearing was being measured. According to Trenhaile (1987) rates generally vary between 0.1 and 35 mm a^{-1}. Such rates are comparable to the cliff erosion rates on limestone and granite presented by Sunamura (1983), i.e. 10^{-3} m a^{-1} for granite and 10^{-3}–10^{-2} m a^{-1} for limestone. This is as might be expected, given that limestone and granite cliffs tend to erode through weathering and marine abrasion, rather than through mass movements. They therefore share the same major sculpturing processes with shore platforms.

On many rocky coasts, predominantly those composed of limestone and those found in tropical and Mediterranean areas, biologically constructed features are important components of platform morphology. Vermetids and calcareous algae produce rimmed pools and small reef-like forms. 'Plates-formes a vasques' is a term of French origin (Guilcher, 1953) used to describe a suite of wide, flat-bottomed pools, forming shallow steps in the intertidal zone. Although the rims between the pools are often made by encrustations of algae and vermetids, they may also be residual features made of platform rock (as noted in southern Madagascar by Battistini, 1981). Thus, they are not always a biologically produced feature. Other French terms, i.e. 'corniche' and 'trottoir', are used to describe ledge-like, or pavement-like features which occur at or around sea level. Characteristically, these too are produced by encrusting organisms growing out from a steeply sloping cliff or platform edge. In other cases they are primarily erosional features colonized by a thin layer of encrusting organisms. The role of encrusting and boring in the production of such platform relief is discussed more fully later in this chapter.

Shore platforms are often characterized by a diverse microtopography as well as the large 'plates-formes a vasques', corniches, trottoirs and notches. On the southern Madagascan coast, for example, a clear zonation is found on platforms cut into calcareous aeolianite. Lapiés occur on the alternately wetted and dried intertidal zone, and flat-bottomed pools in the permanently wet zone. Blue-green algal weathering occurs only in the lapiés zone (Battistini, 1981). Along the Mediterranean coast near Marseilles, there are two similar zones on limestone shores, with an intensely dissected, dark brown supratidal relief produced by biological weathering, and a smooth, yellow zone in the lower intertidal (Le Campion-Alsumard, 1979). Kelletat (1980) has observed coastal limestone features in north-east Mallorca and shows one profile from Cala Guya (*see* Fig. 4.12) which shows an upper, relatively smooth zone, with a deeply dissected zone full of rock pools below, which ends in a notch form which reaches down to mean sea level. Just below the intertidal zone are a whole suite of encrusting organisms and boring organisms producing a small scale, complex topography. Ley's studies of marine karren around the Bristol Channel coast, south-west England show the importance of the amount of wave energy expended in the intertidal zone to the relief (Ley 1977, 1979). Here relief around the high tide mark is pitted with some shallow pools. At or just below mid-tide level the relief is at its maximum, and around low water level the plat-

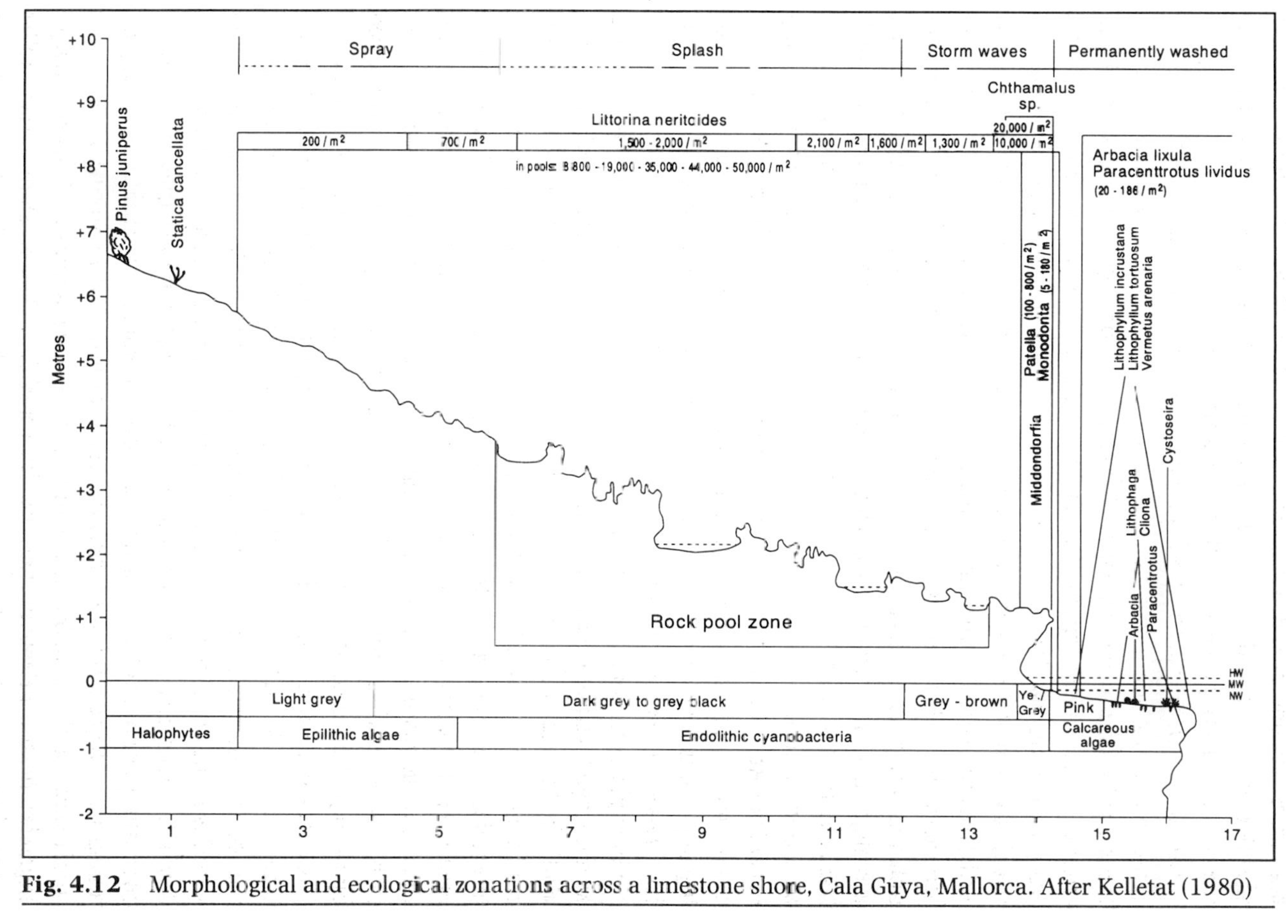

Fig. 4.12 Morphological and ecological zonations across a limestone shore, Cala Guya, Mallorca. After Kelletat (1980)

form surface is generally flat and smooth. Guilcher (1953, quoted in Trenhaile, 1987, p. 241) suggests that the small scale morphology found on platforms and the lower sections of cliffs in limestone areas varies according to temperature and tidal regime.

Geomorphological change on shore platforms

As the preceding sections have illustrated, it is difficult to make a clear separation between platforms and cliffs in many cases — especially in limestone areas. In many ways it is more realistic to talk about the intertidal and immediately adjacent areas regardless of the particular slope angle found there, as the same processes dominate in this zone. The major processes at work shaping the near-tidal areas are weathering (especially through salt and boring microorganisms, and frost in suitable climates), abrasion, bioerosion and encrustation. Biological weathering of near-tidal rocks is of major importance, and some very high rates have been recorded. Blue-green algae, for example, have been found to colonize fresh intertidal surfaces in a matter of days (Le Campion-Alsumard, 1975) and to be capable of boring up to 900 μm into the surface creating a dense network of boreholes which significantly weaken the surface (Lukas, 1979; Schneider and Torunski, 1983). As discussed, there has been debate over the ability of sea water to dissolve limestone. Work by Trudgill along the Irish coast has shown that nearshore waters are usually only undersaturated with respect to calcium carbonate, and therefore able to dissolve limestone, at night and in intertidal pools, and that such chemical weathering is vastly overshadowed by a suite of biological processes (Trudgill, 1987). Salt weathering is an extremely potent force in the splash and spray zones. Alternate wetting and drying allow crystallization to occur with consequent stress effects on the rock surface; hydration may also produce physical stress, and salt solutions may chemically weather many rocks.

Salt weathering can act in a potent combination with frost action. On a chalk shore platform in southern England, with an average gradient $< 5°$, observations following very cold conditions in January 1985 showed widespread cracking and spalling of the upper parts of the platform (Robinson and Jerwood, 1987). Laboratory simulations showed just how efficacious these linked processes could be (Jerwood *et al.*, 1990a and b). Abrasion is undoubtedly important in some rocky near-tidal areas, especially where there is a nearby source of abrading material and high wave energies to transport such material. There are few studies of abrasion in action although Robinson (1977) reports on abrasion of rock platforms in Yorkshire, England. Tjia (1985) uses morphological evidence (smooth surfaces) to infer the action of abrasion, but shows no process observations.

Bioerosion is often spectacular on rocky coasts, especially those composed of limestone. Bioerosional processes may be split into two main types, that is boring (whereby the organism creates a tunnel or hole, often as a dwelling-place) and grazing (whereby the animal produces shallow scars on the rock surface by grazing on plants growing in and on the surface). A huge range of organisms participate in such processes and there have been many studies of the rates at which they bore and graze, leading to estimates of overall erosion rates. The key point about bioerosional processes for our purposes is that they provide a link between the biological and geomorphological worlds. The zoned organisms create an imprint on the rock surface, which therefore tends partially to reflect this zonation. It is not only a morphological relationship, but also a functional one. Organisms participate in erosion, and the rock becomes involved in nutrient cycling. A sudden change in ecology or geomorphology will thus have knock-on consequences for the other components.

Encrusting organisms are those that actively form, or act as passive nuclei for, cement which binds them together to form a relatively hard, but often highly porous, substrate. Calcareous algae, for example, actively precipitate calcium carbonate to form a sort of external 'skeleton'. Vermetids produce tubes which are then often cemented together by calcareous algae. Such encrustations act to protect the underlying rock from weathering and erosion. However, complications arise where encrusting organisms are found side by side with boring organisms. Some organisms, such as some sponges, may be capable of both encrusting and boring.

Geomorphological change has been related to structural controls and zonation of bioerosive processes on a Carboniferous limestone foreshore in County Clare, Ireland by Trudgill (1987). In this area the structural arrangement of benches (which is itself determined by glacial history and geological structure) controls the vertical biological zonation. Thus, there is a clear interdependence between geomorphological form and biological and inorganic processes. Where there are extensive subhorizontal benches in the intertidal zone, there is more space for *Hiatella*, *Paracentrotus* and *Cliona* to colonize and create characteristic bioerosive microlandforms. In general here, boring algae (presumably cyanobacteria) dominate in the mid-intertidal, with boring lichens in the upper intertidal and boring sponges in the lower intertidal. Trudgill's work suggests a hierarchical model of platform development, with structure providing macro-scale controls, interacting with biological zonation to produce the meso-scale platform geomorphology.

Ecology of shore platforms

The diversity and importance of near-tidal ecology on rocky coasts has been hinted at in the preceding sections. Such ecology, in terms of species present, their zonation, and their interactions, has been far better studied than those of cliffs (or, to be more precise, those on the upper areas of cliffs above the splash zone). The major controls on the species present on any one coastal profile are sea and air temperature, water currents, exposure, shore topography, substrate, aspect and tidal range. Thus the exact composition of intertidal communities varies with climate, geology and wave environment. Despite variation in the species found on rocky shores, all such shores are characterized by a zonation down the coastal profile. In recent years, biotic factors have received more and more attention as being a major control on the exact nature of zonation, and being responsible for variations over time. Barnes and Hughes (1988) suggest that for species where the factors that influence zonation have been elucidated it is physiological factors (such as tolerance of desiccation) which fix the upper zonal limit and ecological and behavioural factors (such as competition and predation) which fix the lower limits.

It is vital at this stage of the discussion to establish a useful framework for describing shore ecology. As with many parts of the coastal system, geomorphologists and ecologists view the coast in a slightly different way. In geomorphological terms rocky shores are divided up according to tidal levels, thus producing the supratidal zone, the intertidal zone and the subtidal zone. In most ecological studies very different criteria are used to divide up organisms on the coastal profile in terms of a littoral zone (*see* Fig. 4.13). Thus, Lewis (1964, p. 49) states:

> The littoral zone is here defined entirely in biological terms, and is quite distinct from the 'shore' or the 'intertidal zone' which are physical entities and with which its limits rarely if ever coincide.

This immediately has important ramifications for geomorphological studies of bioerosion and bioconstruction which attempt to relate two things (zonation of coastal topography and zonation of coastal organisms). If the two systems are very different, one might expect problems. We may go on to postulate, however, that in areas where bioerosion and bioconstruction are important elements of the geomorphological system, then the biological and geomorphological zonations will tend to coincide. What is also clear is that the biologically defined 'littoral zone' is a more fluid, changing entity than the hydrologically defined intertidal zone. As Little and Mettam (1994, p. 181) put it 'the distribution of species across a shore may be thought of as a shifting mosaic of species superimposed on a general pattern of zonation'. This implies that any bioerosive or bioconstructive impact on coastal geomorphology may suffer some fluctuations unrelated to those affecting physical processes.

The upper part of the littoral zone is called the littoral fringe, or supralittoral zone. Generally this zone is dominated by lichens and *Littorina* spp., except in

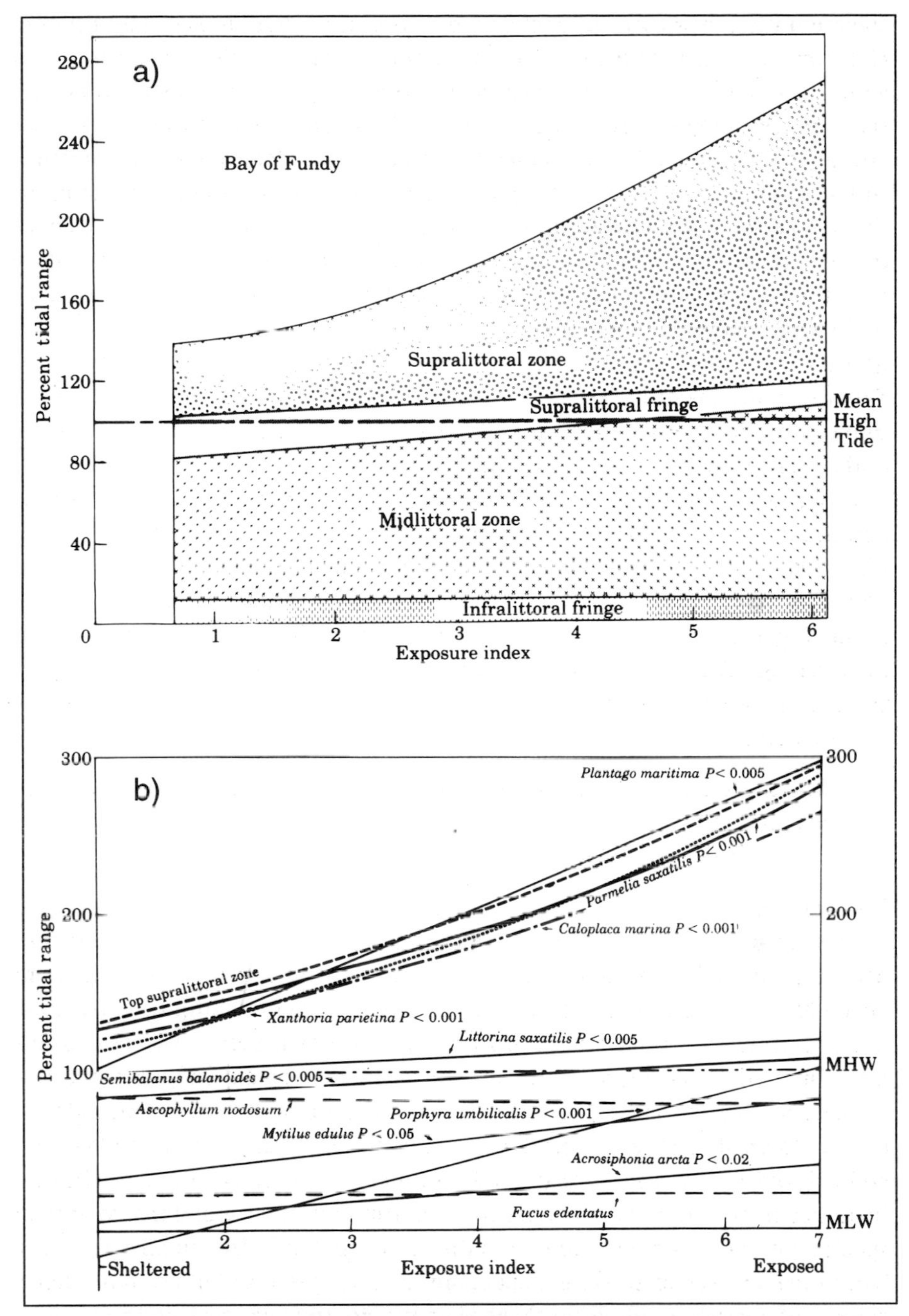

Fig. 4.13 Changes in a) littoral zonation boundaries and b) upper limits of important species, with exposure in the Bay of Fundy, Canada. After Thomas (1994)

calcareous rocks where blue-green and green algae may replace the lichens. The supralittoral zone is poor in species and shows less variability than the lower zones. Lower down in the littoral zone larger algae may be found, as well as barnacles and limpets. On the Massenkalk limestone on the Mediterranean coast of Spain, Kelletat (1985) describes a blue-green algal and *Littorina* (periwinkle) zone, above the notch. In County Clare, Ireland, again on limestone, Trudgill (1987) found lichens present in the splash zone with *Pelvetia canaliculata* (channel wrack) coming in in the upper intertidal zone. Note here that Steven Trudgill is a geomorphologist and, therefore, uses geomorphological zonation terms when describing ecological zonation. On limestone coasts in the northern Adriatic, Schneider and Torunski (1983) found *Littorina neritoides*, *Chthamalus* sp. (barnacle) and blue-green algae present in the supralittoral zone. On tropical coasts similar zonations result, thus Trudgill (1976) found *Littorina glabrata*, and *Nerita* spp., plus the sponge *Cliona* sp. and the chiton *Acanthopleura* sp. in the upper parts of the intertidal. Boring blue-green algae are also present in this zone (Viles, 1987).

In Britain and similar areas, the eulittoral or midlittoral zone is dominated by barnacles, and is generally more species rich and variable than the littoral fringe. Three types of community are found in the midlittoral zone in Britain, largely dependent on exposure. In exposed situations the mussel *Mytilus edulis* and red algae dominate. In moderate conditions limpets (including *Patella vulgata*) and barnacles dominate, with a lower belt of algae. Under sheltered conditions fucoid algae dominate. Schneider and Torunski (1983) illustrate how similar communities may be found in warmer Mediterranean climates with *Monodonta turbinata*, *Patella caerulea*, *Fucus virsoides* and *Mytilus galloprovincialitis* found in the midlittoral zone in the northern Adriatic.

In Durness, northern Scotland, Kelletat (1988) found the midlittoral on limestone dominated by *Patella vulgata*, *Mytilus edulis*, *Semibalanus balanoides* (barnacle) and *Fucus*. On Curaçao, Netherlands Antilles, in the southern Caribbean, Föcke (1978) found a range of encrusting algae in the midlittoral on exposed shores (e.g. *Lithophyllum* sp.). In fact, the midlittoral zone in many areas is the site of major encrustations such as those forming vermetid, serpulid and calcareous algal formations. It is here also that the bulk of the notch-forming bioerosion processes occur, involving the bivalve mollusc *Hiatella arctica* and *Paracentrotus lividus* (sea urchin) in County Clare, Ireland (Trudgill, 1987). Hodgkin (1970) reports that the major notch bioeroders in the Malaysian coastal areas he studied were the sponge *Cliona* sp. and the bivalve molluscs *Lithophaga lima* and *Gastrochaena cuneiformis*.

The sublittoral or infralittoral fringe is dominated around Britain by large algae of the order Laminariales. Exposure controls the species mix found, with very exposed shores dominated by *Alaria* spp. and averagely exposed coasts having *Laminaria* spp. instead, plus *Lithothamnia* spp. The infralittoral zone contains the greatest variety of animals of any littoral zone, but they are usually inconspicuous. The infralittoral zone in some parts of New Zealand is dominated by kelp beds, including *Durvillea antarctica* and *Macrocystis pyrifera*. Calcareous algal accretions often develop in the infralittoral zone, with *Tenarea tortuosa* and *Lithophyllum incrustans* producing platforms or ledges on aeolianites, limestones and granites on the Mediterranean coast of Spain (Kelletat, 1985). Recent work by Norton (1985) has clarified the controls on zonation of seaweeds across the coastal profile on Britain's rocky shores. He concludes that seaweed zonation is determined by events after the settlement of the propagules with physiological stress, desiccation and environmental fluctuations, limpet grazing, water motion and light levels all being important factors in different parts of the littoral zone. According to this work biotic factors (grazing and competition) are dominant in the midlittoral and infralittoral zones.

Within the generalized zonation patterns described above, more detailed work has been carried out on interactions between organisms. The food webs and nutrient cycles in the rocky littoral zone are often quite complicated. One point is particularly important here, that is that ecosystems in the littoral zone are not in any sense closed, but have important links with offshore, onshore and alongshore systems. Thus, some species shed reproductive stages into the sea where growth and dispersal periods may precede re-establishment on the shore (Lewis, 1977). Similarly, many species rely on material in suspension for food — with such material coming from the intermixing of coastal and runoff waters. Bioerosion may be seen as a component of such food webs with some borers (especially microborers such as blue-green algae) extracting nutrients directly from the substrate, and grazers removing rock incidentally while grazing on such algae. An important focus of ecological work on the rocky littoral zone has been the identification of 'keystone species', i.e. species which, if removed, have enormous ramifications for the community as a whole. Such keystone species are often particularly sensitive to environmental fluctuations. Jones and Baxter (1985) report on the importance of such a keystone species, the limpet *Patella vulgata*, in the monitoring of rocky shore pollution. *Patella vulgata* is especially sensitive to oil and dispersants (that is, the detergents used to clean off the oil) which cause its detachment from the rock substrate.

A final important focus of ecological studies of the rocky littoral zone has been to identify natural fluctuations, especially on the small scale, in community composition and zonation. These fluctuations may be seasonal, annual or irregular in timing. Because the littoral zone ecosystems are not isolated, but rely on interactions with onshore, offshore and alongshore systems a disturbance to any one of these may lead to fluctuations in the communities elsewhere. We must see the impacts of pollution in the context of these interactions and fluctuations. Along the Severn Estuary, for example, hot dry conditions during 1975 and 1976 killed off fucoid algae and prevented settlement of young ones. Fucoid algae, mainly *Ascophyllum nodosum*, *Fucus vesiculosus* and *F. serratus*, had previously dominated the extensive shore platform at Watchet (on the southern shore of the estuary). Subsequently, limpets were able to colonize. They are normally affected by 'whiplash' from algal fronds, preventing their establishment. The fucoid algae began to recolonize, but were heavily grazed by limpets during the 1980s. By 1989 full cover of fucoids was restored (Crothers and Haynes, 1994). Many investigations of such fluctuations have shown how physical and biological (predation) disturbances can enhance species diversity, by removing large amounts of the organism which is the dominant competitor for space (Barnes and Hughes, 1988). An intermediate scale of disturbance appears to maintain species diversity, whereas high levels of disturbance can have the opposite effect.

Human influences on rocky coasts

As we have seen in the preceding sections, rocky coasts are characterized by varying stages of geomorphic and ecological change which are often extremely patchy over space and time. We can recognize two contrasting zones of human activity on rocky coasts which may affect such natural processes, i.e. above and below extreme high water mark. Below high tide level usage is constrained by water cover, and is dominated by harvesting of edible or useful organisms, such as seaweed, or quarrying. In Britain, for example, mussels, winkles, crabs and sea urchins are collected for food and anemones are used as bait for line fishing (Lewis, 1977). Large brown algae are sometimes used for manure; whereas other algae are prized for their food and medicinal values. At higher levels, rocky coasts become attractive sites for settlement, communication routes (roads and railways) as well as some rather more esoteric harvesting activities. Goldsmith (1975), for example, describes some commercial uses for cliffs in Britain (now largely stopped) such as picking *Crithmum maritimum* (rock samphire) for pickles, collecting seabirds' eggs for eating, and killing seabirds for eating and their oil. More esoterically he describes useful side products from these activities on the Shetland Islands: 'the skin of gannets necks . . . was once used as a primitive shoe; also their beaks were used as pegs for holding the strings on thatched roofs'.

Such activities may affect ecological interactions, and in extreme cases may affect geomorphological change. Building and quarrying activities produce obvious impacts on cliffed coasts, often also having a major influence on linked coastal sediment systems. Thus, Hudson (1980) mentions the huge changes to cliff profiles, including the creation and destruction of caves and stacks, along the north Yorkshire coast, England, because of centuries of mining and quarrying for jet, alum and ironstone. Railways and roads built at the base of active cliffs can reduce the sediment transport to nearby beaches, especially in cases where the cliffs require armouring to protect the routeways from mass movements. Conversely, building on cliffs can facilitate mass movements and encourage the supply of coastal sediments.

Meeting the challenges on rocky coasts

Engineering around hard rock cliffs

Hard rock, stable cliffs pose a unique series of challenges. Roads and railways may pass through these areas, and they must be protected from rockfalls. Conversely, a solid appraisal of the hazard of rockfalls should enable better decisions over the location of such roads in the future. Engineering roads and railways along the edge of hard rock cliffs often necessitates the creation of a 'ledge' in the cliff profile, in much the same way as in mountainous areas inland. Some spectacular cliff top, or cliff side, roads have been built, despite all the problems, including the road around the southern Cape Peninsula in South Africa. Rockfalls continue to be a serious problem along many such roads, especially under severe weather conditions, such as frost spells in temperate latitudes. A variety of strategies has been developed for reducing or preventing rockfalls. Such strategies are also used in inland areas where cuttings go through hard rock masses with artificially steepened faces (e.g. the M40 motorway in Berkshire, England, where it passes through the Chilterns escarpment). Two main types of approach may be followed, singly or in combination: either the prevention of rockfalls occurring, or the prevention of the debris produced from reaching the road or railway. The second option is probably easier in the short term, but needs long term management. Studies such as the one by Allison (1989) on the Dorset coast provide data on rock parameters and the frequency of rockfall hazards which can be used as a basis for planning and management. To prevent rockfalls in any one area, it is necessary to know what causes them. Where frost is deemed to be a major cause, methods to prevent frost gaining access to fractures must be considered. Armouring the rock with a protective coating might be a possibility.

Rockfall damage is often prevented by covering the rock face with netting to hold rockfall debris. This netting should be emptied of debris from time to time to prevent it from bursting, thereby releasing its stored debris and producing a more serious rockfall. Vegetation can also be used to prevent rockfalls, at least on cliff faces where there is some soil and, preferably, a stepped profile. Essentially, the idea here is to produce a sort of rock garden in which the plants will help provide slope stability. The key factor here is to use the right plants; the limited knowledge about hard rock cliff ecology in many areas implies this might be easier said than done.

Managing cliff recession and failure

As we have already seen, cliff failure is caused by an interplay of wave attack at the cliff base, and water and loading changes within the cliff profile. Both of these can be severely affected by human activities. In San Diego County, California, for example, south of the Santa Margarita River, many cliffs are actively retreating not because of wave action, but due to groundwater attack. Further, storm drains installed to divert water from the cliff face collapsed in 1978, 1980 and 1983 allowing the water to erode the cliff face (Kuhn and Shepard, 1984). Along much of this coastline, groundwater solution has been encouraged by over-watering of non-native vegetation such as the ice plant, and tending garden lawns. Some major failures have resulted from such activities.

In southern Portugal, some workers have ascribed the high erosion rates in the Vale de Lobo area to development on the cliff hinterlands, and coastal modifications (Alveirinho Dias and Neal, 1992). Several examples show the impact of shore protection and development works on cliff failures. On the south-east shore of the Dee estuary, England, cliffs cut into glacial deposits are highly unstable. Protection works undertaken for a local golf course to stabilize the cliff foot in the area have resulted in severe toe erosion and slope instability over a 73 m length of cliff nearby (Pitts, 1983). In Kent, the increased landsliding activity in Gault clays overlain by chalk at Folkestone Warren, on top of which is a vulnerable railway, has been ascribed to a large extent to the progressive interruption of littoral drift by the developing Folkestone harbour (Hutchinson *et al.*, 1980). So what can be done to ameliorate problems on failure-prone cliffs?

Solving the problems of cliffs prone to recession and failure is by no means easy, but several alternative approaches are available, which have also been used in various combinations. Protection of the cliff foot zone from wave attack using solid walls, or baffling devices such as tetrapods, is used where marine erosion is a major problem. In Japan, for example, the 9 km long Byobugaura cliff, Honshu Island, is composed of Pliocene mudstone at the base with Pleistocene sand and gravel overlain by volcanic ash on top (Sunamura, 1982). In the early 1960s tetrapod seawalls were built along two sections of this cliff to reduce wave erosion. These walls were built up to 15 m away from the cliff base at either end of the Byobugaura beach. They have been effective in reducing erosion of the cliff faces immediately behind them (rates of 0.05 and 0.3 m a^{-1}), but the cliff area in between is still eroding rapidly (average rate of 1.0 m a^{-1}) as described by Walker and Mossa (1986). Seawalls themselves can be vulnerable to scour and erosion, and often have an undesirable effect on the beauty of an area. Beach management may provide a better solution, making a natural barrier between waves and the cliff base, but may not be a long-lasting solution. Clearly, the interrelationships between cliff erosion and nearshore sedimentation discussed earlier imply that any cliff protection strategies must take into account local conditions, and must attempt to treat the coast as a whole, and not just focus on protecting individual parts.

Along the Dorset and Devon coasts, southern England, attempts have been made to stabilize mudslides using a combination of drainage and grading of steep cliff faces. Denness *et al.* (1975) made a detailed study of the Higher Sea Lane mudslide complex in this area and recommended draining the overlying landslide material (on top of Lias clay), grading the upper cliff to a slope angle of 33°, and superimposing an artificial shallow drainage system. However, the cliff complex is still unstable, and clearly the 'gigantic conveyor belt' systems found in such areas are not going to be easy to control.

An alternative approach, feasible in many areas, is to use vegetation to aid cliff stabilization. In 1918, Carey and Oliver wrote with some confidence about the future of such approaches:

> As the study of vegetation in connection with the stabilization and protection of cliffs and steeply-sloping ground is only in its infancy, there is no doubt that bold experiments among a wide selection of plants would lead to discoveries of great utility in the treatment of ground of this kind.

Whether this aim has as yet been achieved is debatable, but some successes have been recorded. May (1977) describes the cliff management schemes on the cliffs at Bournemouth, on the southern English coast. Cliffs formed in Eocene sands and clays capped by plateau gravel and blown sand have been prone to movement and failure. The first protective walls were built in 1907, and several schemes using standard engineering techniques such as revetments have been tried since then. Recently, grading and terracing of cliff faces has been undertaken in association with the building of paths and steps, and the seeding and planting of vegetation. Terraces were made, planted with privet hedges at 90° to the cliff face, and seeded with grasses or planted with *Mesambryanthemum* (a ground cover plant). On the cliff faces various shrubs, including *Salix* spp. (sallow) and *Quercus ilex* (evergreen oak) were used to aid stability. Improved drainage, grading and seeding with grass and gorse were also undertaken to prevent deep failures and further stabilize the slopes. Carey and Oliver (1918) recommend the use of willows (*Salix pulchra*) for preventing cliff top cracking and movement, through their matted surface growth and deep

roots. Vegetation by itself is unlikely to tackle major cliff failure problems, which require grading, drainage improvements and sea defences, but it can ensure the continued success of 'rehabilitated' slopes.

Clearly, many failure-prone cliffs are going to require long term management, and the hazards associated with cliff recession in many areas of rising sea level and unresistant cliff materials are not going to go away (Plate 4.4). Where cliffs can be protected from erosion (by seawalls or beach nourishment) they may still require drainage to prevent failure. Often such drainage schemes only last 20–30 years (McGown *et al.*, 1988). Careful planning needs to be implemented along failure-prone coasts to identify areas where future housing developments should be restricted (*see* Case study 4.2) and to monitor and manage protection works to ensure maximum benefit over the longest possible period. As a first step, geomorphologists and geologists should be involved in evaluating and explaining the spatial and temporal variability of failure hazards in key areas (as, for example, Bosscher *et al.* (1988) have done for Lake Michigan, USA, and Grainger and Kalaugher (1987), did for Devon, UK).

Rocky shore pollution and the impacts of tidal power schemes

On rocky shore platforms, which are largely found around the intertidal zone, the problems mainly come from the sea, or from changes in tidal regime. The major problem seems to be that of pollution and the disruption it causes to the ecological and geomorphological systems. Where pollution (such as oil) damages plant and animal communities on limestone shores, for example, their contribution to geomorphological processes (through bioerosion and bioconstruction) will be affected, and sediment production may be dramatically reduced. It has been stated that oil residues become harmless to organisms after they 'age' to tar and wax. Several workers disagree with this, however, arguing that exposure to sunlight can reliquefy the deposits (Zimmerman and Kelletat, 1984). Lewis (1977) argues that most oil has lost its toxic components through evaporation by the time it is washed ashore, and that its major damaging effects are suffocating. He further proposes, as have several workers, that it is the detergents used to remove the oil which cause the most serious damage. Zimmerman and Kelletat (1984) suggest that oil spills which affect the Mediterranean supralittoral zone are particularly damaging, diminishing or even stopping completely the natural bioerosion of limestone coasts here which proceeds at a rate of 0.25–1 mm a^{-1}. The severity of oil impacts on shore platforms depends on exposure: sheltered areas tend to concentrate pollution, whereas exposed headlands facilitate flushing. Where grazing animals are killed by oil, massive growths of fast growing opportunistic seaweeds may result (Dicks *et al.*, 1988).

Following the major *Torrey Canyon* oil spill, involving 117000 tonnes of crude oil in March 1967, there was large scale mortality of many rocky intertidal species along the Cornwall coast, England. After disturbance from the oil and dispersants used to clean it up, it took *c.* 10 years for the communities to return to their previous conditions (Southward and Southward, 1978). Observations following this, and other major spills such as the *Exxon Valdez* (*see* Case study 7.1), have suggested alternatives to the ecologically damaging detergents, such as bacteria which 'eat' the oil.

Tidal power schemes can also have dramatic and long term effects on rocky shores. The Rance Estuary, near St Malo in northern France, has had an operational tidal barrage since 1967. The tidal range at the mouth of the river is 13.5 m at maximum, and within the barrage it varies between 4 and 5.5 m depending on turbine operating conditions. During construction, between 1963 and 1966, the estuary was cut off from the open sea for 3 years, which killed off many organisms. Others, such as *Mytilus edulis*, survived. When rocky shore zonation was studied in 1980 the average salinities in the estuary were higher than before the barrage, but showed less variability. As shown in Fig. 4.14 considerable compression of the midlittoral zone was found in the 1980 survey, with a general replacement of barnacles, such as *Balanus balanoides*, by macroalgae, such as *Ascophyllum nodosum*. The 1980 survey suggests that the shores have recovered from major disturbance, but that irregularities in the artificial tide (related to operating conditions) have perhaps prevented an equilibrium developing (Little and Mettam, 1994; Rétière, 1994). These results are of great interest to other areas, such as the Severn estuary in England, and the Bay of Fundy in USA, where similar tidal power schemes are planned.

Rocky coasts and the greenhouse future

Rocky coast geomorphology and ecology will be affected by several components of global warming. Groundwater conditions in failure-prone cliffs may change as climate changes, and increases in storminess and sea level will increase basal erosion (as foreshadowed by the changes wrought to the Oregon coast by the 1982–3 El Niño event described earlier). Compared with interest in assessing the future of sandy and muddy parts of the coastline, there has been relatively little interest in predicting the future for rocky coasts. Sunamura (1992) presents a simple mathematical model, along the lines of the Bruun rule (Chapter 3) for sandy shores:

$$\frac{dX}{dt} = \frac{dx}{dt} + \frac{dZ}{dt}/I$$

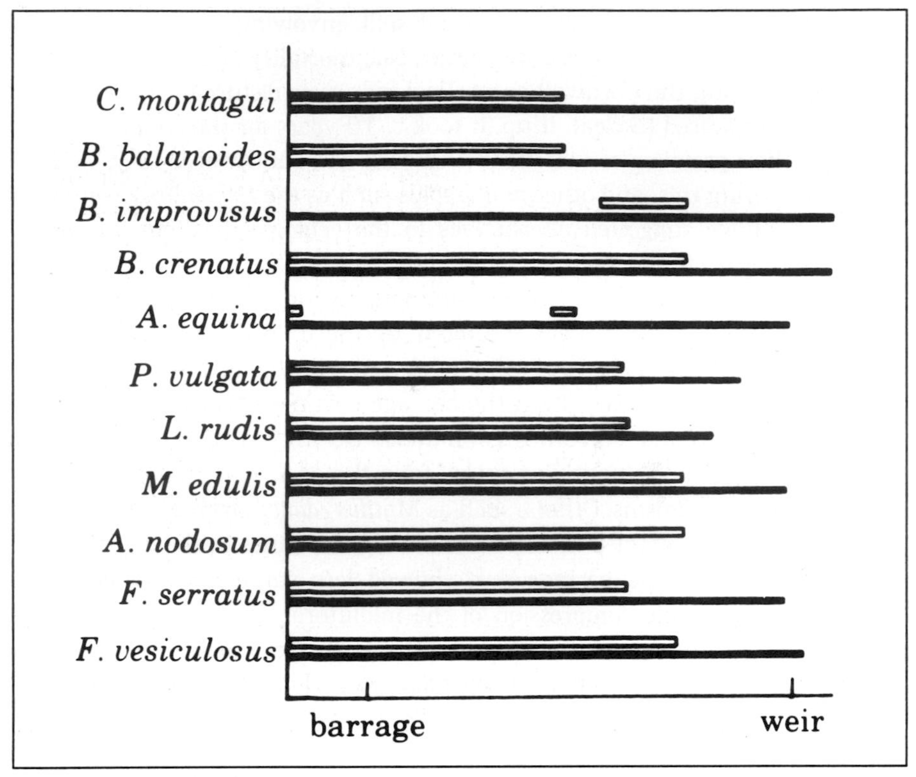

Fig. 4.14 Changes in upstream limits of species in the Rance estuary in 1931 (black bars) and 1980 (white bars). The distance between the barrage and the weir is approximately 15.5 km. After Rétière (1994)

where:

$\frac{dX}{dt}$ is the cliff erosion rate under a rising sea level;

$\frac{dx}{dt}$ is the cliff recession rate under stationary sea level, where x is recession distance in metres; and

$\frac{dZ}{dt}$ is the rising speed of sea level and I is the platform gradient.

Using this equation to predict changes to cliffs on Nii-jima Island, Japan, Sunamura (1988) found that recession rates and distances would be highly influenced by sea level rises of the order forecast by the EPA in 1984 (of course, recent sea level rise estimates are much lower as discussed in Chapter 2). Wave climates are likely to change as well, making predictions much harder to make, and in any areas where cliff erosion is related to a 'conveyor belt' system, as in Dorset, the situation is likely to be much more complex.

Plate 4.4 Coastal erosion and property loss on the mid-Oregon coast, USA: the view north from Roads End near Lincoln City towards Cascade Head. Retreat of cliffed Pleistocene marine terraces is estimated at 10 cm a^{-1} at this locality

Case study 4.1: Llantwit Major — blasted cliffs!

Along the Glamorgan coast on the northern shore of the Bristol Channel, UK, there are extensive cliffs cut in generally horizontally bedded, Lower Liassic sedimentary rocks. Alternating shale and limestone beds occur, with the limestone here being strong and the shales relatively weak. The wave energy environment is high here, with a long wave fetch across the Atlantic, coupled with prevailing winds coming from the west and south-west. Storms are common, and the cliffs are actively retreating through rockfalls induced by basal erosion, enhanced by strong east–west longshore drift in the area which removes protective debris quickly.

This area of coast is an important focus for recreation, and its scenic attraction was confirmed when it was recognized as a Heritage Coast by the British Government in 1973. Heritage Coasts are a form of protected coastline, entitled to special planning attention and funds for management, aimed at reconciling conservation with use for tourism and recreation. Rockfall activity has long been seen as a problem in this area, especially as many of the cliffs have overhanging portions, and in 1969 the coastal protection agency here (Crowbridge Rural District Council) became concerned about the hazard and its possible effects on recreation at Colhough Bay, Llantwit Major (Fig. 4.15a). They employed consulting engineers to undertake a survey and advise on solutions. These engineers advised that both modification of the upper cliff profile and protection of a shale band at the cliff base were necessary to provide stability. However, cost and other considerations led to only cliff top blasting being undertaken in an attempt to both stabilize the cliff profile and protect the basal shale band.

Test blasting was carried out in June 1969, followed by blasting near the beach promenade, and blasting of the central cliff area (Fig. 4.15b). Blasting

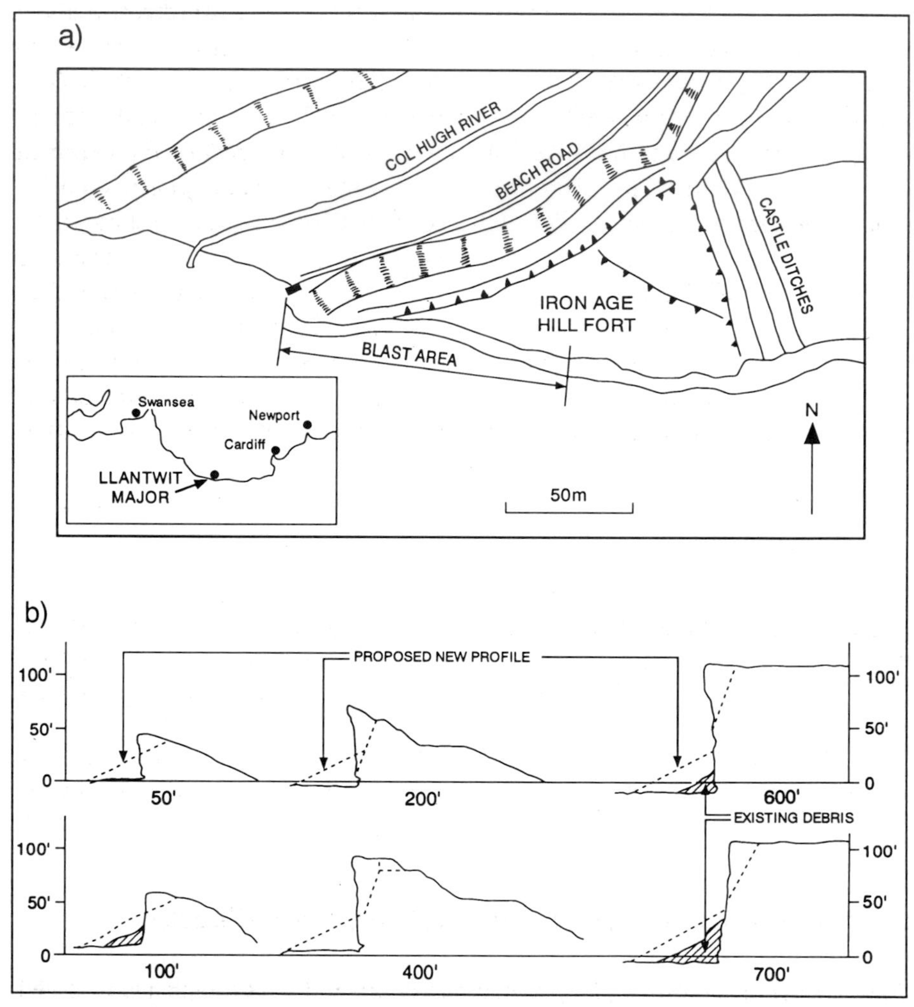

Fig. 4.15 Blasting at Llantwit Major cliffs: a) location and b) old profiles and proposed blasted profiles. After Williams and Davies (1980)

alone cost just under £3000. The blasting programme was carefully planned so as to exploit the highly fractured nature of the rock. Field and photographic surveys 9 years after this blasting showed that the actual debris cone was smaller (about two-thirds of the planned volume) and less permanent than had been predicted (because of wave attack and longshore drifting). Williams and Davies (1980) suggest that this implies 'a lack of awareness of the nature and speed of operation of geomorphological processes in this particular environment'. Even the modified, upper cliff profile has not turned out quite as planned. Weakening of the rock has occurred through blasting and Williams and Davies conclude that 'the danger of rock falls in the upper face has not lessened'.

Thus, it is clear that this cliff modification scheme was ill-conceived, and did not adequately take into account the actual processes shaping the cliff environment. The juxtaposition of limestone and shales here creates unstable cliffs prone to rockfalls even where basal erosion is minimal. Furthermore, the design of the blasting seems to have been flawed, as it did not produce the expected result of a consistent upper cliff profile at 23° to the vertical. A more expensive scheme of stepped terracing might have produced better results, although the combination of geological factors and a highly erosive environment probably prevent any such scheme being successful in the long term. Increasing the awareness of, and adjustment to, the rockfall hazard through notices and regulating recreation at the cliff base seems a more sensible long term option (Williams and Williams, 1988). Recent scientific research underlying the causes and nature of rockfalls in this area provides an important basis for any further schemes (Williams *et al.*, 1993).

Selected references

WILLIAMS, A.T. and DAVIES, P. 1980: Man as a geological agent: the sea cliffs of Llantwit Major, Wales, UK. *Zeitschrift für Geomorphologie Supplementband* 34, 129–41.

WILLIAMS, M.J. and WILLIAMS, A.T. 1988: The perception of, and adjustment to, rockfall hazards along the Glamorgan Heritage Coast, Wales, UK. *Ocean and Shoreline Management* 11, 319–39.

WILLIAMS, A.T., DAVIES, P. and BOMBOE, P. 1993: Geometrical simulation studies of coastal cliff failures in Liassic strata, south west UK. *Earth Surface Processes and Landforms* 18, 703–20.

Case study 4.2: Jump-Off Joe: building on a disaster

The Newport area of the Oregon coast is characterized by high rates of cliff recession because of the presence of vulnerable, seaward-dipping (10–30°) mudstones and siltstones forming low cliffs on uplifted marine terraces within a tectonically active setting (as discussed earlier). The city of Newport was founded in the 1860s, and has attracted settlers and tourists because of its scenic beauty and abundance of natural resources. Jump-Off Joe was the name given to a picturesque headland which suffered rapid erosion into an arch, then stacks, of which only small remnants are visible today. Now Jump-Off Joe refers to this area in general. As shown in Fig. 4.16 there has been considerable coastal recession in the Jump-Off Joe area since the 1860s, including two major

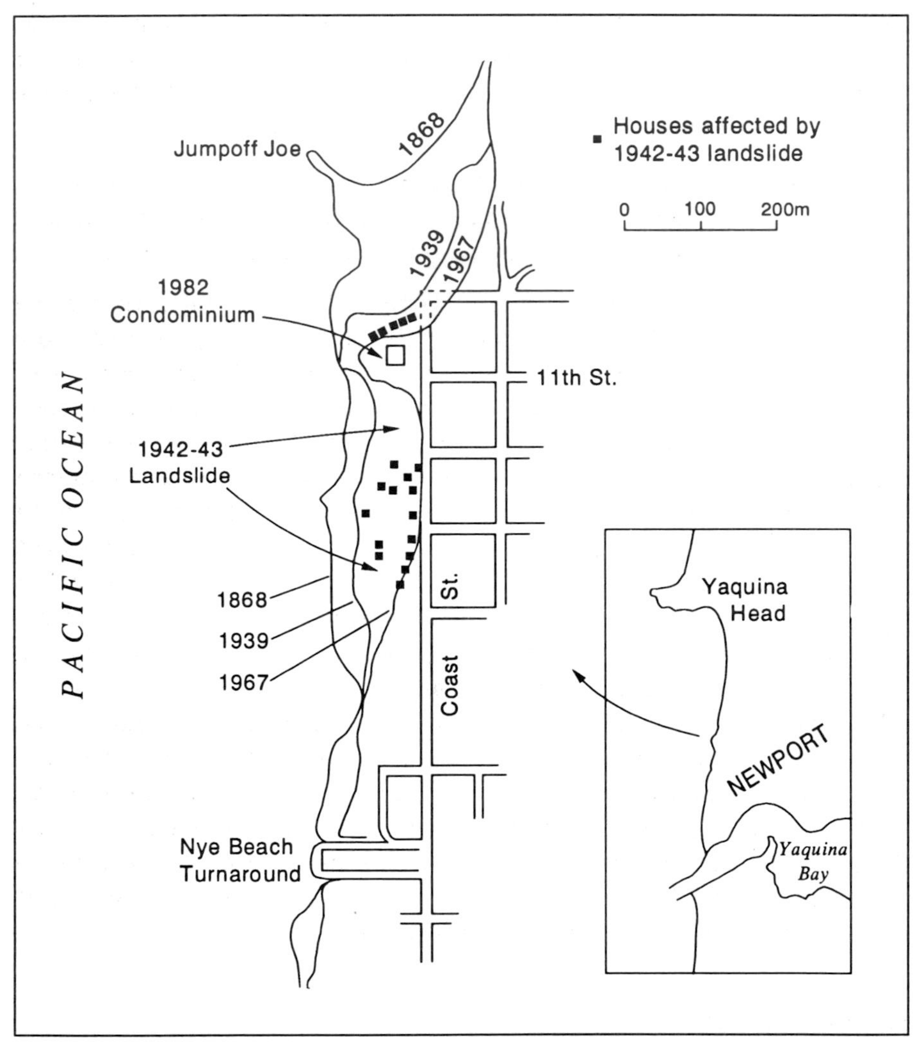

Fig. 4.16 Changing shoreline profile 1868–1967 at Newport, Oregon, and location of Jump-Off Joe site (1982 condominium). Black squares show houses affected by the 1942–3 landslide. Modified from Sayre and Komar (1988)

slumps: one in the winter of 1942–3 which affected fifteen houses, and one 100 years previously. These two slumps left a small uneroded cliff between them where a condominium was built in 1982.

It is surprising that any building at all was permitted on such an area, with a history of active slumping. In 1980 developers started to plan a large development of houses on the 1942–3 slumped block, as well as on the small uneroded cliff. They planned to regrade the site and construct a seawall at the cliff toe in order to reduce the geomorphological hazard there. Considerable legal wrangling ensued between the developers, the Newport Planning Commission and opponents to the scheme represented by the Friends of Lincoln County (a group formed in the 1970s to oppose development of wetlands within the Newport area). As Sayre and Komar (1988, p. 15) put it:

> It was a classic confrontation between developers who thought their project would help a city grow, and environmentalists who wanted to preserve the coastline: in the end, the issues were decided by Nature.

Eventually, a much smaller plan was proposed by the developers for the uneroded cliff area only, based on a geologist's report which gave a very low figure for cliff retreat here and no indication of any slump hazard. This was in direct contradiction to most of the evidence. The scheme was approved, despite great opposition from the Friends of Lincoln County. Construction of the condominium began in March 1982 in a depressed housing market. A drainage pipe to control groundwater saturation was installed under the building to reduce any further slumping hazard. In May 1983 the developers went bankrupt, and by September 1984 the cliff face retreat had undermined the perimeter fence around the condominium. Then the drainage pipe burst, probably because of slippage, and aided a larger slump on the site, which caused the foundations to fail. In 1986 the condominium was demolished.

The area is still zoned for housing development, although there is now a geological hazards overlay which means that any developers must provide a detailed assessment of the hazards, and any schemes which are too risky will be rejected. The geologist whose report encouraged the development of the Jump-Off Joe site has had his certification revoked for gross negligence and incompetence. The history of this failed scheme exemplifies the conflicts that can arise when developing a naturally hazardous coastline.

Selected reference

SAYRE, W.O. and KOMAR, P.D. 1988: The Jump-Off Joe landslide at Newport, Oregon: History of erosion, development and destruction. *Shore and Beach* 56, 15–22.

Chapter Five

COASTAL WETLANDS

Introduction

Coastal wetlands appear to be both important sinks of global carbon (Twilley *et al.*, 1992) and exporters of nutrients (Nixon, 1980) which help sustain the productivity of estuarine and other coastal ecosystems. A nexus of physical, biological and chemical processes is at the heart of wetland dynamics and alterations to any one element can have repercussions for the whole marsh. This chapter discusses the major environmental settings of muddy coastlines in more detail, and then introduces the ecological and geomorphological characteristics of mudflats, marshes, mangroves and sabkhas. How wetlands react to changing sea level is illustrated for salt marsh systems. Finally, this chapter illustrates the types of problems which are affecting these coastal areas, including erosion, pollution, high magnitude–low frequency events and biological invasions, and assesses their future prospects and appropriate management strategies.

Muddy coastlines form in low-lying, relatively sheltered areas with abundant sediment supply, such as behind barrier islands, around estuaries, bays and on deltaic coasts. In general, they are characterized by low wave energy climates and dominated by tidal processes. Some wetlands, however, form on open coasts where low coastal gradients are accompanied by large sediment supplies (e.g. the Surinam coast). In the upper intertidal zone the mud surface is covered with halophytic vegetation but even apparently bare mudflats in the lower intertidal zone are rich environments for algae and benthic fauna. Major muddy coastlines worldwide include the huge mudflats along the west coast of South Korea (Wells *et al.*, 1990), mangrove swamps around the Bay of Bengal in India and Bangladesh (Blasco, 1977) and in many estuaries in northern Australia (Woodroffe, 1990), wetlands on the Mississippi delta (Walker *et al.*, 1987; and Case study 5.1) and along large stretches of the Netherlands coast.

Despite the common perception of coastal wetlands as being 'wastelands' they are in fact subjected to a vast range of human activities. Grazing is commonly practised on marshes in the United Kingdom, with reclamation permitting further agricultural development on former marshes. A range of aquacultural enterprises is also practised, ranging from oyster beds in the Colne Estuary, eastern England to shrimp ponds in mangrove swamps in the Philippines. Salt production is important in many wetland areas, and timber is commonly obtained from mangrove forests for use as a fuel and in the building and charcoal industries. Several wetland areas have been affected by major industrial developments, such as oil extraction (in, for example, the Mississippi and Niger deltas), nuclear power installations and port developments. Furthermore, many wetland areas have achieved important conservation status because they act as temporary resting and feeding stations on flyways for migrating birds and other animals. Finally, wetlands have been found to be important natural agents of coastal protection and purification, helping to protect the landward zone from flooding and acting as a sink for toxic wastes.

The magic of mud

Unlike beaches, where sediments coarsen onshore, tidal environments show a progressive decrease in sediment size from the low tide to the high tide level (Pethick, 1984). Thus mudflats and marshes are dominated by fine sediments and their transport, sedimentation and erosion processes. Mud is fine-grained sediment in the silt and clay fractions, with particle sizes ranging in diameter from 0.0005 mm (0.5 μm) to 0.063 mm (63 μm), with most grains in the range 0.001–0.02 mm (a pin head has a diameter of *c.* 1 mm (1000 μm)). Mud is, however, often found in association with other sediment types. Thus, for example, the Mississippi River deposits nearly 5×10^{11} kg a^{-1} of sediment into the Gulf of Mexico, of which 45 per cent is clay, 36 per cent silt and 19 per cent fine sand (Wright, 1985).

It has been traditionally assumed that the deposition of sediment in wetland environments takes place at times of slack water, when tidal current velocities fall to zero as the tide reverses. However, the period of slack water in tidal systems is typically short and the settling time for small particles very long; thus while a coarse sand particle would take 7 min to fall through a 100 m settling tube it would take a clay-sized particle 2 months to settle through the same water column. The 'periodic settling model' was tested by McCave (1970) for the deposition of fine sediments in the German Bight, North Sea; even generously allowing 30 min of completely slack water and 90 min of water with very low velocities in each tidal cycle, his model only generated a deposition rate of 1.34–5.36 cm per 100 years as opposed to an actual deposition rate of 15.5 cm per century. One answer to this problem is to invoke sedimentation at times other than slack water; another is to generate larger particles and hence more rapid settling times. One method of generating larger particles is through biogenic pelletization. Organisms in mudflat and marsh environments trap suspended silts and clays and strip their surfaces of organic matter. The remaining inorganic sediments are then excreted in the form of large faecal pellets which then behave like sand grains rather than like their constituent grains.

Besides this of organic flocculation, there are also processes of physico-chemical flocculation taking place in estuarine and nearshore environments. This highlights the crucial fact that clay-sized particles are not simply miniature versions of sand-sized grains. Whereas individual particle sizes, shapes and densities are important attributes of sand-sized particles, clay-sized material has important bulk properties which relate to physico-chemical behaviour between particles and interparticle cohesion. Clay grains possess short-range attractive forces which bind neighbouring clays together. In freshwater, this process is prevented by surface charges which repel neighbouring grains but in saltwater the effect of these surface charges is reduced and clay particles can form large aggregations known as 'flocs'. Flocs are loose, open structures typically 10–20 mm across and with water contents of *c.* 90 per cent. The flocs join into floc groups, 50–200, perhaps 800, mm in size and containing 100000–1000000 individual particles (Pethick, 1984). These large aggregations clearly settle much more rapidly than their constituent grains and their presence can, therefore, help explain the sedimentation of fine-grained sediments in mudflat and marsh environments. Physico-chemical flocculation is particularly likely in these environments because of the mixing of water bodies of varying salinity, particularly in estuarine environments where freshwater and saltwater meet and the presence of organic matter helps to bind and stabilize floc structures. Flocs settle out into fluid 'stationary suspensions' which can be re-entrained. In time, however, further sedimentation removes interfloc voids and the sediments evolve towards 'settled muds'.

Interparticle adhesion and organic binding, high water contents and low surface roughness in settled muds makes them more difficult to erode than uncohesive sands and gravels (Nichols and Biggs, 1985). Once entrained, however, cohesive sediments can be carried by very low velocity waters until eventually they become deposited. These differences between entrainment and erosion thresholds have been used to show how in the Wadden Sea fine silts and clays are progressively transported in a landward direction by repeated tidal cycles, so leading to mudflat formation (Postma, 1967).

Environmental settings of coastal wetlands

Back barrier environments

Barrier islands and spits commonly have an associated back barrier zone with salt marshes, mangroves and lagoonal environments (as discussed in Chapter 3). Great swathes of salt marshes are found behind the natural barrier islands of the east coast of the USA, for example, and mangrove forests appear behind barrier islands along the Nigerian delta coast. Back barrier environments are protected from the open ocean by the barrier itself, although seawater may penetrate through overwash, percolation and movement through tidal inlets. If the barriers become altered by coastal protection structures the back barrier environments tend to become more insulated from the sea, allowing a less saline-tolerant vegetation to develop, and any lagoon to become brackish. Wetlands in back barrier environments are relatively impermanent in the long term (10^3–10^4 years) as, at least on transgressive coasts, the barrier moves inland by 'rollover' processes over the wetlands. Marsh sediments are then exposed on the foreshore and broken up by wave action, although 'mud balls' protected by an outer coating of beach gravels may persist for some time on the beach face.

Estuaries

Estuaries are semi-enclosed inlets where saltwater and river water mix. Often, they form in drowned river valleys, although there are other types, such as the fjords in glaciated valleys (see Chapter 7). They are found worldwide although they are perhaps best developed on mid-latitude coastal plains with wide continental shelves and a trend of locally rising sea level. Estuaries are by no means permanent features on a geological time-scale, and their life span depends on the balance between sea level rise and sediment infilling (Nichols and Biggs, 1985). The key feature of estuaries is that they are complex, dynamic interface zones between freshwater and saltwater areas (Fig. 5.1). This produces associated complex morphological responses, and unique biological and chemical characteristics. Estuarine sediments can come from a range of sources, including drainage basin, continental shelf and coastal waters, the atmosphere, erosion of estuarine margin and bottom sediments, and biological activity. Dyer (1986) suggests that marine sources of sediment are dominant for most temperate zone estuaries.

Three basic types of estuaries can be identified on hydrodynamic criteria, i.e. salt wedge (or stratified), partially mixed and fully mixed types, depending on the relative strengths of river flow and tidal influx. Each type produces different kinds of residual currents or circulation. In areas with high river discharge and low tidal range there is little mixing of saltwater and freshwater — with the

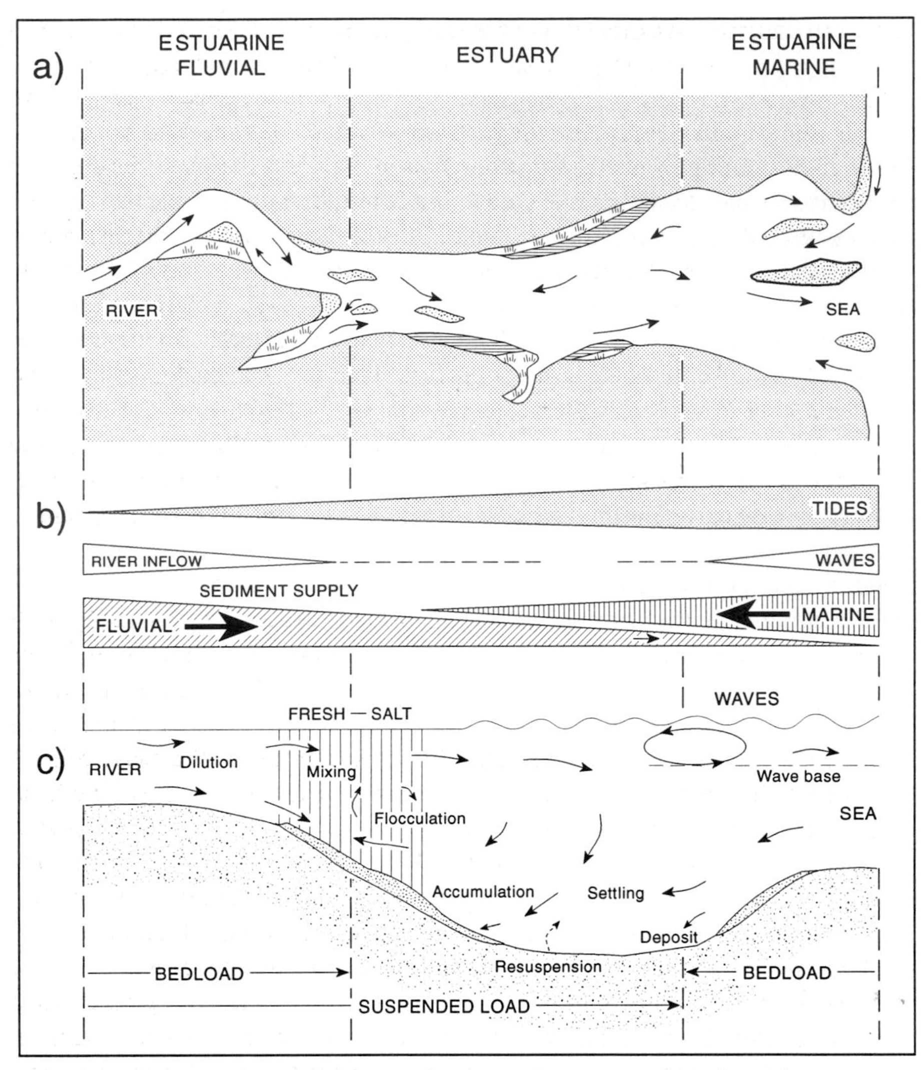

Fig. 5.1 Conceptual model of dispersal zones and routes in a hypothetical estuary. a) Plan view. Arrows show direction of water and sediment movements. b) Relative intensity of dominant dispersal agents (tides, river inflow and waves) along the length of the estuary in relation to dominant sediment supply (marine or fluvial). c) Vertical-longitudinal section of main dispersal routes

river water floating on the denser saltwater and flowing unimpeded all the way to the open ocean. Such flow characteristics have sedimentological consequences, with a delta forming at the mouth of the estuary, and coarser sediment bars upstream at the tip of the salt wedge. In such estuaries, such as the Mississippi, Rhone, Niger and Orinoco, most sediments are derived from river discharge. In the Mississippi the salt wedge penetrates 150 km landward at low river flow periods, and only 1.5 km at high flows (Nichols and Biggs, 1985).

Partially mixed estuaries, with large tidal range and relatively small river inputs, show a clear vertical gradation from saltwater to brackish to freshwater, and much sediment is brought in from the sea. Larger sediment is deposited as bars near the mouth and finer sediment penetrates upstream, forming a 'turbidity maximum' at the limit of saltwater intrusion. Examples of partially mixed estuaries include the Mersey, England and Chesapeake Bay, USA. Fully mixed estuaries occur where tidal range and current velocities are large enough to break down the vertical salinity gradient, to produce totally mixed, homogeneous waters. Where estuaries are wide, however, there is considerable lateral variation as the tidal and river water streams become deflected by the Coriolis force. Such estuaries are wider than *c.* 0.5 km with marine sediments deposited on the left bank, and riverine sediments on the right bank, in the northern hemisphere. Examples include the Firth of Forth, Scotland and the Delaware estuary, USA.

Tidal currents in estuaries

In the deep ocean, tides take the form of a sinusoidal wave (see Chapter 2), with maximum tidal current velocities at high and low tides and equal durations for the flood and ebb tides. As ocean tides are propagated into shallow estuaries they are transformed by frictional damping on the sea floor and the wave becomes deformed. Characteristically, the leading edge — the flood tide — steepens while the trailing edge — the ebb tide — flattens (Fig. 5.2). The flood duration decreases, perhaps to 2–3 hours, while the ebb duration increases to complete the typical semi-diurnal tidal cycle of 12 hours 25 min. Flood velocities increase on the steep crest and ebb velocities correspondingly decrease on the flatter trough, thus favouring the net landward transport of sediment.

However, other processes complicate this pattern. Convergence effects in funnel-shaped estuaries increase tidal amplitudes and the advancing tidal wave may be reflected from the estuary head to the estuary mouth over the course of the tidal cycle. This reflected wave may interact with the incoming wave to produce a standing wave where tidal velocities are zero at high and low water and maximized at mid-tide. Most estuaries, however, show a mixture of progressive and standing waves with the result that flood and ebb velocities typically occur at *c.* 1.5 hours before and after high water. Estuaries also interact with shelf water circulations. Griffin and Le Blond (1990) illustrate, from 19 years' daily salinity records from Juan de Fuca Strait, Vancouver, Canada, how seaward freshwater export is greatest at neap tides, and can be enhanced by favourable winds. Elsewhere, appreciable amounts of sediment have been found to be transported from the Gironde Estuary to Cap Ferret Canyon (2–3 per cent of the total amount) in France (Ruch *et al.*, 1993).

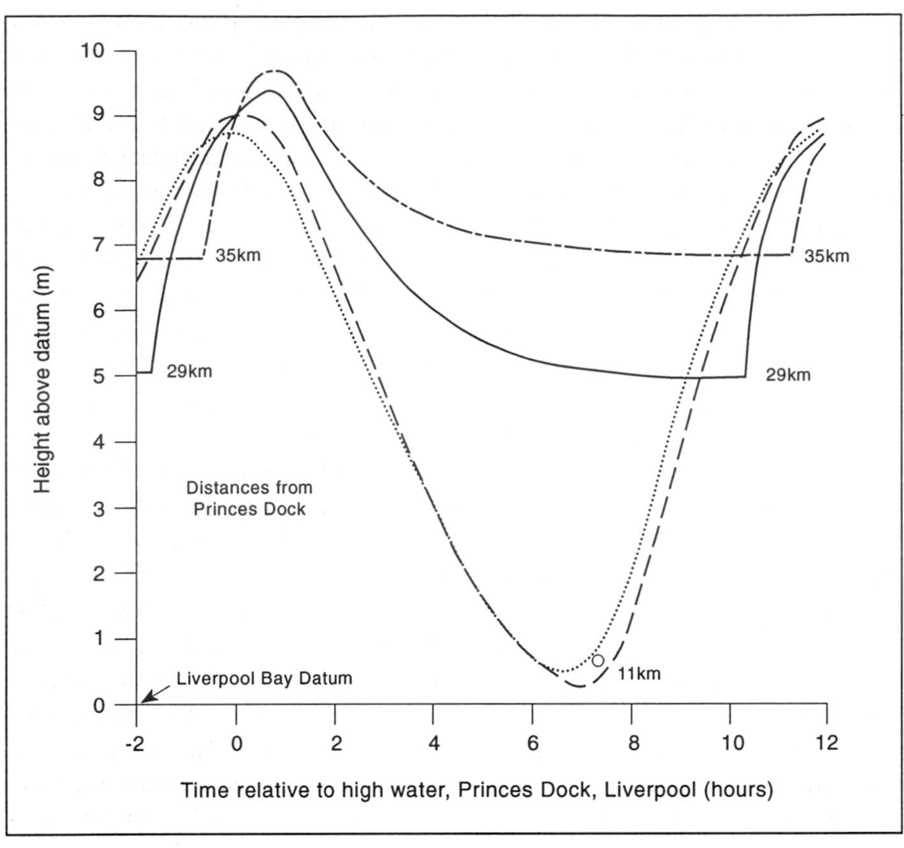

Fig. 5.2 Spring tide recorded in the Mersey Estuary. Note the increasing asymmetry of the tidal wave with distance up the estuary. After McDowell and O'Connor (1977)

Storms

Extreme events such as storms can have a dramatic effect on estuarine sedimentation, producing a pulse of river-borne sediment and discharge. The Susquehanna River, Chesapeake Bay, for example, discharged over 50×10^6 t following Hurricane Agnes in June 1972 which was more than the total discharged in the previous 50 years. During the hurricane, several hundred acres of new intertidal flats were formed in Chesapeake Bay. Calculations suggest that it aged 50 years as a result of this storm, with 17 cm average depth of sediment deposited in the upper reaches (Dyer, 1986).

River deltas

Where rivers draining large drainage basins, producing high sediment concentrations, debouch onto continental shelves a delta may form when the rate of accumulation is greater than the rate of marine removal of sediment, as in the case of the Volga, Ord, Mekong, Amazon, Mississippi and Yangtze deltas. Most deltas form on tectonically inactive, or trailing edge, coasts (Fig. 5.3). Deltas can form at estuarine mouths, or estuaries may be found in some active delta channels. Delta form varies greatly according to the relationship between tidal, wave and fluvial activity, as well as offshore topography. Hoekstra (1993) describes an unusual single-finger delta at the mouth of the Solo River, east Java. This mud-dominated system arises from a drainage basin of 16 000 km^2, carrying a total annual load of *c.* 19×10^6 t. All deltas consist of subaqueous and subaerial portions, and many have active and abandoned parts (which vary in location over time). On the subaerial parts of deltas, complex patterns of distributaries may be produced, and a range of wetland types develop.

The long term history and future of deltas is controlled by the relative rates of submergence (related to tectonic stability) and sea level rise, coupled with sediment availability. Stanley and Chen (1993) used carbon-14 dates to show long term subsidence rates in the Yangtze delta of 0.1–3 mm a^{-1} on land and 1.6–4.4 mm a^{-1} at sea. The great changes in delta deposition possible over relatively short spaces of times are illustrated by the Yellow River delta, China which has a total delta area of 15 000 km^2. Between 1128 and 1855 it formed a large delta, but since then the lower course of the river has moved northwards and now flows into the Bohai Sea. The Yellow River carries an average of 1.6 billion t a^{-1} of sediment (nearly two and a half times that of the Yangtze River), 76 per cent of which is deposited at the mouth. In the last century, erosion has become dominant as no new riverine sediment is being delivered to the delta margin. Consequently, 44 billion m^3 of sediment has been eroded into the ocean, and shoreline retreat has reached 110 m a^{-1} in places (Jin Liu *et al.*, 1987).

Open coast wetlands

Wetlands may form on open coasts where there is a very high sediment supply and a low wave energy environment. Examples include the west Korean coast, the coasts of Surinam and the chenier plain of south-west Louisiana, USA. In England, there are only rare examples because of the prevailing high wave energies, although the Dengie Peninsula marshes in Essex are of this type. In some cases, such wetlands develop in bays, for example, The Wash and Morecambe Bay.

These four environments which act as hosts for wetland development differ greatly, as we have seen, in their complexity, stability and environmental characteristics. Thus, a back-barrier salt marsh in the eastern USA may be influenced by very different factors to a salt marsh in a British estuary, or a Chinese delta setting.

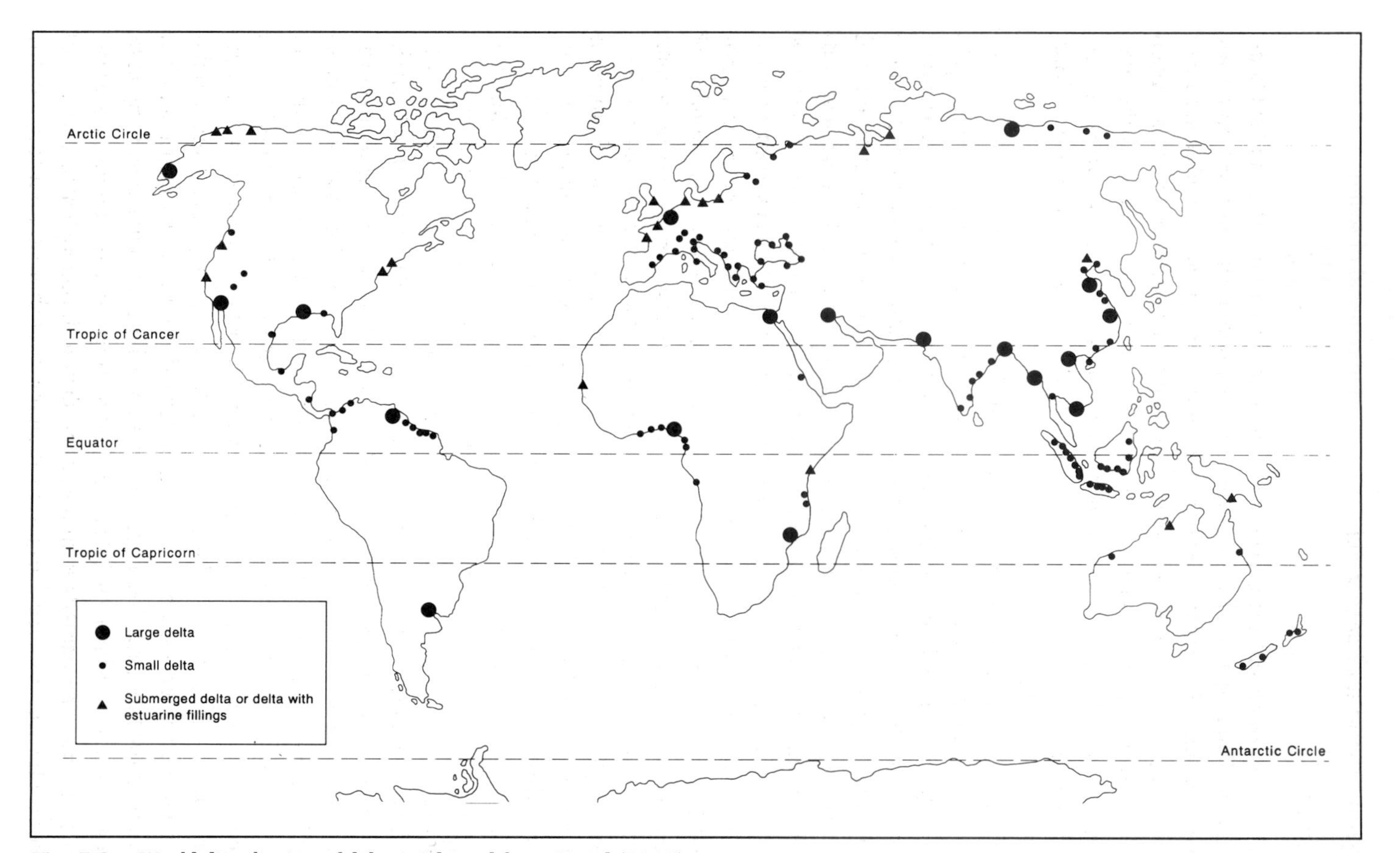

Fig. 5.3 World distribution of deltas. Adapted from Snead (1980)

Types of coastal wetland

Mudflats

In the lower intertidal zone, seaward of mangroves, salt marshes and sabkhas, extensive mudflats (coupled in many cases with coarser sediment units) are often developed. Halophytic vegetation cannot develop because of too-frequent immersion, but there is a wide range of organisms present. Mudflats are often very extensive in macrotidal areas (e.g. 5–25 km wide mudflats on the west coast of South Korea, and similar dimensions at King Sound, north-west Australia where mean spring tidal range is *c.* 9.4 m), but are also significant, if narrower, in mesotidal areas.

Mudflats are far from featureless, with tidal channel networks, terraces, and a variety of 'mud mound' topographies and small scale ripple features. Thus, Wells *et al.* (1990) observed just south of Inchon, South Korea, that the western part of the mudflats was terraced with meandering tributary channels and muddy sediments, whereas the southern part was composed of coarse sediments and had no tributary channels. Mudflats are usually morphologically dynamic, with sediment transferred over tidal cycles. Paterson and Underwood (1990) state that the Severn Estuary, UK, holds an estimated 10 million tonnes of suspended mud, which is thought to cycle between suspension and deposition on the intertidal mudflats. In the Bay of Fundy, Canada, studies on the macrotidal mudflats of Minas Basin have shown that the tidal circulation is produced by the interaction of the tidal wave and the morphology of the intertidal flats in this area of light winds (Perillo *et al.*, 1993). Wells *et al.* (1990) demonstrate how the channels in South Korean mudflats appear to be ebb-dominated, and eroded by storm events; clearly there must be considerable recycling of the sediments thus eroded to account for the maintenance of these mudflats. Other mudflats show clear evidence of erosion. In the Sado Estuary, Portugal, notching and retreat of the upper mudflat surface at 0.8–10.5 cm a^{-1} has recently been observed (Moreira, 1992).

Where mudflats dry out during emersion, mudcracks and runzel marks (which are ripples caused by drying foam) (Klein, 1985) result. Later submersion may produce erosion of sediments in such features. Paterson and Underwood (1990) show how sediment erodibility can change over an emersion–submersion cycle: studies in the Severn and Tamar estuaries, UK, showed that sediment was most easily eroded immediately after emersion. Furthermore, the most stable sediments were found where epipelic diatom populations were greatest, thus suggesting a biological control on sediment stability. Such relationships have been demonstrated in laboratory experiments by both Holland *et al.* (1974) and by Coles (1979); the mucus of the diatoms aids accretion of the mud surface. Other studies have shown the ecological importance of epipelic diatoms. In turbid estuaries diatoms contribute significantly to the estuarine foodweb through their photosynthesis. Furthermore, they act as nutrient sources for birds, mud snails (*Hydrobia ulvae*), and ragworms (*Nereis diversicolor*) (Paterson and Underwood, 1990). In Nauset Marsh, Cape Cod, USA, intertidal mudflats (115 ha, or 12 per cent of the entire back barrier estuary) are dominated by *Cladophora gracilis* with a net primary productivity of 59–63 g m^{-2} a^{-1} (compared with *Spartina alterniflora* here which has an above ground net primary productivity of 664 g m^{-2} a^{-1}), according to Roman *et al.* (1990).

Clearly, mudflats act as an important transitional element for sediments and nutrients between marshes/mangroves to the landward, and the continental shelf to the seaward (Plate 5.1). Indeed, Pethick (1992) draws the analogy between the salt marsh/mudflat system and the dune/beach system.

Salt marshes

In the upper tidal zone, landward of mud and sandflats, salt marshes often form. Found in a wide range of climatic conditions, salt marshes occur from the Arctic to the tropics, where they are mainly replaced by mangrove associations. Salt marshes may occur in the tropics at high levels behind mangroves, according to Chapman (1977a), with common species being *Sesuvium portulacastrum* and *Batis maritima*. Well-studied and impressive salt marshes occur along the north Norfolk barrier coast, England, in Louisiana and Georgia, USA, the Netherlands, and the Severn Estuary, England. The Atlantic coast of the USA contains 589 480 ha of salt marsh (Reimold, 1977). Interesting marshes occur in south-eastern Tunisia (Oueslati, 1992) where vegetation is adapted to high salinity and aridity and thus dominated by *Salicornia arabica*, *S. fruiticosa*, *S. perennis* and *S. radicans*.

Characteristically, salt marshes are of two distinctive types, i.e. low, or pioneer, marsh and high, or mature, marsh. Low marshes, sometimes separated from mudflats by a small mud cliff (e.g. those in the Sado Estuary, Portugal where the mud cliff is 0.5–2 m high (Moreira, 1992)), are usually dominated by a few plant species. In the USA salt marshes, *Spartina alterniflora* is often the dominant species with *S. patens* and *Distichlis spicata* in more landward habitats (Frey and Basan, 1985). In the Sado Estuary, Portugal, *Spartina maritima* dominates. Low parts of the Hut Marsh, Scolt Head Island, eastern England are dominated by *Aster tripolium*, whereas the highest surfaces have a mixed community of *Limonium vulgare*, *Armeria maritima*, *Puccinellia maritima*, *Spergularia maritima* and *Triglochin maritima* (French and Spencer, 1993). High marshes contain a greater diversity of plant species (e.g. species from the genera *Salicornia*, *Suaeda*, *Limonium*, *Atriplex* and *Halimione*). Marshes also vary in the type of sediment found. Many marshes in the USA, for example, have a high organogenic (peat) component, whereas most British marshes are more

Plate 5.1 The transition from mudflat to salt marsh: Cockle Bight, Scolt Head Island, north Norfolk coast, England

minerogenic in content (Allen and Pye, 1992); these differences become important when discussing marsh response to sea level change. Around the UK, salt marsh sediment types vary from muddy (e.g. the Severn estuary) to sandy (e.g. Solway Firth and Morecambe Bay). These different sediments support a range of marsh cliff types (Allen, 1989).

Salt marsh geomorphology

Marsh geomorphology, at the small scale, consists of a range of depositional and erosional features, i.e. channel networks, cliffs, salt pans and creek bank levées. The marsh channel network, often dendritic in nature, is a key component in the functioning of salt marshes, bringing in (and removing) sediments, water and nutrients. Pethick (1992) suggests that the morphology and dimensions of creek networks depend on the tidal prism during the flood phase, and marsh creeks develop as a morphological device to dissipate tidal wave energy. Thus marsh creeks can be seen as another way in which the marsh is linked to the mudflat.

Salt marsh ecology

The flora of salt marshes, although often relatively restricted in diversity, may be extremely productive. In the 330 ha Nauset Marsh, Cape Cod, USA, for example, Net Primary Productivity figures have been measured by Roman *et al.* (1990) as follows:

Marsh surface:	*Spartina alterniflora* (above ground)	664 g C m^{-2} a^{-1}
	Zostera marina	444–987 g C m^{-2} a^{-1}
Creek banks:	*Ascophyllum nodosum* (fucoid algae)	1179 g C m^{-2} a^{-1}
	Fucus vesiculosus (fucoid algae)	426 g C m^{-2} a^{-1}

In the USA, productivity decreases with latitude and increases with tidal range. On a smaller scale in *Spartina* marshes, the grass tends to be taller and more productive on the creek bank levées than in the marsh interiors. This may relate to nitrogen availability or to physiological stress in marsh interiors from higher interstitial salinities (for review see Mann, 1982).

Elsewhere there is usually a clear vegetation zonation in a landward direction across salt marshes. Stress tolerance is the most important quality in vegetation at the lower lying, seaward parts of the marsh, and competition plays a bigger role on drier areas. Salt marsh vegetation also shows clear successional trends over time, as accretion leads to changing soil conditions which permit a different range of plants to grow. The plants themselves play a vital role in the development of marsh surfaces, promoting accretion and improving the shear strength of marsh soils. Van Eerdt (1987) found for marsh cliffs in the Oosterschelde, the Netherlands, that root density was the dominant control on the tensile strength of cliff material, followed by clay content (explaining 62 per cent and 11 per cent of the variance, respectively). Although both Alizai and McManus (1980) and Stumpf (1983) have argued for considerable sediment retention on the reed *Phragmites communis* and *Spartina alterniflora*, respectively, French and Spencer (1993) have found that for inorganic marshes at Scolt Head Island, north Norfolk, UK, direct plant encouragement of accretion is minimal and direct settling is the dominant process.

Field studies in Georgia, USA salt marshes in the 1960s suggested large-scale exports of organic detritus and led to the 'outwelling hypothesis' (Odum, 1971) of nutritive material being supplied by marshes to coastal waters. However, more recently it is clear that patterns of organic production export are more complex: exports of 40 per cent (Bataria Bay, Louisiana, Day *et al.*, 1976) can be compared with negligible export or even seasonal imports elsewhere. Use of carbon isotope tracers has shown the importance of algal pathways in wetland systems. In some marshes it is clear that plant tissues decompose beneath a layer of algae and that it is algal mats and associated diatoms which are exported from the marsh rather than plant material. It is also clear that fungi and bacteria play an important role in converting plant materials to soluble organic matter; some nutrients are then recycled within, or exported from, the marsh in proportions which differ from location to location (Mann, 1982).

Salt marsh fauna include many species of invertebrate, several of which are plant-eating, or phytophagous. Birds either breed in salt marshes (e.g. in Britain, redshank, oystercatchers and lapwings), or are seasonal visitors (e.g. the many species of wildfowl which winter in British salt marshes). On the Avon marshes around the Severn Estuary in England, for example, at least 250 bird species have been recorded this century, including around 140 which are regular visitors. Dunlin breeding here has declined recently, because of the many disturbances to the salt marshes (Rose, 1990). Grazing, or herbivory, is an important component of some marsh ecosystems with insects, crabs and geese as common marsh grazers. The snow goose *Anser caerulescens* is an important grazer in east coast USA marshes, producing 'eat outs' where biomass is reduced over large areas. Erosion may often result from such modifications to the vegetation.

Mangrove swamps

Associations of mangrove trees growing in muddy coastal environments are often called mangrove forests or mangals. They have much in common with salt marshes, often characterized by a tidal creek network, and often showing a clear ecological zonation (Plate 5.2). However, they differ fundamentally in their aerial storage of biomass: the individual plants take the form of shrubs or trees and the community generally has the appearance of a forest. They range from fully saline conditions near mean sea level to brackish water settings in estuarine and other environments with restricted marine exchange. Usually found in sediment-rich, sheltered coastal environments, mangroves can grow on rocky coasts as well. Mangroves extend from Bermuda (32° 23′ N) in the north to Australia (38° 45′ S) in the south (Woodroffe, 1990), and have been estimated as occupying around 20×10^6 ha (Twilley *et al.*, 1992). Snedaker (1993) quotes a figure of 22 million ha. Impressive mangrove areas include West Africa (especially the Niger delta), the Ganges delta and islands in the Bay of Bengal in India and Bangladesh (Blasco, 1977), and the 1600 km stretch of high Avicennia forest along the Guiana coast (West, 1977), as well as large areas in northern Australia and Malaysia (Chapman, 1977b). Mangrove forest floras have been well studied in many areas, and the distribution of mangrove tree species is related mainly to past geographical spread, and thus to plate tectonic history. A major floristic contrast can be drawn between a taxonomically rich Old World of *c.* sixty species and a depauperate New World flora, the latter comprising only four genera and ten species of tree.

Woodroffe (1990) recognizes three major settings in which mangrove forests occur: river-dominated environments (e.g. deltas, such as the Fly River delta in

Plate 5.2 Pioneer *Rhizophora* mangrove, Rufiji delta, Tanzania, East Africa

Papua New Guinea), tide-dominated environments (e.g. estuaries, such as South Alligator River, Northern Territory, Australia) and carbonate settings (e.g. associated with coral reef islands, such as Grand Cayman and south-west Florida (Fig. 5.4). The major difference between these environments relates to sediment supply. The sediment is allochthonous in river- and tide-dominated environments and autochthonous in carbonate settings. Data on long (*c.* 6000

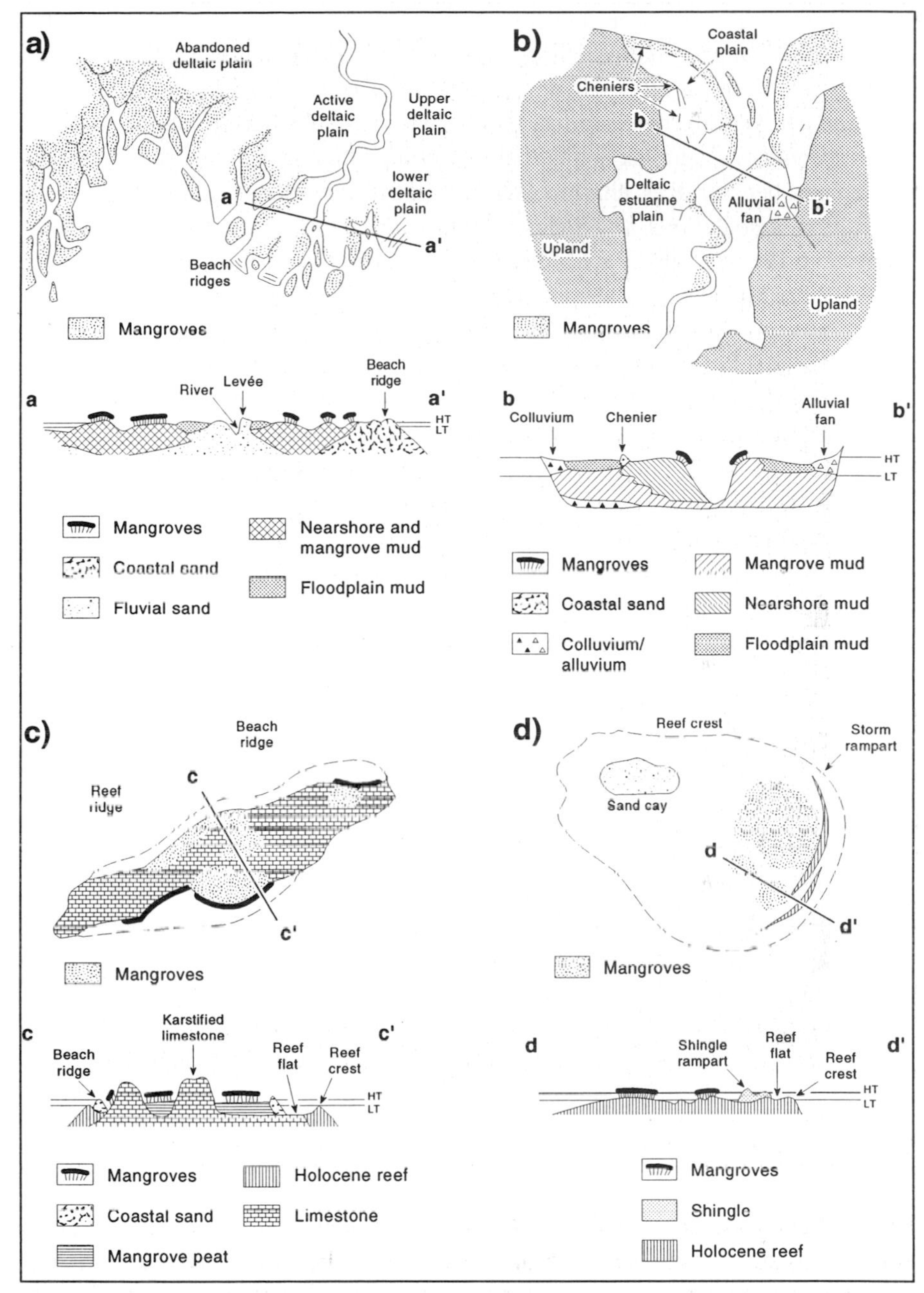

Fig. 5.4 The major mangrove environments a) river dominated, b) tide dominated, c) and d) carbonate settings. After Woodroffe (1990)

years) and short term accretion rates in mangrove forests collected by Twilley *et al.* (1992) suggest that reef mangroves accrete at rates of < 1 mm a^{-1}, basin mangroves accrete at 1–2 mm a^{-1}, and riverine mangroves accrete at > 2 mm a^{-1}. Within these settings there are a range of different mangrove habitats, especially in complex deltas. These include alluvial deposits, salt flats, chenier ridges, beach ridges, channel levées, and they are characterized by different vegetation associations and sediment types. In delta areas, landform assemblages can change rapidly with distributary development and decay. As Thom (1967) noted in Tabasco, Mexico, there is not a strict vegetation succession in deltaic mangroves, but rather different species colonize different geomorphological settings in a more opportunistic patterning as sedimentation foci change over space and time. Most deltaic settings are dominated by river-borne sediments, although extensive studies in northern Australian estuaries have shown that here sediments of marine origin play a vital role (Woodroffe, 1990, 1993). Mangroves, like salt marshes, have developed over the Holocene in relation to an initially fast and then declining sea level rise, as revealed by stratigraphic studies of the suite of sediments under mangroves today. Thus, South Alligator River estuary, northern Australia has been subjected to intensive drilling (131 bore holes), pollen analyses and carbon-14 dating which show the following phases:

1. 8000–6800 years BP: an initial marine transgression and development of mangroves, as sea level rose nearly 12 m
2. 6800–5300 years BP: the 'big swamp' phase of widespread mangrove forest
3. 5300 years BP onwards: sedimentation in mangroves continued, sometimes reaching above high tide levels, and the tidal river developed a sinuous form (Woodroffe, 1993).

Twilley *et al.* (1992) have estimated the global storage of carbon in mangrove biomass as 4.03 Pg C, 70 per cent of which occurs between 0 and 10° latitude. In general, mangrove biomass and tree height decrease with increasing latitude, reflecting the dominant control of solar energy, compounded locally by influences of topography, hydrology, soils and salinity. Average wood production in mangroves also decreases with increasing latitude. Maximum mangrove development occurs where there is high precipitation or freshwater runoff (Snedaker, 1993), because fresh water helps decrease salinity and delivers nutrient-rich sediment. Aridity, hypersalinity, lack of coastal protection and winter frosts all act to inhibit mangrove growth.

Like some salt marsh species, mangroves show adaptations to their environment. Anchorage in mobile muddy substrates is facilitated by extensive shallow root networks (*Avicennia*), prop roots (*Rhizophora*) or buttresses (*Laguncularia*). Many mangroves also show adaptations to physiological stresses, particularly salinity stress and anaerobic conditions. Some mangroves, including *Aegiceras* and *Aegialitis* species have salt-secreting glands on their leaves whilst *Avicennia* produces large numbers of rubbery breathing roots, or pneumatophores, which extend above the swamp surface. The pneumatophores aerate the below-ground tissues which are unable to obtain oxygen from waterlogged soils.

Primary productivity in mangroves is often very high. A study of litter production in mangrove in Guadeloupe found a mean production (measured over 2 years) of 2.37 g dry weight m^{-2} day^{-1} (8.65 t ha^{-1} a^{-1}). Three mangrove species were involved, i.e. *Rhizophora mangle, Avicennia germinans* and *Laguncularia racemosa.* Over 81 per cent of the litter was leaf material, nearly half from *R. mangle.* Leaf litter production increased in the warm, wet season and the main minerals involved were sodium and sulphur (Imbert and Portecop, 1986). Mangroves, like salt marshes, are open ecosystems, with important interchanges of nutrients, sediments and energy between landward and adjacent coastal and offshore ecosystems, but can also act as nutrient sinks (Twilley, 1988). Studies in a *Rhizophora* mangrove forest in Florida showed that *c.* 5 per cent of total leaf production is consumed by terrestrial grazers but that the rest enters the aquatic system as organic debris (Odum and Heald, 1975). Detritus consumers function as either grinders, deposit feeders or filter feeders and include fish, worms, midge larvae and a wide range of crustaceans, including copepods, amphipods and shrimps. Chapman (1977a) lists six faunal habitats in mangroves, i.e. tree canopy (especially for birds, mammals and insects), rot holes in branches (e.g. mosquito larvae), soil surface (e.g. mudskippers, snails and hermit crabs), the soil (nereids, snails and crabs), permanent and semi-permanent pools (e.g. mosquito larvae, crabs and tadpoles) and channels (crocodiles, alligators and fish). As with salt marshes and mudflats, there are important links between plants, animals and sedimentation. Thus the amazing aerial root networks aid the accretion of sediment, whilst crabs play a key role in litter dynamics in many mangroves. Twilley *et al.* (1992) discuss the finding that most leaf litter on the mangrove forest floor in Ecuador is harvested by the crab *Uridies occidentalis* and moved into its burrows.

Several fish and sea food species utilize mangroves for part of their life cycle. A recent study from New Caledonia looked at the interactions between fish in mangroves (262 species collected) and in nearby coral reefs (735 species recorded). Some species were found to use mangroves as nursery sites, but only forty-three species in the St Vincent Bay area occurred in both reefs and mangroves. This study shows that here at least the importance of mangroves for reef fish is less than usually thought (Thollot, 1992). Nevertheless, mangroves

Plate 5.3 Salt flats and mangrove swamps near Derby in tropical semi-arid, macro-tidal (equinoctical spring tidal range: 11.5 m) King Sound, north-western Australia (photograph: H.A. Viles)

offer important habitats for many species, as at Selangor, Malaysia, where 219 species of fish and nine species of prawns were found in the mangrove creeks (Sasekumar *et al.*, 1992).

Sabkhas

Coastal sabkhas are extensive salt flats found in arid and semi-arid settings along tectonically stable coastlines. They are characterized by very low gradient slopes. Good examples are found in the Arabian Gulf (Fig. 5.5), Baja California, Mexico and Sinai, Egypt. Smaller salt flat areas are found in King Sound, north-western Australia (Semeniuk, 1981 and Plate 5.3), near salt marshes in the Gulf of Gabes, Tunisia (Oueslati, 1992), and behind a barrier lagoon coast in Mediterranean Libya (Jelgersma and Sestini, 1992). There are also sabkhas on the Red Sea coast of the Yemen Arab Republic, with salt pools in the lower parts (Youssef, 1991). The hot, arid conditions coupled with high sea temperatures and a sheltered environment, lead to the formation of displacive and replacive evaporite minerals in the capillary zone above a saline water table. The Arabian Gulf possesses a series of sabkhas which stretch

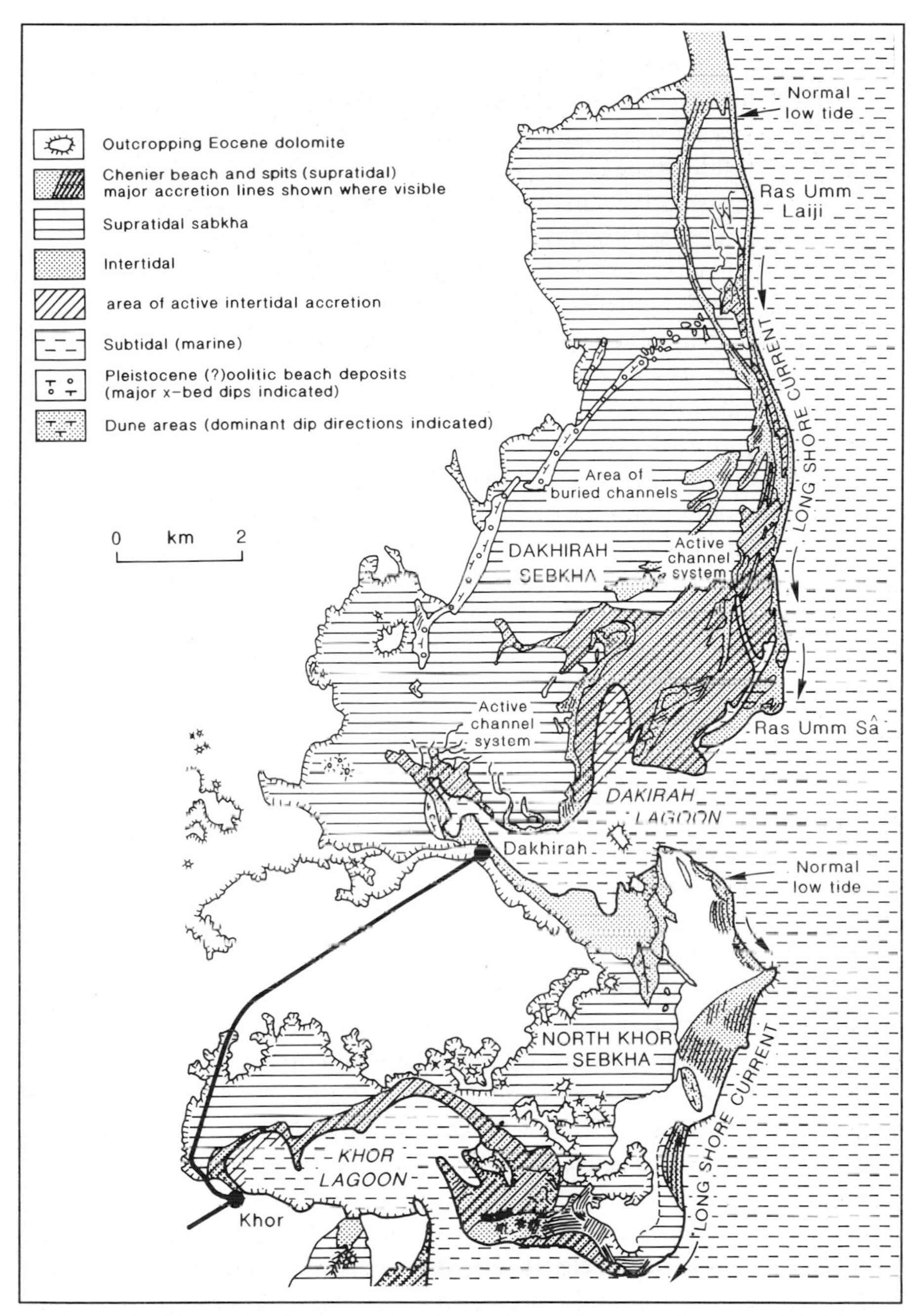

Fig. 5.5 Sabkhas and other coastal sediments in north-eastern Qatar. Modified from Shinn (1973)

approximately 320 km along the coast of the United Arab Emirates (UAE), reaching up to 24 km wide. The sabkhas here are found to the landward of coral reefs and beach ridges, and have formed over *c.* 7000 years (Warren, 1989). The seaward edges of the sabkhas, in the intertidal zone (1–2 km wide), have algal mats forming stromatolites, with patches of *Avicennia marina* (Purser and Evans, 1973) and burrowing crabs. The supratidal zones (which are < 8 km wide) are characterized by gypsum-, anhydrite-, dolomite- and halite-rich sediments. The sediments are generally muddy carbonate materials, but contain coarser materials including siliciclastic sand. Shinn (1973) identified relict tidal channels on supratidal sabkha surfaces in the Qatar Peninsula. In Bahrain, there are two large sabkhas in the south and south-west of the island. The south-western one is approximately 60 km^2 in extent and most of its surface, apart from dunes, is < 3 m above sea level. There is a beach ridge complex to seaward. Both sabkhas are largely devoid of vegetation, and extensive areas are inundated after extreme high tides and as a result of heavy rain. The surfaces are salt encrusted, with gypsum precipitated through the evaporation of groundwater brought to the surface by capillary rise (Doornkamp *et al.*, 1980). Plants such as *Sporobolus arabicus*, *Zygophyllum qatarense* and *Suaeda vermiculata* (which is used locally as a medicine) are found in some parts of the Bahrain sabkhas on soils where salinities (estimated from conductivity measurements) range from 7.6 to 45.5 mS cm^{-1} (Abbas and El-Oqlah, 1992). The sabkhas here often grade into mangrove and salt marsh vegetation, with stunted forms of *Avicennia marina* (only 1 m high), and *Salicornia europaea* and *Halocnemum strobilaceum*, as at Sitrah Island.

There is much potential for crust formation on sabkha surfaces. Algal mats become cemented with aragonite, for example, halite crusts form sporadically on supratidal flats and supratidal aragonitic tufas (called conialites by Purser and Coreau, 1973), coat cliffs and beach rocks. Groundwater in sabkhas is often highly saline. In some parts of the Umm Said sabkha, Qatar, it was found to be saturated with sodium chloride (de Groot, 1973). Evaporative pumping leads to concentration of brine in groundwater and the precipitation of evaporative minerals in sabkha soils (Akili and Torrance, 1981). Thus although sabkhas are not a major world coastal environment, they clearly present many challenges for coastal development in the rich Middle Eastern oil states.

Sea level rise and wetland growth and destruction

The preceding reviews show that the development of wetland environments is controlled by the changing balance between tidal regime, wind–wave climate, sediment supply, relative sea level and wetland vegetation (Reed, 1990; Allen and Pye, 1992). Marshes and swamps grow through processes of vertical accretion of sediment on the wetland surface and progradation at the edges of the marsh or swamp. Rates of marsh growth can be impressive: a study using old maps, and air photographs from the late 1700s over a 190 year period, revealed the growth of an 8 km^2 salt marsh behind a barrier spit at Wells, Maine, USA (Jacobson, 1988). Furthermore, impressive flat and marsh progradation occurred over almost 7000 years in the Holocene along the north-east coast of Argentina, where an average extension of 30 km seaward occurred behind developing barrier islands (Codignotto and Aguirre, 1993).

Models for the growth of salt marshes

Are there characteristic responses of wetlands to sea level change? It is possible to calculate a measure of salt marsh accretionary performance by comparing vertical accretion rate with estimates of sea level rise in common units of millimetres per annum (mm a^{-1}): if positive, the marsh will be in 'accretionary surplus', if negative in 'accretionary deficit' (French, 1994). Studies in the USA (Stevenson *et al.*, 1986; Craft *et al.*, 1993) have suggested that accretionary balances become more positive with increasing tidal range (Fig. 5.6) as internal organic sedimentation is topped up by inorganic inputs from tidal flows. However, in marshes characterized by externally derived, tidally introduced inorganic sediments, patterns of sedimentary performance are more complex and there is no simple relationship between accretionary balance and tidal range (Wood *et al.*, 1989; French, 1994).

However, the data from Scolt Head Island, UK (Pethick, 1981; French, 1994) are instructive: North Cockle Bight (NCB) is a young marsh, at the point of transition from mudflat to vegetated surface; Missel Marsh (MM) is *c.* 60 years old and Hut Marsh (HM) is *c.* 100 years old. Thus, tidally dominated marshes develop rapidly at early stages in their history but accretion rates slow through time as progressively higher surfaces are flooded less and less frequently. Eventually, the marshes stabilize at an 'equilibrium level'; typically in north Norfolk this is reached after *c.* 300 years. Interestingly, the 'mature' marsh surface in this area is found at an elevation of 0.6–0.8 m below the height reached by the highest astronomical tide (Fig. 5.7). This level represents the point at which declining sediment inputs from infrequent high spring tides just manage to offset the regional sea level rise (due in equal measure to regional tectonic subsidence and recent eustatic sea level rise) of 2 mm a^{-1}. This model is just one of a suite of potential models concerning the relations between wetland response and sea level change. Such models have been explored by both French (1991, 1993, 1994) and Allen (1990b, 1990c), using variants of the following equation:

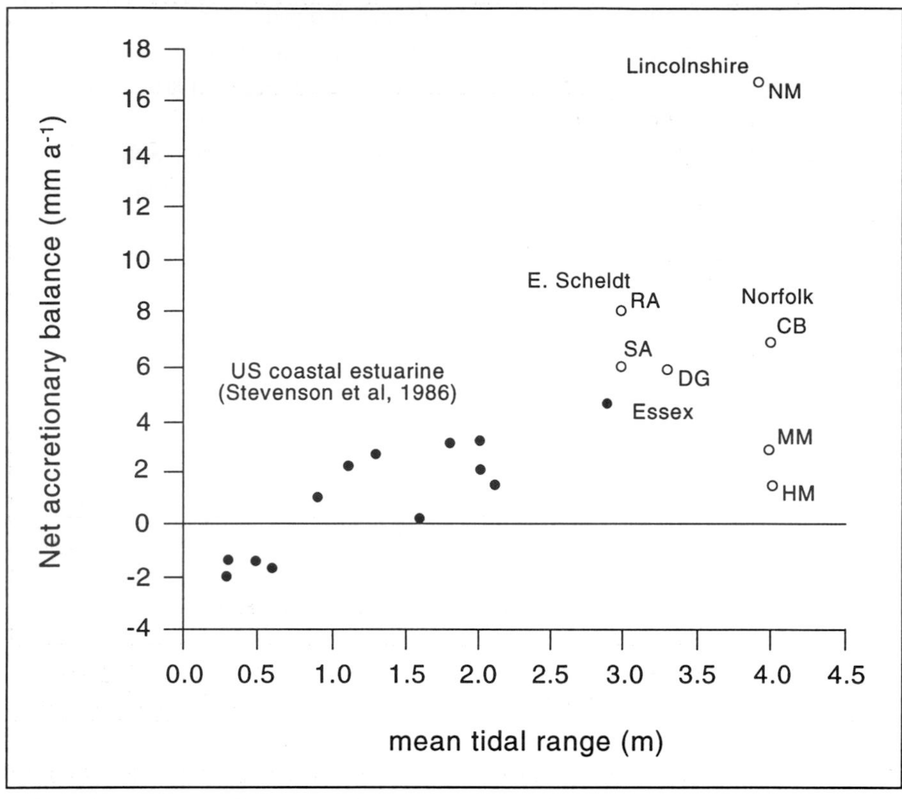

Fig. 5.6 Plot of accretionary balance (mean annual accretion − local relative sea level rise) versus mean tidal range. After French (1994): CB = Cockle Bight, Scolt Head Island; MM = Missel Marsh, Scolt Head Island; HM = Hut Marsh, Scolt Head Island; DG = Dengie, Essex; NM = New Marsh, Gibraltar Point; RA = Rattekaai, East Scheldt, the Netherlands; SA = St Annaland, East Scheldt, the Netherlands

$$\frac{dE}{dt} = \frac{dS_{ex}}{dt} + \frac{dS_{int}}{dt} - \frac{dM}{dt} - \frac{dP}{dt}$$

where:

- dE = change in elevation (relative to a local tidal datum)
- dS_{ex} = elevational change due to externally derived, largely inorganic sediments
- dS_{int} = elevational change due to internally derived, largely organic sediments
- dM = relative sea level rise (positive upwards)
- dP = surface lowering due to compaction (positive downwards)
- t = time

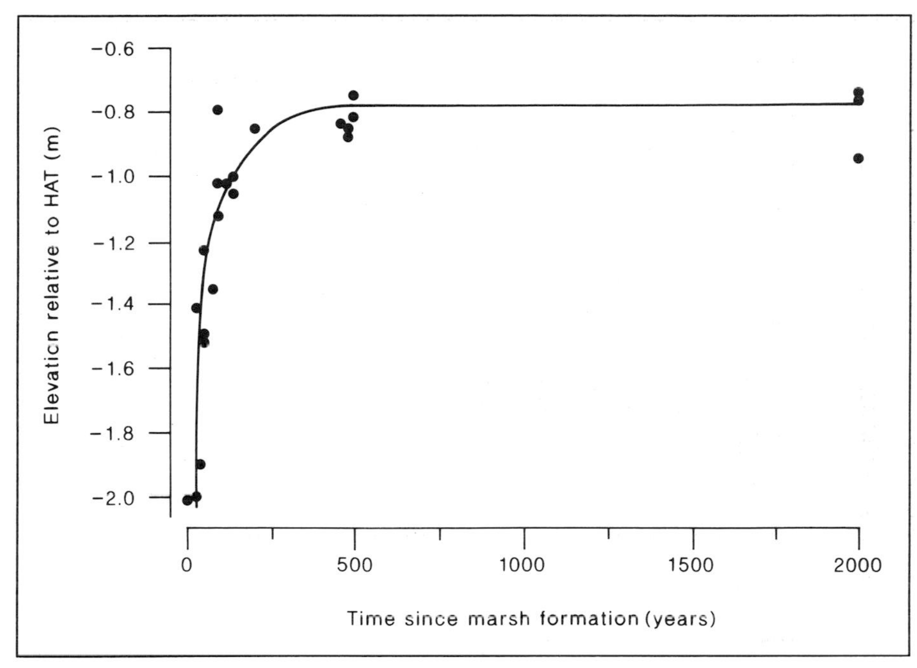

Fig. 5.7 Asymptotic increase in mean surface elevation, relative to HAT (Highest Astronomical Tide) with increasing marsh age in north Norfolk. After French (1993)

Various pathways of marsh development are shown in Fig. 5.8. Marshes may, of course, alternate between these different modes of operation: thus the marshes of the macrotidal Severn estuary, UK show sequences of alternating inorganic- and organic-dominated sedimentation during the Holocene (Allen, 1990b). The Norfolk example described above corresponds to trajectory ABCD, where compaction effects are likely to be minimal and there is a strong feedback loop between elevation, tidal submergence and sedimentation. By comparison, other trajectories show that systems dominated by high rates of organic peat accumulation are not constrained by sea level controls and may reach the upper limit (JKLM) or even grow above the tidal frame (EFGH). However, these linkages are more problematic than those associated with tidal inputs because changes in tidal submergence have implications for the vegetation communities which provide the organic debris to raise wetland surfaces. In the rapidly subsiding wetlands of Louisiana (see Case study 5.1), the marshes are dominated by organic inputs from the marsh grass *Spartina*. Calculations suggest that a 10 mm rise in sea level in this region requires a carbon input of 2–4 kg m^{-2} a^{-1}, a figure corresponding to most of the annual production, to prevent marsh submergence. In fact, under actual rates of sea level rise of > 1.2 cm a^{-1} surface waterlogging leads to reduced *Spartina* productivity and

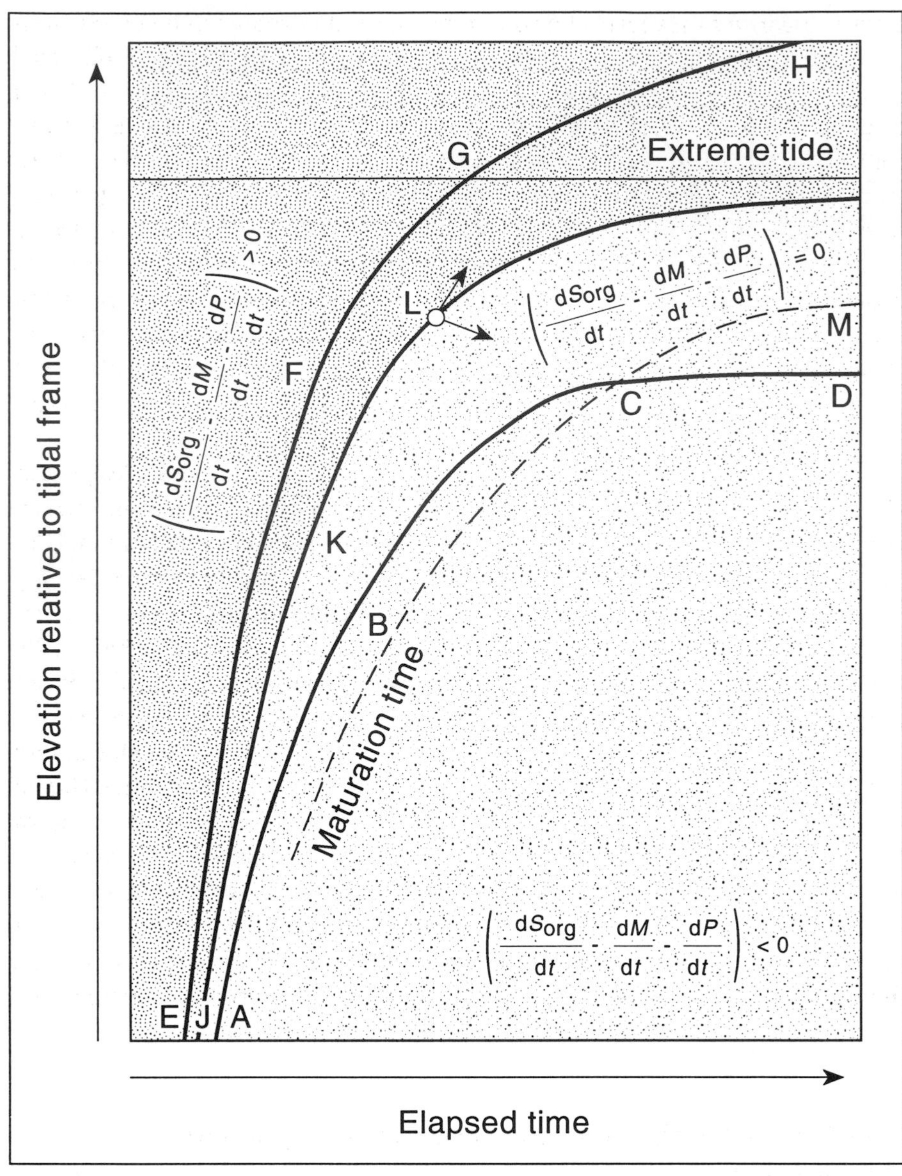

Fig. 5.8 Salt marsh growth models. See text for detailed explanation. After Allen (1990b)

plant die-back, probably from inadequate nitrogen uptake and sulphide toxicity (Koch *et al.*, 1990). Thereafter the wetlands disintegrate and convert to mudflats or open water.

Useful as modelling exercises are, they only give a one-dimensional view of marsh response to sea level change. In fact, accretion rates can be highly variable over a marsh surface. Five-year monitoring at Hut Marsh, Scolt Head Island, north Norfolk, UK, has shown accretion rates to vary between 1 and 8 mm a^{-1} (Fig. 5.9). Although elevation, through numbers of submergences, controls the general pattern of accretion, with the lowest rates on the highest marsh surfaces, this relationship is complicated by sedimentation patterns related to marsh surface flooding pathways and thus distance from feeder creeks, to such an extent that, on the small scale and on individual tides, the usual height–sedimentation relationships can be reversed (Fig. 5.9; French and Spencer, 1993). Complex patterns of sedimentation mean that simple averages of accretion rates from marsh sites may not accurately measure the volumetric sediment sink and thus misrepresent the actual degree to which marsh accretion is able to keep pace with sea level rise (French *et al.*, 1994b).

Sea level rise also leads to changes in wetland margins. Floristically rich upper marshes may disappear under the landward retreat of enclosing barriers and erosion may cause the retreat of creek banks or seaward marsh margins. Open-coast marshes in Essex, eastern England are currently experiencing retreat rates averaging 0.7 m a^{-1} (Harmsworth and Long, 1986) and this erosion is coupled with an average surface accretion rate of 9 mm a^{-1} (Fig. 5.10b; Reed, 1988). Similarly, the marshes of the eastern Scheldt, the Netherlands are being eroded back at *c.* 1 m a^{-1} with this erosion loss being accompanied by surface accretion rates of up to 15 mm a^{-1} (Oenema and DeLaune, 1988). Oueslati (1992) illustrates how erosion at the seaward side of salt marshes in the Gulf of Gabes, Tunisia (which proceeded between 1983 and 1988 at a rate of up to 1 m a^{-1}) was accompanied by onshore extension of the marshes on to sabkha surfaces at rates of *c.* 0.42–3.28 m over the same period. Such erosion–accretion relationships have been investigated more generally by Phillips (1986) in the context of a modified Bruun rule (Fig. 5.10a ; *see* Chapter 3). This analysis shows that recycling of fringe erosion is not sufficient to maintain marsh area except under very low landward slopes.

Problems affecting coastal wetlands — upsetting the balance

As we have seen in the previous sections, coastal wetlands are in a dynamic balance between acting as a source or a sink for sediment, organic matter and nutrients. Natural events, such as cyclones, abrupt tectonic movements and plant die-back, may upset the balance in the short term. Similarly, human activities can have important results in the short and longer term. Thus, for example, Indonesia (which contains the world's largest area of mangroves) had

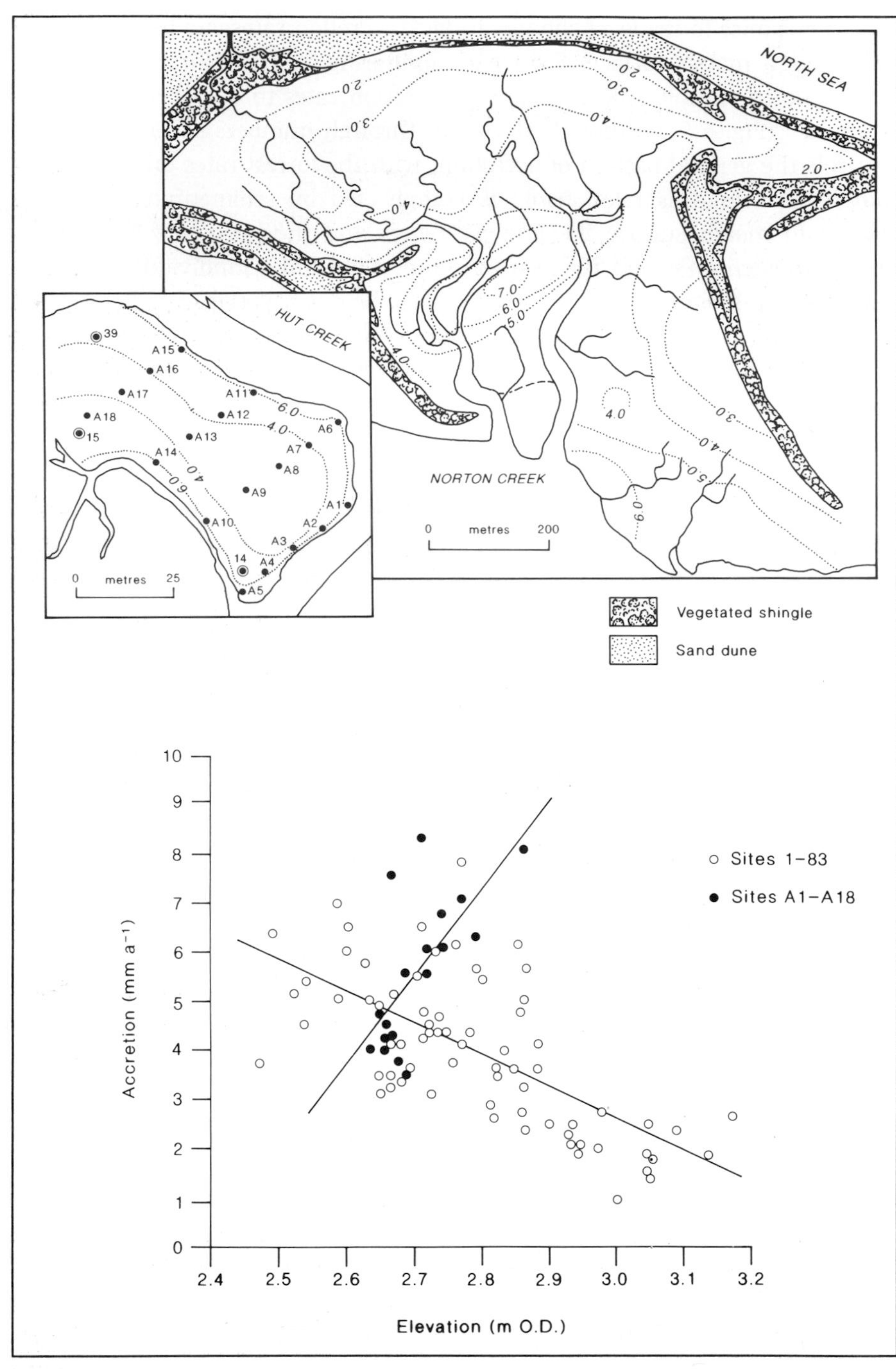

Fig. 5.9 Patterns of salt marsh accretion at Hut Marsh, Scolt Head Island, England, at varying scales; marsh-wide sites 1–83 (top) and local sites A1–A18 (inset). After French and Spencer (1993)

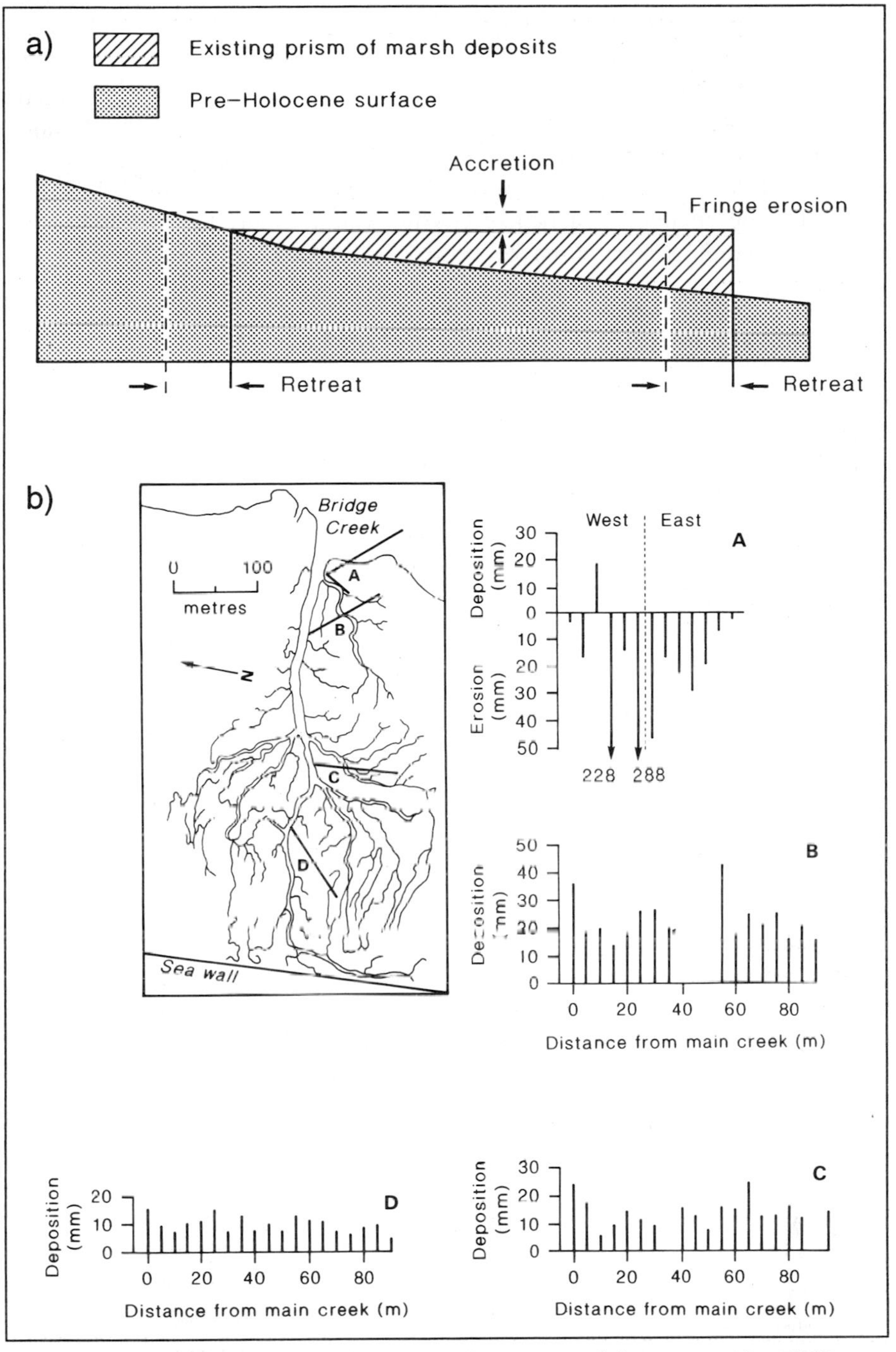

Fig. 5.10 a) the Bruun rule applied to marsh erosion and deposition. After Phillips (1986). b) Marsh edge erosion and marsh surface deposition, Dengie Marshes, Essex. After Reed (1988)

lost *c.* 40 per cent of its mangroves by the late 1980s because of logging, transformation to agriculture and creation of brackish water aquaculture (Giesen, 1993). Over the longer term, changes in the rate of sea level rise (perhaps in the future because of increased atmospheric carbon dioxide) are a major control of coastal wetland development. In the rest of this section, we discuss the human impacts affecting the sedimentary balance, the hydrology, the biological balance and the chemical balance of coastal wetlands, before considering the future confounding effect of sea level rise.

Upsetting the balance between erosion and accretion

Human activities on and around many coastal wetlands have direct and indirect impacts on erosion and accretion and water movements. In many cases the exact contribution of any single activity can be very difficult to identify, as the interaction of a range of processes with natural changes may be involved. Erosion may be encouraged by: removal of protective barriers, removal of wetland vegetation, direct removal of wetland sediment, reduction of sediment inputs, and reclamation of land to the landward of wetlands, thereby restricting their onshore migration and growth. In areas where erosion is naturally enhanced, such changes may be catastrophic. In the western Niger delta, Nigeria, for example, spectacular rates of erosion have been recorded, with 32–39 m a^{-1} average loss between 1973 and 1981 (from air photograph evidence), leading to 487 ha of coastal plain and > 2.5 million m^3 of soil lost to the Atlantic Ocean. Deforestation of mangroves was found here to contribute to the erosion, which was mainly caused by subsidence of the continental margin following oil and gas extraction (Ebisemiju, 1987).

Large areas of salt marsh have been reclaimed, including low-lying areas around the Netherlands (Fig. 5.11) and the UK. Early reclamation was dominantly for agriculture, but more recently land has been reclaimed for industrial, urban and waste disposal uses: thus Doody (1992) describes how over 2000 ha of intertidal land in the Tees estuary, UK was reclaimed up to 1979 for port facilities, oil refineries and a power station. Mangrove swamps have suffered a similar fate throughout the tropics, for example in Sierra Leone, where reclaimed land is used for rice production, Puerto Rico, for sugar cane, and Cambodia for coconut palms (Walsh, 1977). Many coastal wetlands are now threatened by development schemes such as barrages and marinas. In Britain, the Cardiff Bay barrage is planned to create a freshwater lake instead of tidal mudflats, which are home to 5000 dunlins and redshanks, at an estimated cost of £150 million.

In many areas changes in wetland extent and quality may be difficult to predict. Thus, Doody (1992) discusses how large areas of the upper Dee estuary

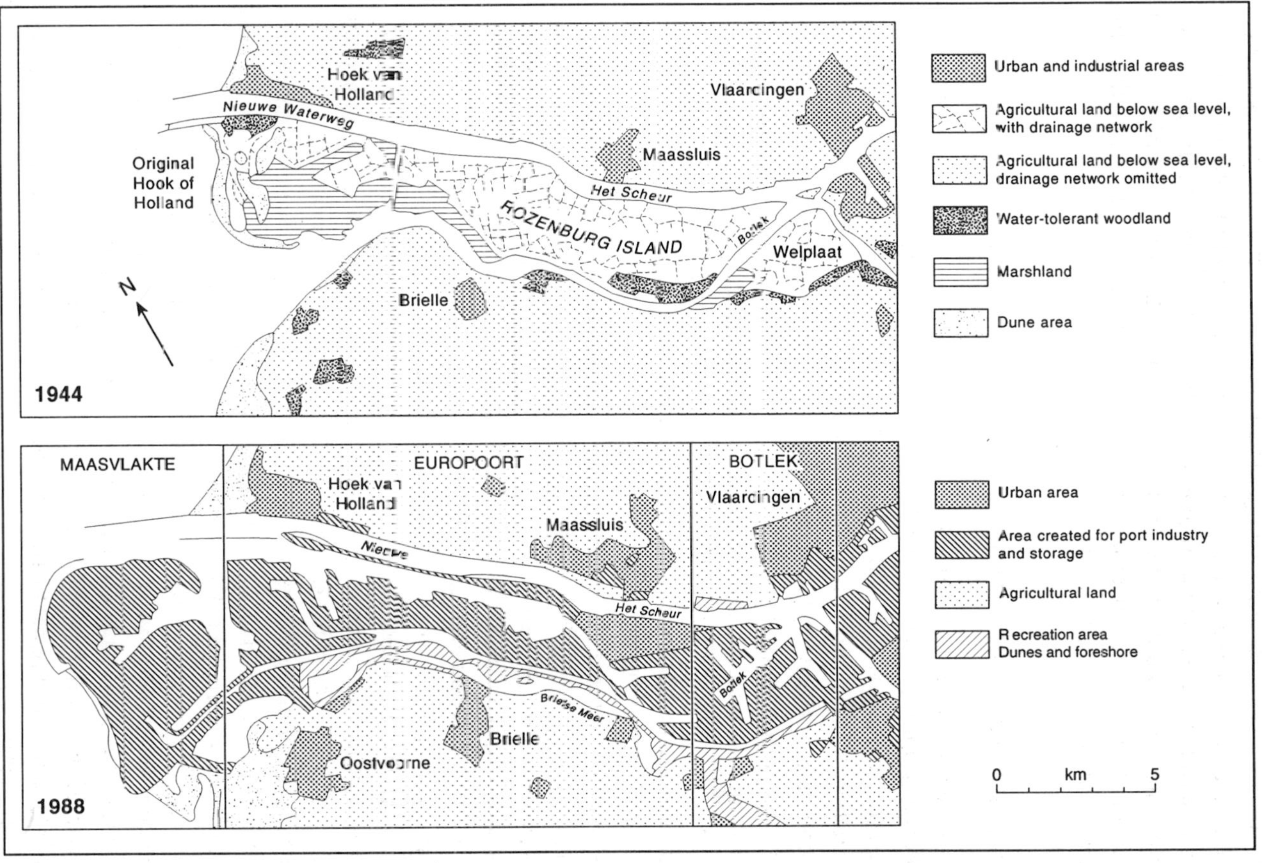

Fig. 5.11 Port expansion and wetland loss in Rotterdam. After Pinder and Witherick (1990)

were reclaimed from *c.* 1730 to 1986, but as the estuary decreased in size, so the area of salt marsh actually increased. The probable cause of this increase was the reduction in tidal volume which reduced scour and promoted sedimentation. Even where the landward fringe of marshes is itself unreclaimed, the building of a dyke or a seawall to prevent coastal flooding can have a similar impact on wetland sediment dynamics. Moreira (1992) describes how 80 per cent of the length of the marshes around the Sado estuary, Portugal is now backed by a dyke.

Decreased sediment input to wetlands has occurred commonly through damming of rivers, and the reduction of soil erosion in drainage basins, as well as from coastal protection schemes which reduce alongshore and onshore sediment movements. In the Po River, Italy, for example, sediment load decreased from 16.9 million to 10.5 million t a^{-1} over the period 1965–73 (Sestini, 1992). Official figures show that 100 million t of sand were dredged from the Po River bed between 1958 and 1981, although this is probably an underestimate with the real figure being up to six times higher (Sestini, 1992). The Ebro delta in north-east Spain has suffered a similar reduction in sediment inputs. River dams have resulted in sediment discharge decreasing from *c.* 4 million t a^{-1} before 1965, to less than 400 000 t a^{-1} now (Mariño, 1992). Such changes in sediment inputs have led to remarkable changes in the wetland shorelines as the location of major deposition centres has shifted.

Kamaludin (1993) illustrates the complex human impacts on wetland sedimentation in a study of the changes in mangrove shorelines in peninsular Malaysia. Maps, air photographs and soil auguring surveys reveal that net progradation occurred between 1914 and 1969, with accretion rates of up to 18–54 m a^{-1} in some areas. A net increase of 26.7 km^2 of mangrove swamp resulted. Kamaludin ascribes this rapid progradation to extensive development of agriculture, especially sugar cane, and to tin mining. Agriculture and mining are assumed to increase sediment yields to rivers by increasing erosion. Since 1969, the trend has reversed, with an estimated 0.5 km^2 of wetland lost to erosion. Damming of rivers here, as well as a reduction in tin mining activity has been seen to be at least partially responsible. Similar complexities arise in the mangrove coasts of the UAE, Gulf of Arabia, where dredging to aid access to ports, coupled with sewage dumping from burgeoning cities, has acted to promote mangrove growth. On the other hand, dumping of aggregates into Abu Dhabi lagoon for land reclamation and building killed mangroves near Um An-Nar and Musaffah industrial areas (Embabi, 1993).

In Chesapeake Bay, the largest estuary on the US Atlantic coast, marshes have shown a variable pattern of erosion and accretion. In five estuaries tributary to the bay significant changes have been recorded since the first European settlements in the mid-seventeenth century. Agriculture led to increased sedimentation and a general increase in tidal marsh area (Froomer, 1980). However, such findings need to be set in the local context of rising sea levels and decreased river discharge and sediment loads consequent upon up-estuary damming. A recent sediment budget study by Stevenson *et al.* (1988) showed that estuarine marshes trap only 5–11% of the annual sediment input (only half the amount previously thought to be trapped in this area). According to Leatherman (1989), over a third of the total marsh area (*c.* 2000 ha) at the Blackwater Wildlife Refuge on the eastern shores of Chesapeake Bay was lost between 1938 and 1979 because of sea level rise.

Construction works which affect tidal water flows can also have a serious impact on wetlands. In the Dampier–Cossack region of north-western Australia, for example, of 70 km^2 of mangroves surveyed, 12 km^2 were dead, mainly because of permanent ponding of seawater within a salt evaporator near Dampier, and the construction of a road which restricted tidal exchange. This led to an increase in soil salinity above the level tolerated by the seven species of mangrove found in this arid tropical setting (Gordon, 1988). Table 5.1 shows the full range of factors affecting mangroves here. Similar problems have been found on the Farasan Islands in the Red Sea where blocking of an inlet increased salinity and decreased waterflow killing *Avicennia marina* here (Alwelaie *et al.*, 1993).

Direct human interference with wetland sediments has been a common local source of disruption in many areas. Mariculture in mangrove swamps, for example, often requires the creation of brackish water ponds dug into wetland soils. In Indonesia, brackish water fishponds (called tambak) now extend over *c.* 269 000 ha, or 6.5 per cent of the former mangrove area (see Table 5.2 for further details). In several places such conversion has been unsuccessful, or severely damaged by storms, because the potential for conversion had not been properly evaluated (Giesen, 1993). In Hong Kong, much of the Mai Po marshes (a combination of marshes, mangroves and tidal flats) was converted to 'gei wais' (areas enclosed by earth dykes and sluice gates to allow fish and shrimp farming in tidal waters) between the 1920s and 1960s. Since the 1960s the gei wais have been gradually converted to rice production and new housing, following the urban expansion of Hong Kong (Nelson, 1993).

Various ideas have been put forward to minimize the ecological and geomorphological damage caused by aquaculture. In Indonesia, the Directorate of Fisheries suggest that 10–20 per cent of mangrove areas can be converted to tambak without problems, but this is not a definite 'safe limit' and more precise controls need to be established. Giesen (1993) suggests that, as a minimum, tambak conversion should be preceded by assessment of soil suitability, tidal range, green-belt requirements, infrastructure, market and seed stock arrangements, existing tambak and the environmental impact of tambak. In Kenya, where prawns and shrimps are cultured in brackish ponds, damaging con-

Table 5.1 Sources of disturbance to mangroves in the Dampier–Cossack region, Western Australia

Source of disturbance	*Effect*	*Impact/extent*
Road construction	Altered drainage rates Diminished tidal recharge Increased salinity Chronic flooding Altered soil structure Erosion	Mass mortality
Dust from unsealed roads/ore stockpile	Heavy dust coating on leaves of mangroves	Unknown but widespread; possible effect on stomata
Site construction:		
dredge spoil	Asphyxiation of roots	Localized mortality
salt evaporator ponds	Permanent ponding Hypersalinity	Mass mortality
salt bitterns ponds	Increase salinity (?) Altered tidal gradient (?)	Localized mortality
harbour	Interference with tidal exchange Removal of mangal	Unknown Localized mortality
jetty	Removal of mangal	Localized mortality
Effluent:		
release of freshwater used as dust suppressant at ore stockpile plant	Not examined	Unknown: possible input of ore dust to sediments
nutrient release from sewage pond	Not examined	Unknown: enhanced growth of mangroves apparent
Recreation:		
off-road vehicles	Erosion of salt flats Ponding of tidal water Scouring/removing of blue-green algal mats	Mortality of mats

Source: modified from Gordon (1988)

version could be prevented by shifting to pump-fed ponds on higher ground instead of tide-fed ones in the mangroves (Rasowo, 1992).

Adam (1990) discusses the example of salt marshes in the Medway estuary, Kent, England, where large amounts of clay were removed during the nineteenth century for brick and cement manufacture. This has led to a dramatic reduction in marsh area and also the development of a very complex surface topography. Salt production in salt marshes (using evaporation in shallow pans) may have also left a small scale geomorphological impact, with pans

Table 5.2 Declining mangrove areas in Indonesia

Province	*Former mangrove area (ha)*	*Remaining mangrove area, 1986–90 (ha)*	*Tambak area, 1986–90 (ha)*
Aceh	60 000	<20 000 (<33%)	39 476
North Sumatra	95 000	30 750 (32%)	1826
West Sumatra	11 000	1800 (16%)	0
Riau	259 500	184 400 (71%)	192
Jambi	18 500	4050 (22%)	40
South Sumatra	354 500	231 025 (65%)	325
Bengkalu	2000	<2000 (<100%)	94
Lampung	56 500	11 000 (19%)	2939
West Java	66 500	<5000 (<8%)	50 330
Central Java	46 500	13 577 (29%)	30 497
East Java	57 500	500 (<1%)	47 913
Bali	1000	<500 (<50%)	626
West Nusa Tenggara	9500	4500 (47%)	4996
East Nusa Tenggara	29 000	20 700 (71%)	550
East Timor	100	<100 (<100%)	59
West Kalimantan	213 000	40 000 (19%)	32
Central Kalimantan	84 000	20 000 (24%)	0
South Kalimantan	115 000	66 650 (58%)	1405
East Kalimantan	680 000	266 800 (39%)	6107
North Sulawesi	30 500	4833 (16%)	590
Central Sulawesi	43 000	17 000 (40%)	861
South Sulawesi	110 000	34 000 (31%)	73 088
South-east Sulawesi	89 000	29 000 (33%)	6636
Malaku	197 500	100 000 (51%)	65
Irian Jaya	>1 500 000	1 382 000 (92%)	95
TOTALS	4 129 100	2 490 185 (60%)	268 743

Source: modified from Giesen (1993)

being dug or extended artificially. More extreme effects may occur in heavily used wetlands where canals, pipe-lines and roads are constructed (e.g. Louisiana, as discussed in Case study 5.1).

Wetland vegetation removal can upset sedimentation processes. In salt marshes a range of species may be affected, such as *Phragmites australis*, used for thatching, *Festuca rubra* cut for turf for bowling greens etc., as well as several edible species such as *Salicornia europaea* (marsh samphire), but the disruption caused is usually small scale and short lived (Adam, 1990). As Reed (1991) has shown, however, natural vegetation die-back may produce ponds in the Louisiana marshes, and thus human removal of vegetation in patches may produce similar, more long lived features. In mangrove forests, exploitable timber species are common and important. Thus, in Thailand *Rhizophora mucronata* is used for firewood (Walsh, 1977) and *Bruguiera gymnorrhiza* is used for poles in the Andamans. On Zanzibar, *c.* 12000 ha of mangroves (with nine species of tree) on Pemba Island are utilized with 350–1937 cut plants ha^{-1} (Ngoile and Shunula, 1992). The mangrove environment here is put to many uses, including charcoal, lime and salt production, as well as timber and fish.

Upsetting the biological balance

The introduction and/or removal of species from wetlands can have serious impacts on species composition and succession as well as on sedimentation. Along much of the British coast salt marshes have been affected by *Spartina anglica* which now covers *c.* 10000 ha of intertidal flats. According to Carey and Oliver (1918) it was first recorded at Hythe, Southampton Water in 1870. Twenty years later it was spreading rapidly around Southampton and the Isle of Wight. In the early years of the twentieth century it was regarded as an invaluable aid to coastal protection, and enquiries were made into its economic value (for forage etc). However, concern was already being expressed:

> [I]ts marked increase in navigable waters raises the question of whether its spread in some of the localities already occupied may not develop into a serious nuisance (Carey and Oliver, 1918, p. 183).

Interestingly, Carey and Oliver (1918) report on a similar use of *Spartina brasiliensis* in land reclamation on the foreshore at Demerara River, Guiana. There, a Mr John Junor planted the foreshore with *Spartina* and when this was established, and a soil developed, *Rhizophora* mangroves were planted 'with the result that a forest springs up and no further attention is required' (Carey and Oliver, 1918, p. 183). Apparently, once established the *Rhizophora* shades out and kills the *Spartina*, although it is unclear whether the scheme was successful in the long term.

Along the Avon coast marshes *Spartina anglica* was planted in 1913, leading to a great spread in the extent of marshes, but bringing a concomitant decrease in plant species diversity and the number of birds present (Martin, 1990). Die-back has occurred here because of the development of extreme anaerobic conditions in the mud surrounding the *Spartina* root system. On the island of Lindisfarne, *S. anglica* was introduced in 1929 to control erosion, and its spread was encouraged by the construction of the causeway to Holy Island in 1954. The main problem of *S. anglica* on Lindisfarne is its competition with *Zostera* sp. (eel grass) which is the major food here for wigeon and pale-bellied brent goose. Since the early 1970s a control programme has been in place involving digging and chemical control (Davey, 1993), although such chemical control may have unforeseen effects elsewhere in this Northumbrian ecosystem. In China, which has some 2 million ha of tidal flats, *Spartina anglica* was introduced from the UK in 1963, and subsequently *S. alterniflora* from the USA has also been used. Chung (1985) reports that planting has largely been successful in promoting accretion here.

Grazing by stock animals is an important, long term component of salt marsh ecology in Britain and elsewhere. Heavily grazed marshes tend to have an impoverished flora, with the grasses *Festuca rubra* and *Puccinellia maritima* being favoured. Conversely, where marshes are only grazed by birds, rabbits and hares they are often diverse and include *Halimione portulacoides, Limonium vulgare* and *Artemesia maritima* (which are grazing-sensitive species) (Doody, 1992). Heavily and lightly grazed marshes favour different animal populations and therefore have advantages for maintaining species diversity, and thus for nature conservation. Calculations suggest that moderate grazing conditions (i.e. *c.* 0.33 cattle or ponies or 2 sheep ha^{-1} a^{-1}) are best for nature conservation purposes. Dijkema (1990) shows how marshes around the Baltic Sea have a long history of grazing and hay-making, but such practices have now stopped with the result that reed beds, grasses and woodland have spread at the expense of halophytic plants.

Upsetting the chemical balance

Domestic sewage, industrial effluent, airborne dust and oil from offshore installations can all end up in coastal wetlands. Heavy metal pollution is of particular concern as found, for example, in the Severn Estuary, where it comes from sewage sludge, industrial sewers, airborne dust from smelting and land drainage from mineralized areas such as the Mendip hills (Martin and Beckett, 1990). In this area Zn, Cd and Pb are especially significant pollutants. Sediment cores from Chittening Warth marsh reveal high concentrations down to *c.* 35 cm, related to the 1928 founding of the Avonmouth smelting complex. At about the same time *Spartina anglica* invasion altered sedimentation patterns here. Interestingly, marsh plants show less contamination than the sediments, a finding backed up by the work of Ott *et al.* (1993) from Dutch salt marshes who found that plant and soil concentrations of Cu, Cd and Zn did not follow the same trends. In the Humber estuary, eastern England, recent sediments show 3.5–6 times the contents of P, As, Pb, Cu and Zn of sediments dating from 5000 BP (Grant and Middleton, 1990).

Similar results have been found from other parts of the world. In Georgia, USA, where pollution of marshes from land drainage is a major concern, a study compared the heavy metal concentrations of *Spartina alterniflora* and sediment from marshes near ports and cities, and those in unpolluted areas (Alberts *et al.*, 1990). Cr, Cu, Hg, V and Zn concentrations in sediments were found to be up to ten times as great in marshes near ports or urban areas than in the unpolluted ones. Concentrations of these elements in *Spartina* plants varied little between sites. These findings provide at least some good news, in that the plants do not seem to be accumulating dangerous levels of metals.

Very high concentrations of heavy metals have been recorded from salt marshes near Wollongong, Australia. Here, sediments contain up to 200 times the concentrations of heavy metals found in rural areas, and the plants *Sarcocornia quinqueflora*, *Suaeda australia* and *Triglochin striata* all contain elevated levels (although not as high as those in the sediments) of Cu, Zn, Pb, Cd, Cr and Mn (Chenhall *et al.*, 1992). Pollution here comes from the Port Kembla industrial complex, urban pollution and airborne dust from steelmaking, fertilizer production, power generation and copper smelting.

In many cases, where sediments can be dated, a clear association can be made between metal contents and specific pollution sources. In Tasmania, for example, Wood *et al.*, (1992) dated sediments from Lindisfarne Bay with ^{137}Cs and found that the highest concentrations of Cd, Cu, Pb and Zn occurred at 0.8 m depth, related to high recorded emission in the 1970s from the Electrolytic Zinc Co., which was granted exemption to exceed state discharge levels in 1973. In the Dee estuary, UK, the level of iron contamination of salt marshes from the Shotton steelworks has been investigated by mineral magnetic methods. Dating by ^{137}Cs (from the nearby Sellafield nuclear fuel reprocessing plant) helps to show the spatial and temporal distribution of pollutants in the marsh (Hutchinson, 1993).

The danger that metals may become concentrated in edible organisms within wetlands spurred the study by Niolne *et al.* (1992) of oysters from Senegalese mangroves. The oysters are eaten locally as well as exported, but encouragingly the study found that heavy metal pollution here 'has not yet reached alarming limits.' More worryingly, along the Taiwanese coast in 1986 *Crassostrea gigas* were observed to turn green and then die 3 months later, because of copper pollution from industrial and domestic waste discharged into rivers (Hung and Han, 1992). Oil pollution can be a serious nuisance in wetland areas, as it gets absorbed into the soil and may take years to be removed. In Panama, mangroves were seriously affected by an oil spill in 1986 which occurred during seasonally very low tides, leading to the death of bivalves and barnacles living on mangrove roots (Garrity and Levings, 1993). The impacts have been felt over several years, as contaminated mud becomes resuspended and transported. Polychlorinated biphenyls (PCBs) have been shown to be transported with suspended particulate material in the Seine estuary, France, and their distribution is related to freshwater discharge and tidal amplitude (Abarnou *et al.*, 1987). Pentreath (1987) notes that a large echiuran *Maxmuelleria lankestri* in the Irish Sea feeds on surface sediment and leaves faecal pellets at the bottom of its burrow. This bioturbation process acts to mix radioactive Pu from Sellafield into the sediments.

All these examples show that chemicals introduced into coastal wetland settings can affect the flora and fauna, but that the nature of the problem relates to hydrological transfer routes, reactivities, sedimentation, erosion and bioturbation patterns.

Engineering problems posed by sabkhas

Sabkha sediments, like all coastal zone sediments, can be eroded, and thus their potential role as a coastal protection agent can be reduced. The south-western sabkha in Bahrain is undergoing erosion, according to Doornkamp *et al.* (1980) because of waves generated by shamal winds. The more immediately serious problems are those related to development on and near to sabkhas. The high concentration of salts may cause engineering problems as foundations are prone to salt attack. Similarly, if sabkha sediments are used as aggregates for the construction industry excessive salt concentrations may cause problems. Buildings and roads in the coastal plain of Saudi Arabia, for example, have deteriorated rapidly because the construction materials were derived from sabkha deposits with high evaporite contents (Orhan, 1989). On the other hand, salts can be an advantage when constructing unmetalled roads as they help to bind the surface together (Fookes, 1978). The extensive development in the Middle East associated with the exploitation of oil resources has led to an expansion of road networks and airstrips. On sabkhas, roads need to be built on embankments, to avoid the dangers of periodic flooding. Sabkha soils make good embankment material, although in the UAE, for example, only soils with $<$ 10 per cent salt concentration are used (Tomlinson, 1978). In Libya, construction of sewage systems for Benghazi, Tripoli and other coastal towns in the 1970s was constrained by the presence of sabkhas. Culverts had to be buried in sheet pile cofferdams to avoid heave processes across soft sabkhas with high water tables (Newberry and Siva Subramaniam, 1978). There have been many studies of the geotechnical properties of sabkhas which enable best use to be made of them while limiting the environmental damage caused.

Future sea level rise — towards a new balance?

As wetlands are finely tuned systems whose state is highly dependent upon sea level and sediment availability, they are likely to change in the future, just as they have responded to sea level changes over the Holocene. Much work has been carried out using information on the Holocene development of coastal wetlands and their present day dynamics to predict what will happen under possible future global warming conditions. All such studies have been hampered by a lack of clear predictions of the rate and magnitude of sea level change likely, as well as a lack of knowledge about allied regional climatic changes. Mathematical modelling offers one way forward and French (1993) has simulated adjustment of the north Norfolk marshes to a range of future sea level scenarios. Encouragingly, these simulations show that these marshes are unlikely to reach an ecological 'drowning point', where the marsh is lowered (relative to the highest astronomical tide) to the level of an unvegetated tidal flat, except under the most severe of future sea level rise scenarios (Fig. 5.12).

Empirical studies from areas currently undergoing relatively high rates of sea level rise are also useful. Pethick (1993) discusses the future of British estuarine marshes given the recent experience of the Blackwater estuary in Essex which contains 680 ha of salt marshes and 2640 ha of mudflats. Here, over the past few decades there has been a relative sea level rise of 4–5 mm a^{-1}. Between 1973 and 1988 *c.* 23 per cent of the total salt marsh area has been eroded as the mudflat area has increased. Allied to these changes has been a dramatic increase in the width (and a decrease in depth) of the main estuary channel over the past 150 years (Fig. 5.13). As sea level rises, vertical accretion on the

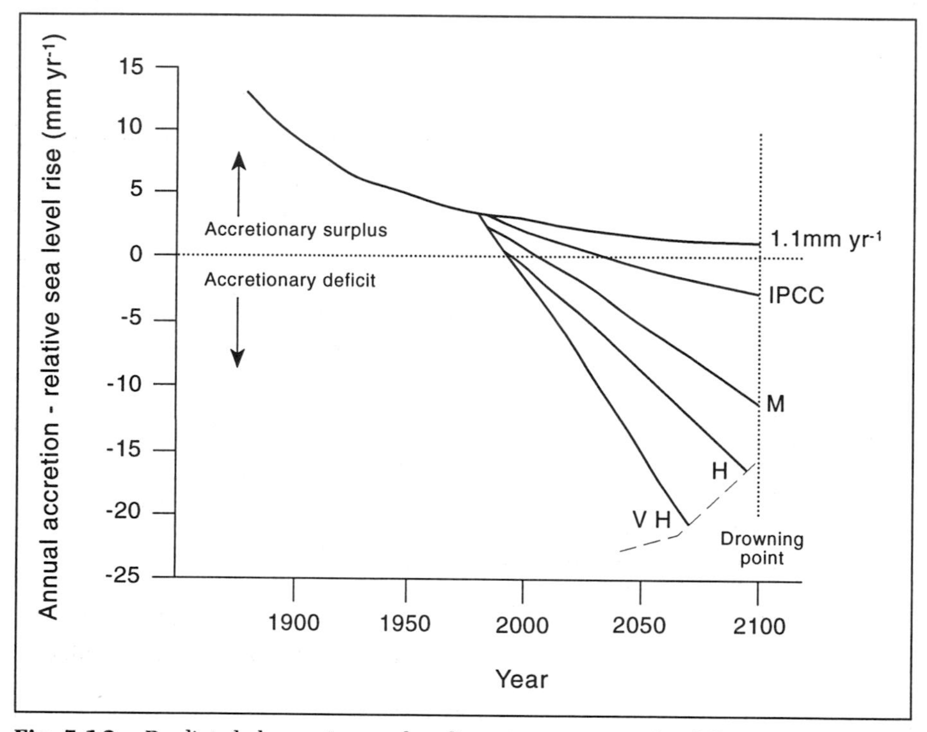

Fig. 5.12 Predicted change in marsh sedimentary status under different sea level scenarios. IPCC = mean best estimate of Warrick and Oerlemans (1990); M, H, VH = moderate, high and very high scenarios of Hoffman *et al.* (1983), respectively. After French (1993)

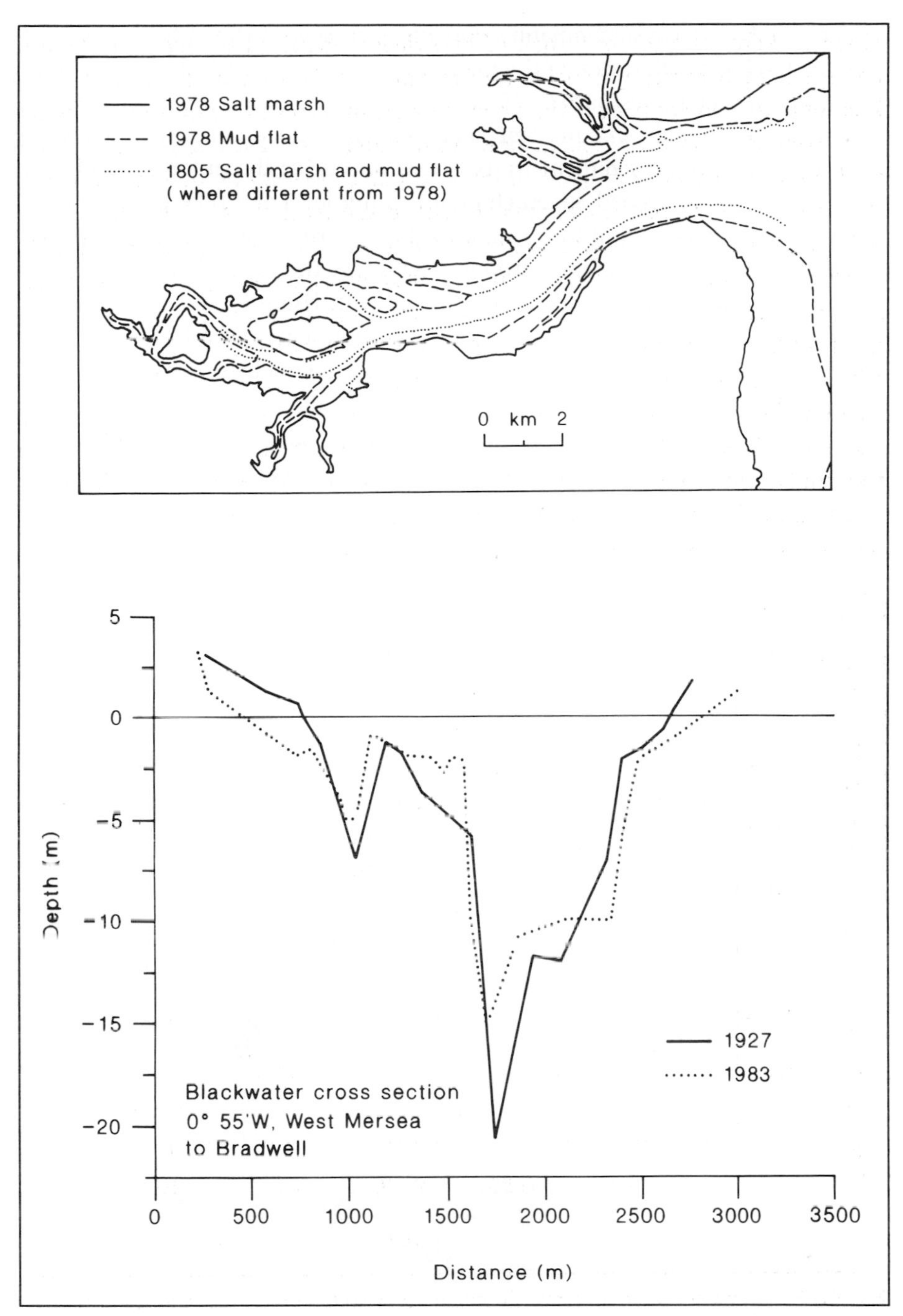

Fig. 5.13 The Blackwater estuary, Essex, England showing (a) changes in salt marsh and low-tide level since 1805 and (b) changes in estuary cross-section. After Pethick (1993)

upper marsh surfaces accompanies horizontal erosion at the outer edges, leading to a landward migration of the marsh. However, in the Blackwater estuary, as in most other UK estuaries, 95 per cent of the coast is protected by flood embankments which will prevent such onshore relocation. Increased wave action and longshore sediment transport will occur as the outer parts of estuaries become more eroded, given changing water level and tidal conditions.

Mangroves will respond in a similar fashion to rising sea level, although in some mangrove settings (i.e. carbonate settings) the lack of sediment availability will make them less likely to keep up with rising levels. Ellison (1993) confirms this from a study of mangroves on Bermuda where Holocene stratigraphic records show that they can keep pace with a sea level rise of up to 9 cm per 100 years (0.9 mm a^{-1}). Recent mangrove die-back on the island has caused the loss of 2.24 acres of mangroves over the last few hundred years, especially over the periods 1900–47 and the last decade. Sea level rise now in this area is 28 cm per 100 years, and the seaward margin of many of the mangrove swamps is now below mean sea level.

Plate 5.4 Mangrove swamps as biological nurseries: shrimp netting, Rufiji delta, Tanzania, East Africa

In river-dominated systems the response will depend on fluvio-tidal characteristics, and tidally dominated systems are likely to experience some seaward erosion. Once again, the landward constraints on onshore migration of mangrove swamps provide important limitations even where sediment is freely available (Woodroffe, 1990).

Balancing the possibilities — effective coastal wetland management

The challenge facing those involved in the management of coastal wetlands is to conserve the sedimentary and biological resources, whilst tapping their economic potential (Plate 5.4), in the light of an uncertain environmental future. Each area requires individual consideration, although there is a danger that badly coordinated small scale schemes will end up being mutually damaging.

In the Blackwater estuary, Essex, England (whose erosional problems have already been discussed) a new scheme for coastal protection and marsh management has been promoted by conservationists. The 'set-back' scheme on Northey Island aims to get round the problem of building ever more large and expensive seawalls behind a suite of eroding marshes to protect the farmland behind. Northey Island, *c.* 146 ha in size, is owned by the National Trust and is used for pasture. The site is important for dark bellied geese in winter, and the salt marshes are an important high tide roost site for wading birds. After 1903 much of the north and east of the island returned to salt marsh when the seawall was breached. In the south of the island the salt marsh had shrunk and the seawall was badly eroded. The cost of a new wall was prohibitive, and so a scheme of managed retreat, which would create a new small (*c.* 0.8 ha) marsh, was planned. A 200 m length of seawall was lowered and a 20 m spillway created, to allow inundation by *c.* 100 tides a year. The scheme cost £25 000 in comparison with the £55 000 estimated for renewing the seawall. Over 2 years the scheme seems to have been successful, with salt marsh plant colonization, sediment accretion and algal mat development on the former pasture surfaces (Turner and Dagley, 1993). However, although locally successful, this set back scheme may have a negative effect on the overall estuary, by affecting the hydrodynamics, and thus may promote enhanced erosion elsewhere (Pethick, 1993).

Integrated conservation schemes in estuary, bay and back barrier areas are clearly required to reconcile varying land uses such as wildlife conservation, coastal protection, grazing, military use and economic development. These schemes are operating against a historical background of land claim (conversion of coastal marshes to agricultural, industrial and urban land), sea level changes and biological changes (such as the introduction of *Spartina anglica*). In Britain over 80 per cent of coastal marshes are found in Sites of Special Scientific Interest (mainly because of their role as bird habitats) and in 1992 sixteen sites were National Nature Reserves. The Royal Society for the Protection of Birds (RSPB) has twenty-two reserves which include areas of salt marsh, and the National Trust manages many salt marsh areas. In most of these conserved areas a range of uses, such as cockle dredging and wildfowling, are allowed, but there is an ever-present threat from projects such as estuarine barrages and other major developments. In southern California development threatens many salt marshes (over 75 per cent of coastal wetlands here have already been destroyed) and many rare birds are at risk. Various development plans threaten these salt marshes further, and although many schemes include wetland restoration as part of the project these must be well planned to avoid loss of species (Zedler, 1988).

Case study 5.1: The disappearing wetlands of south Louisiana, USA

Louisiana is experiencing the highest rates of coastal erosion and wetland loss in the United States. Coastal retreat here averages 4.2 ± 3.3 m a^{-1}; by comparison Atlantic coasts show mean erosion rates of 0.8 m a^{-1} and Pacific coasts are effectively stable. This state suffers 80 per cent of all US wetland losses. These dramatic processes have become well documented in recent years but are still quite poorly understood, not least in disentangling natural and human-induced effects. These difficulties are reflected in the debate over the correct strategies for coastal protection and restoration.

Over the last 7000 years the Mississippi River has built a series of delta complexes; each has passed through stages of active growth, consolidation and then abandonment and decay, with subsidence of deltaic deposits and the growing dominance of marine processes over fluvial ones. Five delta complexes have completed this cycle (the Plaquemines delta of the Modern complex, and the Maringouin, Teche, St Bernard and Lafourche complexes) and only two are active, over *c.* 20 per cent of the delta plain, the 'bird's foot' Balize lobe of the Modern complex and, to the east, the Atchafalaya delta complex (Fig. 5.14). Under natural distributary-switching, the Atchafalaya should by now be the main channel but a series of control structures north of Baton Rouge maintains the position of the Mississippi River and the viability of the Balize delta. The natural sequence of delta abandonment, and associated land loss has been well explained by Shea Penland and his associates (Penland *et al.*, 1988; Fig. 5.15). In Stage 1, the active delta front becomes an erosional headland, with long-

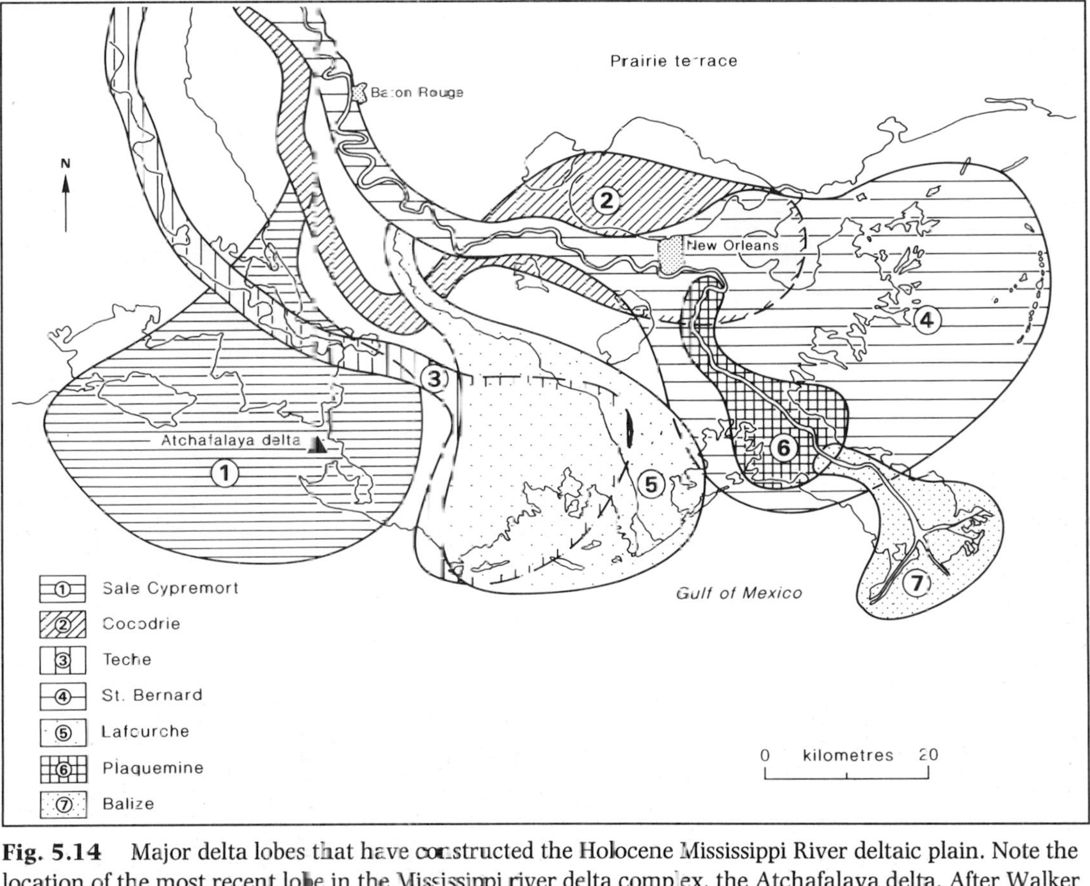

Fig. 5.14 Major delta lobes that have constructed the Holocene Mississippi River deltaic plain. Note the location of the most recent lobe in the Mississippi river delta complex, the Atchafalaya delta. After Walker *et al.* (1987)

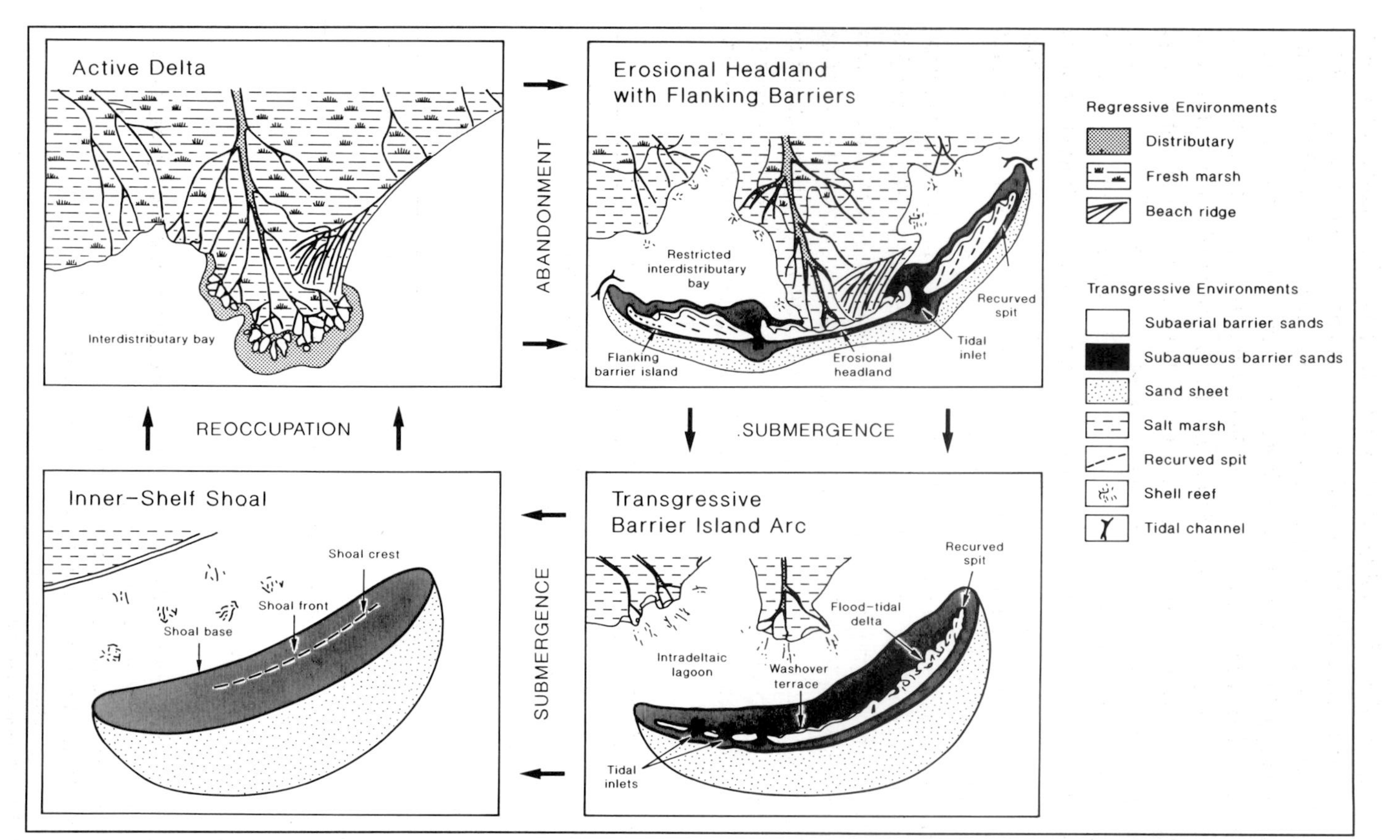

Fig. 5.15 The genesis and evolution of transgressive depositional systems in the Mississippi River delta plain shown by a three-stage model. Stage I, erosional headland and flanking barriers; Stage II, transgressive barrier island arc; Stage III, inner shelf shoal. After Penland *et al.* (1988)

shore transport of sand forming flanking barrier spits. The abandoned Bayou Lafourche headland, eroding at 20–50 m a^{-1} is the most recent example of an eroding headland and the Timbalier Islands to the west and Grand Isle to the east are embryonic barrier systems. With continued subsidence, the barrier becomes detached from the mainland and forms a barrier island arc (Stage 2), such as in the Isle Dernieres (associated with the Lafourche delta complex) and the Chandeleur Islands (associated with the St Bernard complex). Ultimately, starved of new sand supplies, the barrier islands are reduced by overwash processes to inner-shelf shoals (Stage 3). Studies of historical maps and aerial photographs have been used to show that the present barrier shorelines have been experiencing accelerated rates of shoreline retreat in the last two decades, with current mean shoreline retreat of 14.0 m a^{-1}. Projections suggest that most of Louisiana's barriers will have disappeared by 2036.

The removal of the barrier systems means that wetlands not only have to cope with the effects of accelerated subsidence but also are then subject to the full force of open marine processes. These natural processes are further exacerbated by human impacts (Table 5.3). The massive levées which control the course of the Mississippi River prevent the river's sediment load from replenishing the bordering wetlands but rather deliver it far out to sea, to the edge of the continental shelf. Extensive systems of periodically dredged canals and waterways,

Table 5.3 Primary factors affecting Louisiana coastal wetlands

Origin and evolutionary development
Channel switching cycles and high rates of riverine sediment deposition
Slow-to-modest rates of relative sea-level rise
Sheltered low-energy environment
Low-to-modest storm activity
High biological productivity
Degradation and loss
Natural processes
Sediment starvation due to channel switching
High rates of subsidence (sediment compact/consolidation)
High rates of eustatic sea-level rise
High storm activity
Erosion of protective barrier islands
Human activities
Sediment starvation due to building dams, levées, flood control engineering structures
Canal and waterway dredging; saltwater intrusion
Introduced species increasing herbivory activity
Local subsidence due to extraction of minerals and fluids
Forced drainage, land reclamation

Source: Williams *et al.* (in press)

constructed to aid hydrocarbon exploration and production and commercial and recreational water traffic, have allowed saltwater intrusion into freshwater and brackish marshes. Extraction of minerals and fluids, and drainage schemes associated with residential developments, generate local subsidence 'hotspots'. Furthermore it has been estimated that between 1938 and 1964 urban groundwater extraction in New Orleans caused over 0.5 m of ground lowering. Improved mapping programmes are now underway to identify spatial patterns in wetland loss and thus gain a better idea of the causative factors.

The result of all these varied processes is that the current coastal land loss rate for south Louisiana is 12 000 ha a^{-1}. It is clear that the rate of land loss has accelerated over the last 75 years, although with perhaps some deceleration in

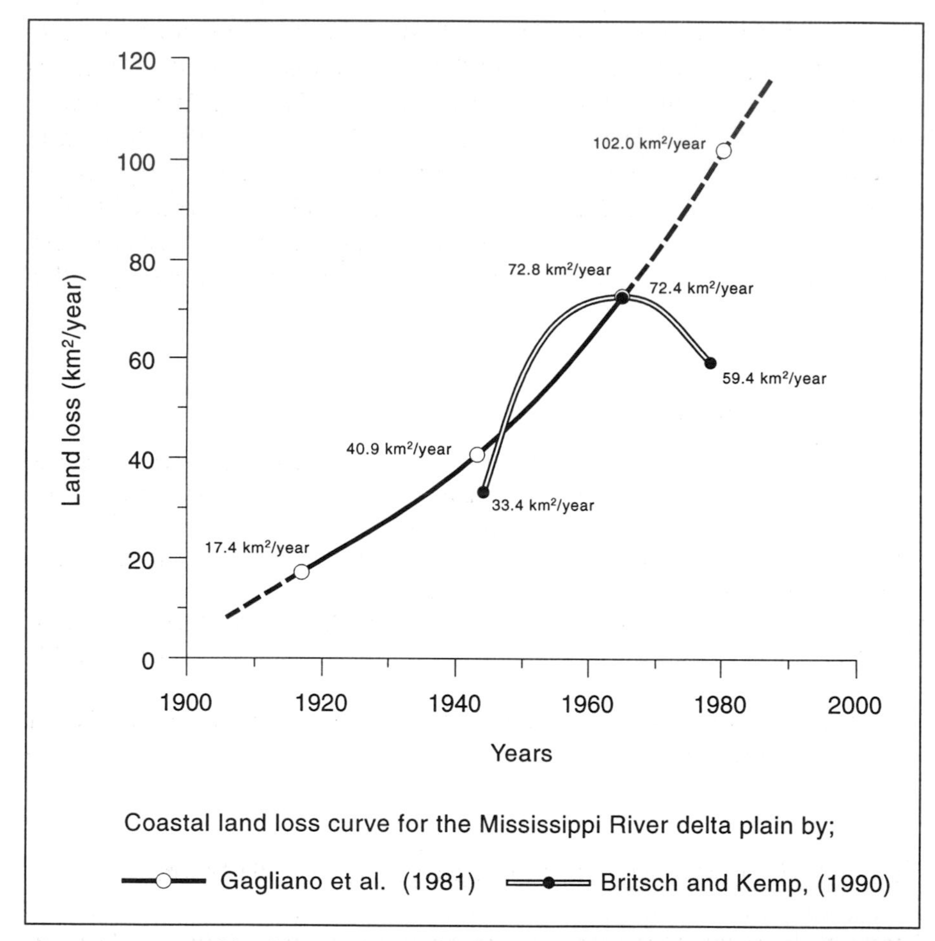

Fig. 5.16 Coastal land loss curves for the Mississippi River delta plain. After Gagliano *et al.* (1981) and Britsch and Kemp (1990)

recent years (Fig. 5.16). The problems of subsidence and land loss are of great concern to the private landowners who own 75 per cent of Louisiana's wetlands. This is not just because of fisheries, wildfowl and hunting resources, which are valued at US$1 billion per year, but also because when an area of marsh becomes open water and a navigable waterway, ownership reverts to the State or, if navigation is also an issue, the Federal Government. In particular, landowners lose control over, and income from, any gas or oil reserves which might have been found beneath the now eroded land. One response by private landowners has been to build up levées around their properties and close waterways with weirs in an attempt to prevent saltwater intrusion; in the period 1980–9 owners of 9 per cent of all coastal marsh habitats were granted permits for this kind of water management. Whilst such schemes may benefit the individual it is not clear how they affect vegetative growth and the functioning of wetland hydrology at larger scales. Preliminary studies on marsh sedimentation, however, show that by reducing tidal exchange the weirs do not promote vertical marsh accretion inside the weirs, raising the possibility that the control structures may in fact promote the deterioration of the marshes that they are supposed to be protecting. Clearly, more scientific studies are required to find the best ways in which hydrological, sedimentological and ecological processes can be manipulated to encourage marsh accretion and wetland recovery in this region.

Selected references

PENLAND, S., BOYD, R. and SUTER, J.R. 1988: The transgressive depositional systems of the Mississippi delta plain: a model for barrier shoreline and shelf sand development. *Journal of Sedimentary Petrology* 58: 932–49.

REED, D.J. 1992: The effect of weirs on sediment deposition in Louisiana coastal marshes. *Environmental Management* 16: 55–65.

TEMPLET, P.H. and MEYER-ARENDT, K.J. 1988: Louisiana wetland loss: a regional water management approach to the problem. *Environmental Management* 12: 181–92.

WALKER, H.J., COLEMAN, J.M. ROBERTS, H.H. and TYE, R.S. 1987: Wetland loss in Louisiana. *Geografiska Annaler* 69A: 189–200.

Case study 5.2: War and the mangroves of Vietnam

The southern coast of Vietnam is dominated by the Mekong Delta. The Mekong river basin occupies over 800000 km² of land in Laos, Cambodia and Vietnam. On the delta coast Quaternary and Holocene sediments provide a prograding blanket around the mouth of the river (Eisma, 1985). Sediments brought down to the coast by the Mekong River are mainly moved southwards, round into the Gulf of Thailand, under the influence of the north-east monsoon between November and March. To the north of the delta are the mouths of several smaller rivers, including Song Saigon on which Ho Chi Minh City (Saigon) is located. This whole coastline would have been dominated by mangrove swamps under natural conditions.

During French rule between 1862 and 1907 there was considerable development of the coastal fringe of the Mekong delta for rice cultivation, with a network of canals dug from 1870 onwards to drain the land and provide navigation routes. The low-lying land here is prone to flooding, and dykes were built during the nineteenth and early twentieth centuries to provide some protection. Large floods in 1926 prompted the reinforcement of dykes, and the modernization of irrigation works in the delta (Fisher, 1964). During the Second World War much of the delta land given over to rice cultivation reverted to swamp.

In 1977 Chapman estimated that there were 2800 km^2 of mangroves in Vietnam, dominated by *Avicennia marina* as pioneer species, with *Bruguiera parviflora* and *Ceriops tangal* behind, *Bruguiera gymmnorhiza* on high land and the palm *Nypa fruticans* along streams where the water is brackish. Orians and Pfeiffer (1970) also recorded *Avicennia intermedia, Rhizophora conjugata* and *Ceriops candoleana.* Crabs dominate the fauna, often in huge numbers.

Concern about Vietnamese wetlands arose during the Vietnam war in the 1960s when the US Army used chemical defoliants and napalm bombing to 'flush out' the Vietcong from the Mekong delta. Over 2×10^6 kg of 2,4,5-T (in Agent Orange) was dropped on South Vietnam during the period 1962–70. According to Vo (1994, personal communication) about 50 per cent of all Vietnamese mangrove forest was destroyed by the war, and about 40 per cent of the entire land area is now classified as unproductive wasteland. Observations made in 1969 show the scale and the severity of the problem. On the mangroves along the Nha Be River, in the estuarine area just south of Saigon, which had been sprayed particularly seriously with defoliants 'several years earlier' there was almost no regrowth of mangroves, or of the saltwater fern *Archrosticum aureaum* which often invades mangrove areas (Orians and Pfeiffer, 1970). In defoliated mangrove areas there were found to be almost no insect- or fruit-eating birds, and very few fish-eating birds, such as egrets, herons and ospreys.

It was estimated in the early 1970s that, given favourable conditions, it would take at least 20 years for the dominant mangrove association to re-establish itself, but the long term storage of defoliants in the mangrove soils and problems with seed germination have made recovery very slow. Several reasons have been proposed for the absence of regrowth since then, including lack of a source of seedlings, soil surface compaction, soil chemistry, or crab predation of any seedlings which did establish themselves. Many of the soils in the Mekong delta, especially those in seasonally inundated *Melaleuca* forest areas are now high in sulphur because of attempts to dry them out. This renders them highly acidic and difficult to recultivate. Very large crab populations have been recorded within the area, and so crabs may have played an important role in preventing mangrove re-establishment. Whatever the exact cause, it is clear that many mangrove areas subjected to wartime defoliation have shown almost no natural regrowth since then, and the once fertile coastal wetlands have remained barren and unproductive.

In recent years, there have been attempts to reforest the mangroves, to provide a source of timber and other products, as well as to encourage the recolonization of wildlife. Around 30000 ha were successfully replanted in the 1980s and early 1990s (Vo, 1994, personal communication). Special reforestation techniques have been developed to encourage the successful regrowth of sensitive species, by reproducing as closely as possible their natural environmental conditions — for example, using foreign shade-giving species to provide temporary shelter for understorey plants. The mangroves now supply wood for local people, fisheries are returning to pre-war conditions, and the wetlands provide tree roosts to birds such as herons, which are now returning to their old homes.

Clearly, the Mekong Delta mangrove swamps suffered an acute stress through wartime defoliation, which drew world attention to the long term damage caused. The mangrove ecosystem has proved to be both highly sensitive to, and show low resilience to, the extensive use of defoliants over a timescale of over 20 years. However, it is encouraging to note that the replanting schemes are proving successful so far in returning the mangroves to their important roles as coastal protection agents, conservation areas and sources of revenue for the Vietnamese people.

Selected references

CHAPMAN, V.J. 1977: Wet coastal formations of Indo-Malesia and Papua New Guinea. In Chapman, V.J. (ed.), *Wet coastal ecosystems*. Amsterdam: Elsevier, 261–70.

EISMA, D. 1985: Vietnam. In Bird, E.C.F. and Schwartz, M.L. (eds), *The world's coastline*. New York: Van Nostrand Reinhold, 805–11.

FISHER, C.A. 1964: *South East Asia*. London: Methuen.

ORIANS, G.H. and PFEIFFER, E.W. 1970: Ecological effects of the war in Vietnam. *Science* 168, 544–54.

Chapter Six

CORAL REEFS

Introduction

Coral reefs are the largest biological constructions on earth (Plate 6.1 and 6.2) and are visible from outer space; the Great Barrier Reef, for example, is over 2000 km long and contains over 2500 individual reefs. The complexity of reef systems involves interactions between physical, chemical, biological and geological factors over a range of temporal and spatial scales. Reefs create their own environments for efficient nutrient supply and cycling by building a large,

Plate 6.1 One of the remotest atolls in the world: Ducie Atoll, Pitcairn Group, south Pacific Ocean

complex, wave-resistant structure within which environmental conditions are optimized. Fundamentally, structures like the Great Barrier Reef are the product of the *in situ* constructive abilities of a few simple organisms, most notably scleractinian, or stony, corals, assisted by hydrocorals and alcyonarian corals, coralline algae and other calcareous algae. How do these biological systems of corals create these geological structures?

The scleractinian, or stony, corals are a highly successful class of sea-anemone-like organisms which first appeared in the Mid-Triassic, showing rapid diversification since that time (Dubinsky, 1990). Although corals are common throughout the world's oceans and form, for example, extensive carbonate banks in the fjords of Norway, the reef-building, or hermatypic, corals are restricted to tropical or sub-tropical latitudes (Rosen, 1981), where water temperatures do not fall below 20 °C in the coldest month of the year, and to relatively shallow (< 100 m) waters. The biogeography of the reef-building corals shows that they are organized into two provinces, the high biodiversity (80 genera, 500 species) Indo-Pacific province, centred on Indonesia and the Philippines, and the impoverished (20 genera, 65 species) Atlantic province (Fig. 6.1; Stehli and Wells, 1971).

Some system level characteristics

Pioneering studies by American reef researchers in the late 1940s and early 1950s regarded coral reefs as ecosystems (e.g. Sargent and Austin, 1954; Odum and Odum, 1955). Viewed in this way, some remarkable characteristics emerge.

First, oceanic coral reefs are foci of high gross primary productivity. Community metabolism in reef systems can be inferred from diurnal and nocturnal changes in oxygen and carbon dioxide concentrations in seawater flowing over and through reef structures and their communities (e.g. Smith, 1973). These studies show remarkable uniformity between coral reefs from different latitudes, reef types and settings, and shallow water provinces, suggesting typical community-level gross photosynthesis of 8 g C m^{-2} d^{-1} for reef flat environments (Table 6.1a). These figures are high, but not exceptionally high, by comparison with other ecosystems around the world. What is remarkable, however, is that this productivity is achieved on open ocean reefs surrounded by waters with very low concentrations of key nutrients (dissolved inorganic phosphate, nitrate, nitrite and ammonia). Thus gross production at Rongelap Atoll, Marshall Islands, Central Pacific has been estimated at 1800 g C m^{-2} a^{-1} yet the atoll is bathed by oligotrophic Pacific Ocean waters with a productivity of 28 g C m^{-2} a^{-1} (Mann, 1982). Comparable measurements of reef phytoplankton productivity (Table 6.1c) are one to two orders of magnitude lower than for atoll reef-flats, and open ocean values are even lower. It is not surprising, therefore,

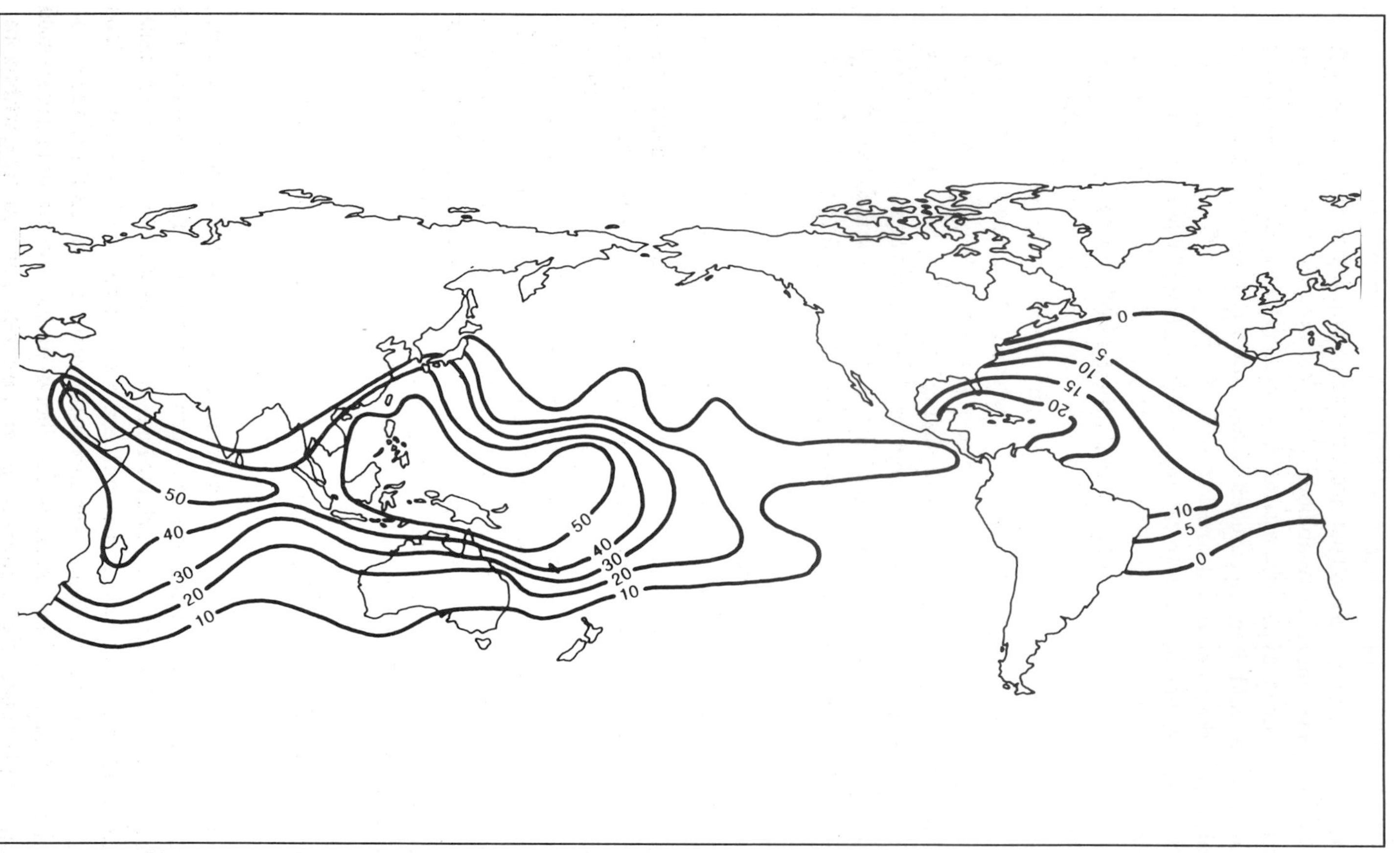

Fig. 6.1 World patterns of the abundance of hermatypic corals; the figures are numbers of genera. After Stehli and Wells (1971)

Plate 6.2 Reef margin island, or 'motu', Tetiaroa, Society Islands, south-central Pacific Ocean

that open ocean reefs have been described as 'an oasis in a desert ocean' (Odum, 1971). Second, in both reef flat (Table 6.1a) and 'whole reef' systems (Table 6.1b), rates of photosynthesis and respiration are roughly in balance, indicating that what is produced is also consumed within the community. This means that coral reefs attain production levels to meet their consumer (i.e. respiratory) demands. This type of behaviour is known as autotrophism and is characteristic of photosynthesizing plant ecosystems which meet their needs by the efficient conversion of solar energy. Third, reef productivity is typically about 12 times reef biomass.

Nutrient cycling, nutrient sources and nutrient sinks in reef ecosystems

All these characteristics point to very rapid and highly efficient nutrient cycling and re-cycling within coral reef ecosystems. This re-cycling takes place at a range of trophic levels: at the cellular level of within-organism symbiotic associations, at the food chain level, and at the ecosystem level between various geomorphic components and their communities within the reef structure.

The best described symbiosis is that between the reef-building corals and small single-celled algae, known as zooxanthellae, which live within the

Table 6.1a Published values for community metabolism in reef flat environments

Location	*P*	*−R*	*E*	*P/R*	*G*
	(g C m^{-2} d^{-1})				(kg $CaCO_3$ m^{-2} a^{-1})
Rongelap Atoll	4	3.5	+0.5	1.1	
Enewetak Atoll	10	10	0	1.0	
Kauai (Hawaii)	7.9	6.6	+1.3	1.2	
Puerto Rico (various)	5–16	5–19		0.8–1.6	
Puerto Rico				0.7	
Guam	4.5	7	−2.5	0.7	
Oahu (Hawaii)	14	24	−10	0.6	
Florida			0.8–2.0		
Indian Ocean (various)	4–9				
Laccadives	6.2	2.5	+3.7	2.5	
One Tree Is. (Australia)					4.6
Enewetak II	6	6	0	1	4
Enewetak III	11.6	6.0	+5.6	1.9	4
Enewetak II	6	6	0	1	
Enewetak III	10.4	6.0	+4.4	1.7	
One Tree Is. (Australia)	7.5	6.8	+0.7	1.1	
Guam	7.2	6.6	+0.6	1.1	
Lizard Is. (Australia)					4
Moorea (Tahiti)	7.2	8.4	−1.2	0.9	
One Tree Is.	7.2	7.4	−0.2	1.0	4.6
Lizard Is.					3.7
	7.9	7.9			4.2
	2.7	5.0			0.4
	(16)	(15)			(6)

Note: data reported as published. Conversion to g C made if necessary but other corrections not applied
Source: Kinsey (1983)

Table 6.1b Published values for community metabolism of 'complete' reef ecosystems*

Location	*P*	*−R*	*E*	*P/R*	*G*
	(g C m^{-2} d^{-1})				(kg $CaCO_3$ m^{-2} a^{-1})
Fanning Is.			0.0	1	1
Canton Is.	6.0	5.9	0.06	1	0.5
Takapoto (Tuamotos)	4	4	0.0	1	
One Tree Is. (Australia)	2.3	2.3	−0.06	1	1.5
Lizard Is. (Australia)	3.2	3.2	0.0	1	1.8

* It should be stressed that all systems included here are relatively shallow with considerable lagoon patch reef development. It seems reasonable to anticipate that large open atolls such as Enewetak will exhibit much lower overall activity
Source: Kinsey (1983)

Table 6.1c Published values for planktonic metabolism in the water overlying reef systems

Location	*P*	*−R*	*P/R*
	(g C m^{-2} d^{-1})		
Rongelap	0.2		
Florida	0.04–0.06		
Palau	0.08		
Enewetak Atoll		0.003–0.03	
One Tree Is. (Australia)	0–0.06	0.1–0.3	0–0.03
Fanning Is.	0.5		
Moorea (Tahiti)	0.004–0.03		
Vairao (Tahiti)	0.1–0.4		
Tuamotos	0.1–0.3		
Takapoto (Tuamotos)	0.002		
Gulf of Elat	0.003–0.01		
Lizard Is. (Australia)	0.1		

Note: ^{14}C estimates tend to lie somewhere between gross and net primary production. The tabulated values are therefore a low estimate for *P*
Source: Kinsey (1983)

innermost layer of the three-layered coral tissue (Fig. 6.2). Similar arrangements are characteristic also of other reef organisms, including clams, sponges and foraminifera. In return for protection, the zooxanthellae photosynthesize in a nutrient rich microenvironment within the host coral, producing large amounts of carbon which are translocated to the host for, in part, construction of the skeleton (see below). Typical rates of gross photosynthesis in Florida reef corals have been calculated at 2.7–10.2 g C m^{-2} d^{-1} (Kanwisher and Wainwright, 1967). Comparable figures for coralline algae range between 2.4 (Curaçao) and 1.5 g C m^{-2} d^{-1} (Enewetak Atoll). This internal nutrient cycling takes place without release and loss of nutrients to surrounding waters and is, therefore, highly efficient; it is well known, for example, that reef-building corals excrete less inorganic phosphorus than non-reef-building corals.

Outside such internalized systems, tight nutrient cycling is encouraged by the presence in close proximity of producers and decomposers. Thus, for example, the porous surfaces of calcareous algae are covered with filamentous bacteria and other organisms which recycle the organic nutrients excreted by algal cells. However, this level of recycling is clearly less efficient than at the cellular level because of 'downstream' losses associated with wave action and current flows through the reef. Nutrient losses, therefore, have to be compensated by nutrient sources. One important process is the extraction of dissolved organic

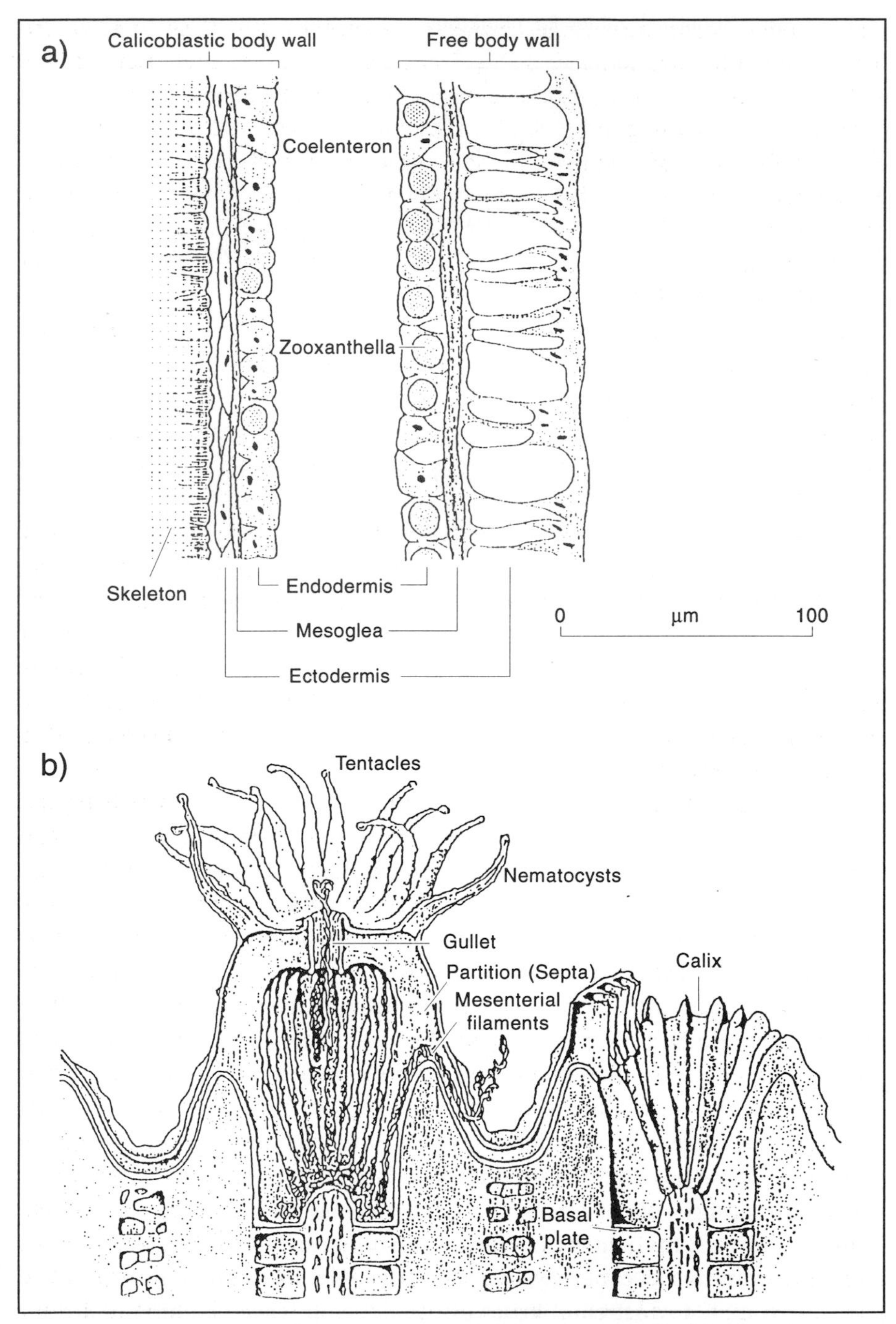

Fig. 6.2 a) Relationships between tissue and skeleton in a reef-building coral; skeleton outside the animal, to the left. After Barnes and Chalker (1990). b) Anatomy of the coral polyp and skeletal architecture. After Goreau *et al.* (1979)

matter, phosphorus and organic nitrogen — perhaps 20–40 per cent of the total annual nitrogen demand at One Tree Island, southern Great Barrier Reef — in turf carpets composed of filamentous algae on reef flats. The turf structure results from the grazing pressure of sea urchins, molluscs and particularly herbivorous fish, a good example of the interaction of biological and chemical processses on the reef. Such turf is highly productive: Johannes *et al.* (1972) have shown that at Enewetak Atoll, Marshall Islands, gross primary productivity on an algal-dominated transect was, at 10 g C m^{-2} d^{-1}, much higher than on coral-dominated transects.

Recently, it has been suggested that nutrients might be further supplied and fixed at the reef margin in association with a process which has been called 'endo-upwelling' (Rougerie and Wauthy, 1993). Nutrient-rich ocean water enters atoll bases and, warmed by geothermal heat flux from volcanic basements, rises to the surface (at < 1.0 cm h^{-1}) to supplement ocean nutrient inputs. Nutrients streaming off the productive nutrient source areas on windward reef margins proceed towards burial and removal in the largely nutrient sink regions of atoll lagoons. Sournia (1976), for example, has shown that the foraminifera and other organisms with algal symbionts of the sandy lagoon floor of Takapoto Atoll, Tuamotu Archipelago show gross production rates of only 0.4–1.3 g C m^{-2} d^{-1}. However, seagrasses in sheltered lagoon environments make important contributions to detrital food chains and can be highly productive: estimates of net productivity of 8–18 g C m^{-2} d^{-1} have been reported (McRoy and Helfferich, 1977).

A second set of important nutrient pathways is the way in which particulate organic matter is trapped by the biofiltering activity of a diverse range of reef organisms: corals, sponges, polychaete worms, bivalves and many elements of the zooplankton which themselves are then scavenged by larger predators. It appears that the nitrogen budget of corals is strongly dependent upon the uptake of particulate organic nitrogen from zooplankton. Many of the major framework-building 'massive' corals appear to be dominantly carnivorous, capturing zooplankton with the aid of batteries of specialized stinging cells on the tentacles surrounding their oral disc; using fine hairs covered in sticky mucus to waft bacteria and detritus into the gullets; and extruding mesenterial filaments through the mouth to aid food capture (Yonge, 1930). Specialized carnivores have large polyps to capture large and active elements of the plankton but have small surface area to volume ratios. Although zooplankton densities in open ocean waters are low, recent sampling within reef communities has revealed the presence of a distinctive and diverse reef zooplankton, much of it of small size (< 2 mm) and with typical densities of 7900 organisms m^{-2} of substrate, with local aggregations of up to 3.3 million individuals m^{-3}. This fauna includes a cryptic component which migrates up the reef at night, when it is exploited by the carnivorous reef corals which expand their polyps at night, unlike the more autotrophic corals which expand their polyps by day. One particularly efficient means of energy transfer appears to be the breakdown of coral mucus secretions by grazers, filter-feeders and reef bacteria — whose biomass may exceed that of the phytoplankton by an order of magnitude — into particulate organic matter which can then readily be assimilated, along with algal detritus, by zooplankton and, in turn, by reef macroinvertebrates and fish. Bacteria, algae and corals also sweep up organic and inorganic nitrogen released into the water by animal excretion.

Fig. 6.3 shows that only a few of the many reef nutrient pathways have been described here and confirms the complexity of the reef ecosystem. Clearly disruption of such a system — by overfishing for example — might well have unforeseen consequences elsewhere on the reef.

Nutrient status and reef survival

Coral reefs manage, clearly through a variety of means, to flourish in nutrient-poor environments. However, this ability is also a necessity: in natural oceanographic settings where upwelling brings nutrients into shallow waters, reefs and thus benthic-algal food sources may be replaced by phytoplankton-based food chains. In such circumstances, corals may be outcompeted by macroalgae, sponges, bryozoans and coral bioeroders which do well under such nutrient-rich conditions (Hallock *et al.*, 1988). Continental shelf upwelling and riverine inputs from South America contribute nutrient- and phytoplankton-rich waters to the Caribbean current which rises along the margin of the Nicaraguan shelf. In this setting, normal Caribbean reef-building processes are suppressed and reef growth is replaced by the development of bioherms constructed by the calcareous green alga *Halimeda*. This example shows the danger of eutrophication of reef waters by human activity.

The biological scale: calcification, coral growth and reef accretion

The previous section indicates that the remarkable metabolic performance of the reef system is an internal one and that this results from the presence of a self-constructed framework — the reef — which provides an immensely complex three-dimensional structure and modified environmental conditions within which these interactions take place. How, therefore, do reef dwellers build this structure?

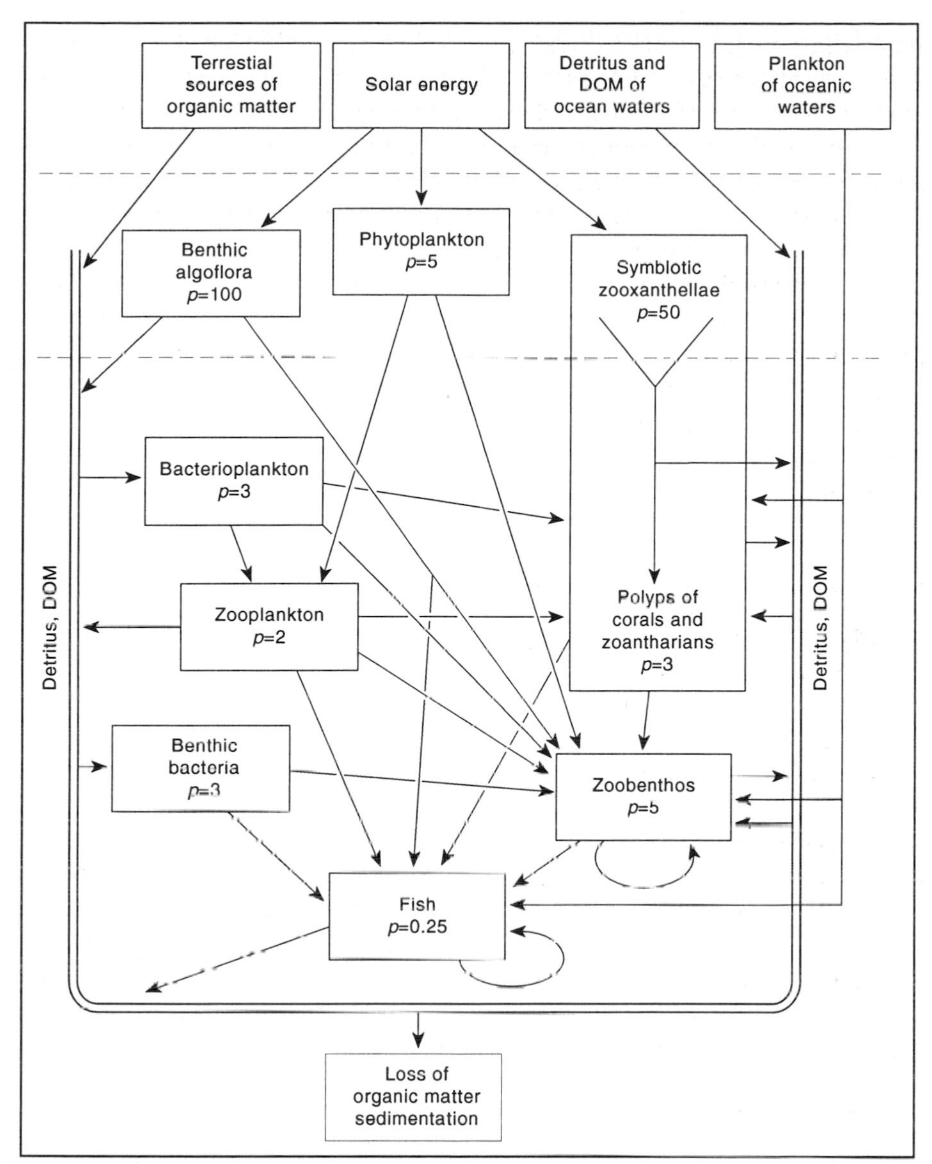

Fig. 6.3 Scheme of energy flow in a coral reef. Numbers in squares: food ratio of subsequent component of ecosystem. After Sorokin (1990)

The coral polyp, the zooxanthellae and skeletal growth

The basic unit of the coral is the polyp (Fig. 6.2). This sits within an external skeleton of calcium carbonate which is secreted by the outer layer of the three-layered coral tissue. Over time as the coral grows, new skeletal material is laid down so that while the structure of the coral may be considerable only the outermost layer is living tissue. Sexual reproduction in corals leads to the release of coral larvae, or planulae, to the ocean plankton. Coral growth in a new location starts by the settlement of planulae and the secretion of a basal plate of calcium carbonate followed by the development of vertical ribs, or septa (Fig. 6.2); this explains why corals needs need stable, rocky surfaces for initial settlement and growth. Once established, coral growth proceeds by asexual budding and/or splitting and even detachment of groups of polyps as polyp balls or individual 'polyp bail-out' from adult colonies. This activity produces a collection of polyps, a coral colony, which are clones of one another. Although there are some 'solitary' corals, the vast majority of corals are colonial as this allows an organism which is sedentary great flexibility for subsequent growth and the potential to rapidly exploit changing resources in a crowded ecosystem.

The role of the zooxanthellae in efficient nutrient cycling on the reef has already been noted. They also, however, play a key role in the promotion of skeletal growth, for the zooxanthellae aid the fixation and translocation of carbon within the organism, a process known as 'light-enhanced calcification'. The details of this process are still unclear but it is indirect: the rapidly growing branch tips of some corals have few zooxanthellae and radioisotope tracer experiments have shown that branch tip calcification is related to photosynthetic activity further down the branch. Corals which are kept in the dark show lower growth rates than corals in the light (Goreau, 1959); if corals are held in the dark for several weeks they expel their zooxanthellae and cease to grow, regaining only some of their growth performance on re-exposure to full light. Furthermore, this process is remarkably sensitive: corals grow twice as fast on a sunny day as on a cloudy one (Goreau and Goreau, 1959) and even show variations in growth rate during the day with the passage of the sun. In some cases, photosynthetic activity may be so efficient that under optimal light conditions and in particular species of coral, the organism may meet all, or most, of its respiratory needs. This kind of behaviour is typical of the multilayered, small-polyped branching corals where surface area to volume ratios are high. As crystal growth rates may vary according to varying light conditions, corals can show banding in a way analogous to tree growth. This has resulted in the growth of scientific studies in 'sclerochronology', akin to dendrochronology.

Coral growth and reef accretion on ecological time-scales

The highest ratio of surface tissue area to skeletal mass, and hence the most efficient form for food capture, is a hemispherical colony and under optimal conditions massive corals adopt this form. Rates of coral growth appear to vary considerably because of the style of growth. The branch tips of open-structured branching corals may extend at 100 or even 200 mm a^{-1} whereas hemispherical corals typically grow by radial expansion over their entire surface area at 4–20 mm a^{-1}. However, when growth is standardized to the performance of a solid stucture it is clear that all corals grow at the same mean rate of 5–15 mm a^{-1} (Buddemeier and Kinzie, 1976). It is misleading, however, to equate the gross growth performance of individual corals with the performance of a reef ecosystem as a whole. The calculation of vertical reef accretion for a complete reef complex requires the assessment of the role of other reef-building organisms, particularly coralline algae, the evaluation of variations in the cover and growth rate of reef-builders by reef environment, and the estimation of framework destructive processes (commonly bioerosion) as well as growth additions. These processes are clearly highly variable over space and time, as has been shown by Le Campion-Alsumard *et al.* (1993) at Tiahura reef flat, Moorea Island, south Pacific, and, although full-scale budgetary calculations have been made (Stearn *et al.*, 1977; Scoffin *et al.*, 1980), the approach remains problematical. Thus an alternative, indirect method of assessing 'whole reef' performance has been preferred. This is the 'alkalinity depression' method (see Smith and Kinsey, 1978 for details), whereby calcium carbonate gain in the reef community depletes total alkalinity in reefal waters. Field studies have identified three modal rates of calcium carbonate ($CaCO_3$) production; a 'fast' rate (10 kg $CaCO_3$ m^{-2} a^{-1}) characteristic of coral thickets, an 'intermediate' rate (4 kg $CaCO_3$ m^{-2} a^{-1}) associated with typical Indo-Pacific reef flats, and a 'slow' rate (0.8 kg $CaCO_3$ m^{-2} a^{-1}) typical of lagoon floor sand and rubble substrates (Kinsey, 1983). If 1–2 per cent of the reef calcifies at the fast rate, 4–8 per cent at the intermediate rate and 90–95 per cent at the slow rate, then a typical reef system will produce 1–1.2 kg $CaCO_3$ m^{-2} a^{-1} (Smith, 1978). If certain assumptions are made as to porosity of the reef framework then this community gain of calcium carbonate can be seen as equivalent to a vertical accretion rate of *c.* 0.8–0.9 mm a^{-1}.

Environmental constraints on coral growth and reef accretion in the vertical plane

Coral reefs provide excellent indicators of past sea levels as they are strictly shallow water constructions being strongly restricted in their position relative to contemporary sea levels. Light and exposure are fundamental controls; other controls such as salinity and turbidity are considered by Hopley (1982).

Corals and light

Clearly the studies in coral physiology outlined above show that light is a fundamental control on coral growth performance. Different measures of light levels all change dramatically with depth: illumination at 25 m depth in Jamaica is only 1 per cent of the surface illumination. Radiant energy falls also (Fig. 6.4), and daylength at Funchal (33° N) is 11 hours at 20 m depth but only 15 min at 40 m (Wells, 1957). These changes are accompanied by reductions in coral diversity from depths as shallow as 10 m; below 20 m the number of coral species falls rapidly (Fig. 6.4). In general, corals are restricted to depths of <100 m, although corals are known to occur at depths of up to 145 m in the Red Sea and to 165 m in the Pacific Ocean. There are photoadaptive changes in pigmentation characteristics of the zooxanthellae with water depth as they attempt to harvest the remaining light more efficiently, and as light intensity decreases on the deep reef, so calcification rate declines. As food supply is relatively invariant with depth, corals attempt to maintain their surface area but cannot continue to fill a hemispherical form: thus massive hemispherical colonies reorganize their growth to overlapping sheets and ultimately thin plates, growing only at their leading edge, with increasing depth (Fig. 6.5). The typical limiting depth for coral growth in the Caribbean Sea is 90–100 m. However, a coral presence at these depths does not imply the construction of a structural reef framework. Besides a lack of light, deep reefs are a difficult environment for corals, with steep, and in places unstable, substrates and high sediment rain from more productive reefs above. In particular, bioerosional activity, especially by boring sponges, undermines coral skeletons. The limit for reef construction is therefore closer to half this depth limit and deep reefs are veneered with encrusting sponges rather than high living coral cover.

Corals and subaerial exposure

If corals are depth limited then they are also strongly constrained by emergence, particularly if their exposure corresponds to high mid-day temperatures or to heavy rainfall. Studies of large coral colonies in tropical lagoons (micro-atolls) have suggested that growth is restricted to a tidal level constrained by mean low water springs. Thus, for example, low water levels associated with ENSO events (*see* Chapter 2) in 1972 lowered water levels in Guam by 44 cm below mean sea level, exposed reefs and caused mass mortalities in corals and associated reef dwellers (Yamaguchi, 1975). Loya (1976) documented an exposure of corals at Eilat, Red Sea to noon time temperatures over a series of unusually low tides lasting a week; 15 years later recovery from this event was still not complete.

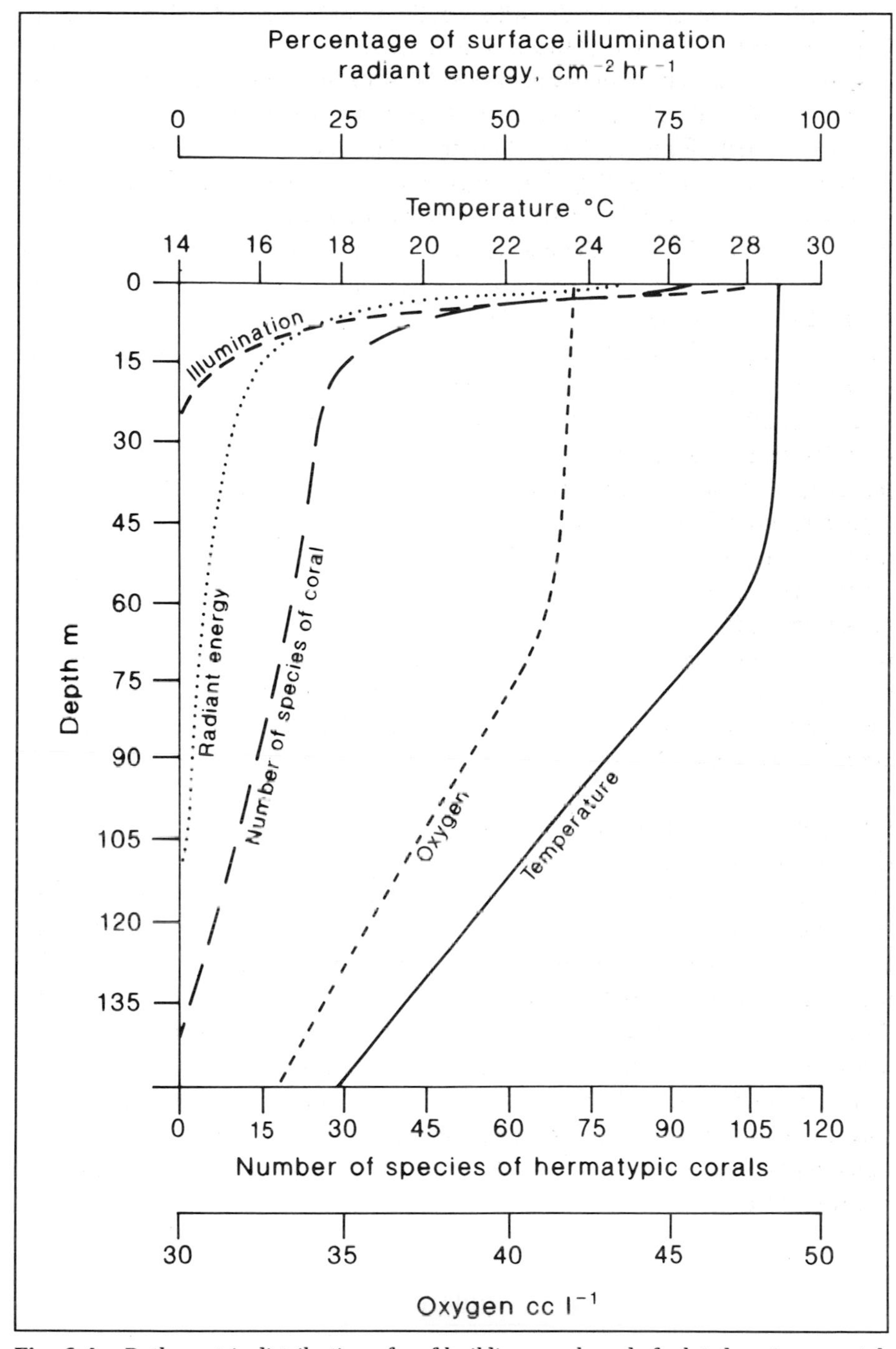

Fig. 6.4 Bathymetric distribution of reef-building corals and of related environmental parameters. After Stoddart (1969)

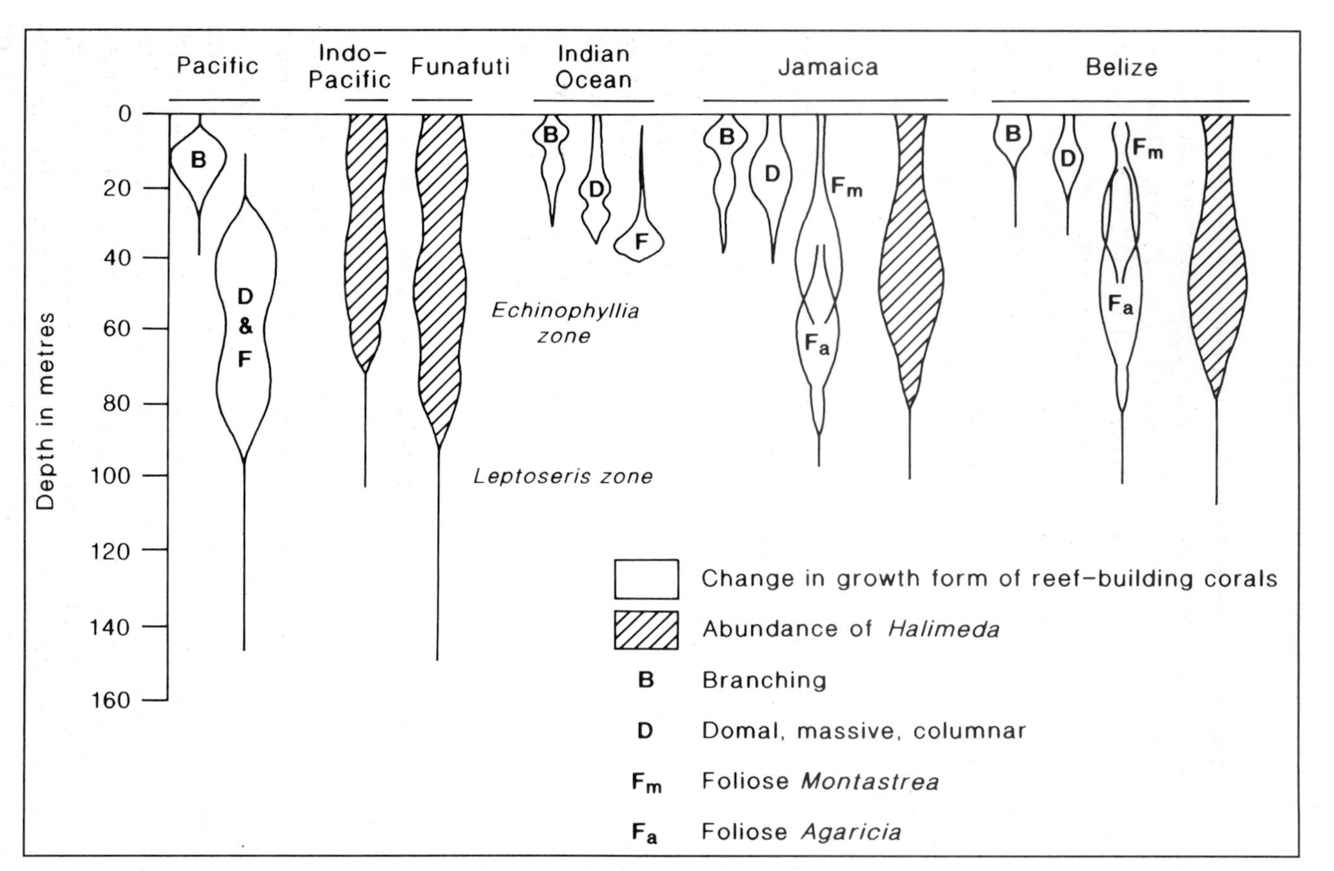

Fig. 6.5 Depth ranges and growth forms of reef-building corals in four different areas. After James and Ginsburg (1979)

The geological scale: reef construction, destruction and net accretion

Construction

The basic framework of the reef is a porous structure provided by corals and coralline algae, cemented and sealed by coralline algae, bryozoans and encrusting sponges and infilled by reef-derived sediments from calcareous algae, molluscs, foraminifera and echinoderms. This open matrix is infilled by secondary constructors and sediments whilst the primary framework becomes subjected to bioerosional processes which create both new cavities and sediments to fill interstices of various ages. The secondary framework itself becomes cemented and lithified and may itself then become subject to alteration by bioerosion. In time, the whole of the original framework may be lost or replaced.

Destruction

Constructional activity on the reef is always offset by the activities of organisms living on the reef which destroy carbonate substrates. A number of processes are involved which go under the general title of bioerosion. Reef grazers browse for blue-green algae which burrow into reef surfaces and the hardened radulae of these grazers remove carbonate grains as well as algae. Other organisms, like echinoderms, excavate burrows on the reef and a wide range of bivalves, sponges and worms bore into coral skeletons. These processes effectively limit reef framework construction where calcification rates are low in deep water or under adverse temperature or other environmental conditions and produce vast quantities of sediments on reef fronts and reef tops.

Net reef accretion on geological time-scales: responses to sea level change

The processes of construction and destruction outlined above combine over geological time to result in varying rates of upward growth of complete reef systems. To what extent, therefore, are contemporary 'biological' rates of coral growth characteristic of reef behaviour over much longer periods of time? This question can be addressed through the estimation of reef growth rates from radiometrically dated carbonate sequences obtained from reef top coring, a method which effectively integrates reef accretion through time. A comparison of rates of reef growth from drill cores, by tectonic setting and reef type suggests a wide range of reef response over the last 8000 years of reef growth during the Holocene (Table 6.2; Spencer, 1994). This variation is due to the different response modes exhibited by reefs to differing rates of sea level rise during the post-glacial trans-

Table 6.2 Rates of reef growth from radiocarbon dating of vertical sequences of reef carbonates recovered by reef-top drilling

Plate-tectonic setting	*Locations*	*Rate of coral growth* (mm a^{-1})
Plate-margin, subduction zone	US Virgin Islands	1–2
Plate-margin, continental shelf	Atlantic Panama	0.3–12
	Alacran, Gulf of Mexico	12
	Belize barrier reef	1.1–8.3 (max. 12)
	N. Great Barrier Reef	1.3–14.2
	C. Great Barrier Reef	1.9–10.2
	S. Great Barrier Reef	0.6–8.3
Mid-plate fringing reefs and barrier reefs	Réunion, Indian Ocean; Society Islands and Hawaiian Islands, Pacific Ocean	7–10 (range 1.6–50)
Mid-plate, coral atolls	Enewetak and Bikini Atolls, Central Pacific Ocean	1–2 (max. 10)
	Mururoa Atoll, S. Pacific Ocean	2–3
	Tarawa Atoll, Central Pacific Ocean	5–8

Source: Spencer (1994)

gression: 'keep-up' reefs were able to keep pace with sea level rise, 'catch-up' reefs were left behind by rapidly rising sea levels but regained sea level on rapid vertical growth as sea level rise slowed, and 'give-up' reefs were terminally drowned through being unable to keep pace with sea level rise and then being at too great a water depth for significant vertical growth to be activated on sea level stabilization (Davies and Montaggioni, 1985; Neumann and MacIntyre, 1985). Whilst recognizing the role of regional and local factors in influencing reef growth, Spencer (1994) reported that the literature suggests that the 'give-up/catch-up' threshold lies near a rate of sea level rise of 20 mm a^{-1}, from well-documented failure of reef banks around the island of Barbados at 13.5 and 11.0 ka BP (Fairbanks, 1989; Bard *et al.*, 1990), and the 'catch-up/keep-up' threshold at 8–10 mm a^{-1}, although this threshold is less well constrained by the available dataset. Relating these thresholds to key water depths suggests that Caribbean platforms at depths of 15–21 m have been able to keep up with sea level rise whilst those at 20–25 m have not been able to do so (Adey and Burke, 1977).

Responses to sea level change and coral reef morphology

Rapidly growing 'catch-up' reefs may accentuate the morphology of the underlying basement. Purdy (1974), in a classic study of the morphology of the Belize

Barrier Reef, showed how Holocene reef growth has magnified the underlying limestone solutional depressions (doline karst) in the drier northern part of the barrier system and highlighted the karst towers and limestone 'eggbox' topography, with lagoonal shoals and the reef islands of the barrier respectively, in the wetter, southern region on a karst marginal plain created by runoff from the non-carbonate Maya Mountains. Elsewhere, however, other authors have argued for modern reef topographies to be related to reef growth on sedimentary accumulation features, either on siliciclastic deposits, such as river levées or on barrier island-related flood and ebb deltas in inlets or storm washover fans. Over time, reef growth is liable to obscure the underlying topography. Thus, on the Great Barrier Reef while shelf edge reefs have just reached sea level and show reefs surrounding relatively deep lagoons, inner shelf reefs growing from shallower pre-Holocene basements have filled in their reef tops. Where windward storm ramparts have been formed and subsequently lithified to protect these platforms, sand cays and mangrove woodland have developed on the surface to give distinctive 'low wooded islands'.

Atlantic versus Indo-Pacific reef morphologies: differences and their potential significance

Fig. 6.6a and Plate 6.3 show a typical Caribbean reef crest of the thickly branched coral *Acropora palmata* and the bladed stinging hydrocoral *Millepora complanata* (Milliman, 1973; Roberts *et al.*, 1988). At best a poorly cemented reef pavement is developed in the back-reef region. This section can be compared with a schematic transect through the windward margin of an Indo-Pacific atoll (Fig. 6.6b). The development of a ridge of crustose coralline algae of the genus *Porolithon*, perhaps a metre above the reef flat, tens of metres wide and showing lateral continuity for tens of kilometres, backed by a reef flat, perhaps 100 m wide and composed of a wedge of competent rock 1–4 m in thickness, is striking. Early views ascribed these differences to the 'low growth rates and immaturity' of Atlantic province reefs. Although algal ridges have been described from high energy reefs in the Virgin Islands, (e.g. Adey, 1975), the Lesser Antilles and Atlantic Panama (Glynn, 1973) they nevertheless do not form such well developed or laterally contiguous structures as in the Indian and Pacific Oceans (Plate 6.4). It is now clear that these differences are not due to differences in growth potential between the two reef provinces — in fact some of the highest growth rates, of 12 mm a^{-1}, are recorded from *Acropora* reefs in the depauperate Atlantic reef province (Table 6.2; Stoddart, 1990) — but are at least in part due to sea level history. Owing to the varying importance of different contributors to sea level change (*see* Chapter 2), the two reef provinces have experienced very different sea level histories over the last 6000 years. In the Caribbean Sea the record

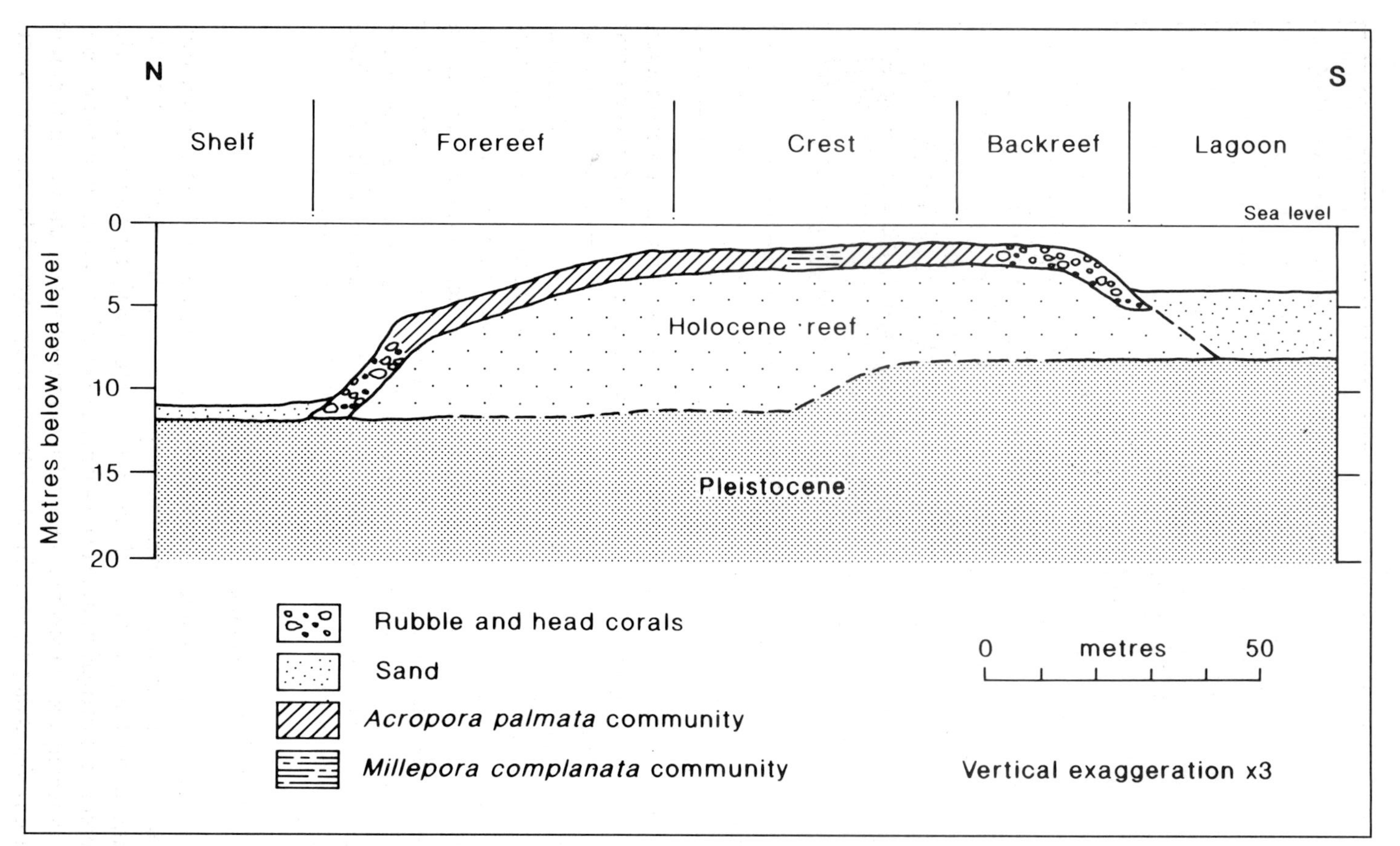

Fig. 6.6a Coral reef morphology in the Atlantic reef province. Note lagoon. After Roberts *et al.* (1988)

Plate 6.3 Caribbean reef crest: *Millepora* sp. and staghorn coral (*Acropora palmata*) exposed by abnormally low tide, Grand Cayman Island, West Indies

has been one of a progressive slowing down of sea level rise with present levels only being reached in the last 1000 years. By comparison, present sea level was attained 6000 years BP in the Pacific and followed by a period of sea level at 0.5–1.0 m above present sea level before falling to its present level. Caribbean reefs, therefore, have been dominated by 'catch-up' behaviour whereas Indo-Pacific reefs have been able to fill in the reef frame over late Holocene time. These differences have a significance in the context of possible future changes in sea level. In the Indo-Pacific region greenhouse gas-induced sea level rise will have to overtop fossil topographies before impacting unconsolidated sediments and reef islands but in the absence of such protective structures in the Caribbean reef, islands will be immediately more vulnerable to sea level rise.

The coral reef community

Reef corals initially capture space by settlement of planulae from the plankton or by the attachments of coral fragments or 'buds'. They then expand in the plane across the substrate and vertically into the overlying water column. The combination of rapid growth of newcomers next to long-lived, large corals leads to crowding and competition for light and nutrients. The exact composition of these reef-building communities is determined by physical factors, which sort

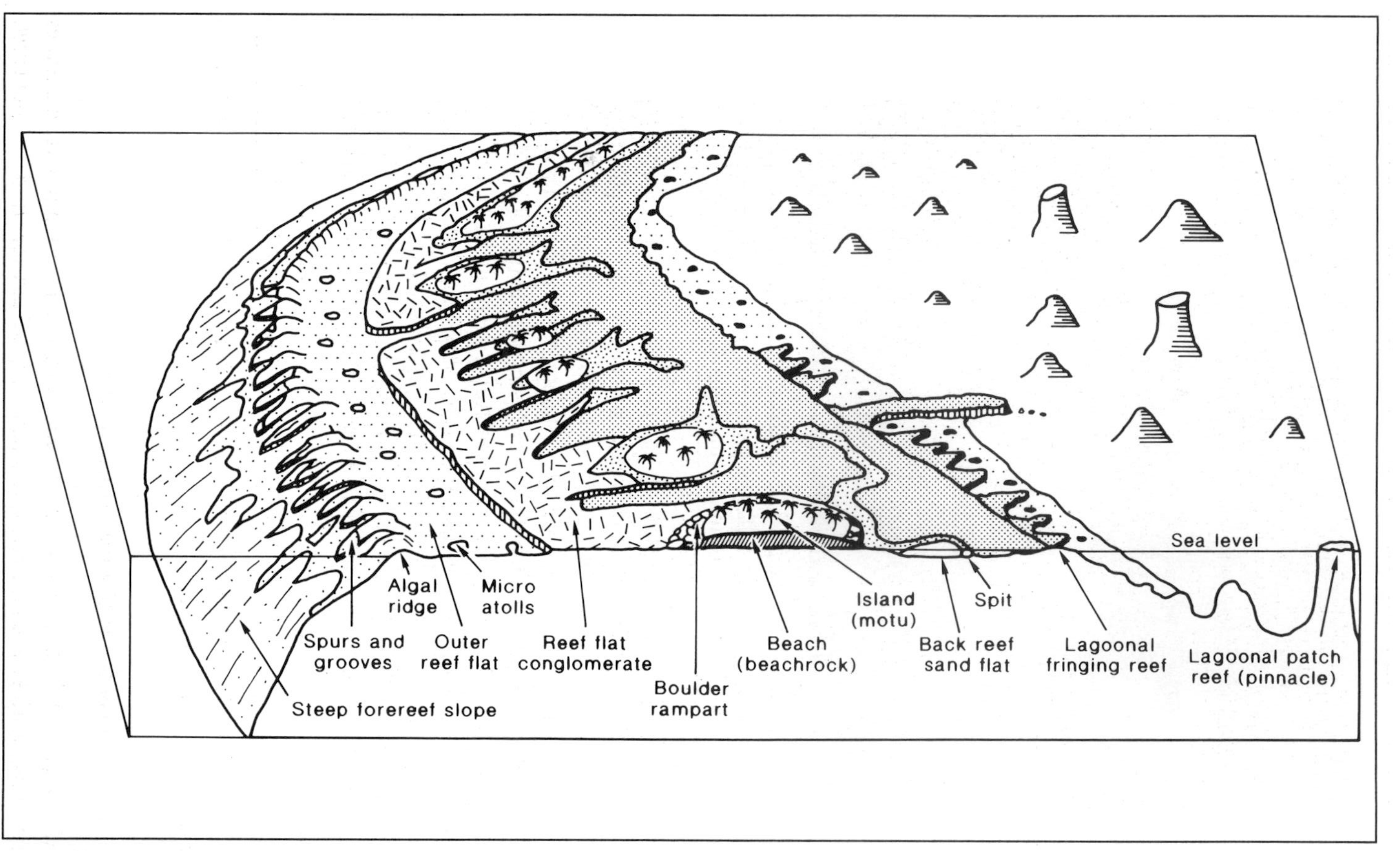

Fig. 6.6b Coral reef morphology in the Indo-Pacific reef province. Note reef flat

Plate 6.4 Algal ridge (top) and reef flat, Niue Island, south-west Pacific Ocean

out species by different process controls and energy levels and by biological interaction between species.

Physical controls in the reef community

On Caribbean reefs, the relative importance of wave forces and current forces varies between the deep fore-reef and shallower reef terraces and the reef crest.

Deepwater waves interact with fore-reef shelves, through refraction, reflection and wave scattering, before breaking at the reef crest. Here typical trade wind-generated waves lose 72–97 per cent of their energy (Roberts *et al.*, 1992). The dominance of wave action at the reef crest means that this environment is dominated by encrusting corals and thickly branched (e.g. *Acropora palmata*) and bladed (e.g. the hydrocoral *Millepora* spp.) forms. As energy levels change, so different coral assemblages migrate to, or are displaced from, the reef crest (Geister, 1977).

Many reefs are characterized by a repetitive pattern of coralgal buttresses separated by coral-free, sediment-filled gullies known as 'spur-and-groove' topography. Such features can be found characterizing deep and shallow fore-reef terraces of Atlantic reefs and elsewhere, and the coralline algal margins of Indo-Pacific atolls. Early observations at Bikini Atoll, Marshall Islands (Munk and Sargent, 1954) argued that spur-and-groove topography and its continuation in the form of 'surge channels' through the algal rim on to the reef flat, act as a dissipating baffle to high wave energy and these ideas have been confirmed by more recent work in the Caribbean, primarily on the island of Grand Cayman. Grand Cayman is aligned in an east–west direction into the typical trade wind field of the northern hemisphere; thus there are considerable differences in wave power around the island (Fig. 6.7a). Spectral analysis of shore-parallel echo-sounding transects shows that high energy environments are characterized by closely spaced and well developed coral spurs separated by deep grooves on both shallow (*c.* 8 m) and deep (*c.* 20 m) fore-reef terraces whereas low energy coasts have widely spaced spurs and a poorly developed shelf topography (Fig. 6.7b). If these differences are pronounced in the microtidal, moderate wave energy environment of the Greater Antilles what role might physical processes play on the macrotidal sections of the Great Barrier Reef or in the storm swell environments of the Central Pacific?

On Grand Cayman, currents are typically weak and variable in shallow water but on the deep fore-reef stong, unidirectional currents predominate. As the relative importance of wave and current forces change with position on the fore-reef, so does the composition and form of the coral communities in these two environments (Table 6.3).

Biological interactions in the reef community

Within these physical constraints is a further layer of controls imposed by biological interactions between species. The different growth styles of branching and massive corals ought to result in branching corals overtopping massive forms and complete domination of the reef environment. However, on most reefs these two types of corals co-exist. One reason is that some corals have

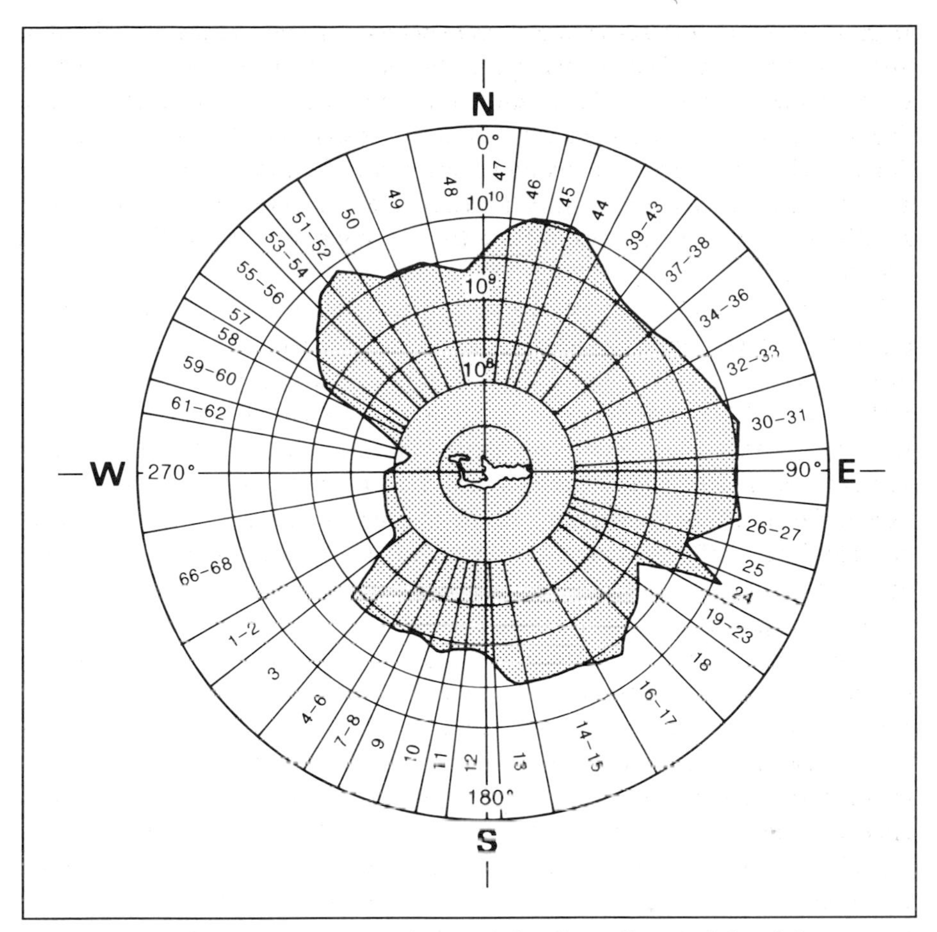

Fig. 6.7a Variation in wave power (ergs m^{-1} shoreline a^{-1}) around Grand Cayman Island, Caribbean Sea

mechanisms for defending or extending their space on the reef: this aggression, over distances of up to 10 cm, can be achieved through the deployment of mesenterial filaments, sweeper tentacles, sweeper polyps or mucus secretions, carrying 'nematocysts', or stinging cells, on to the neighbouring victim (Fig. 6.8). Manipulative field experiments (e.g. Lang, 1973) have been used to construct 'digestive hierarchies' for different species and to show, in some instances, how slower growing forms can be digestively dominant over faster growing opponents. However, this now seems too simplistic. Rather, many reef corals may select from a portfolio of different defensive and aggressive mechanisms which become operative at different spatial (millimetres to metres) and temporal scales (minutes to decades), giving competitive linkages of great complexity. Thus, for example, in a relatively simple interaction a coral may

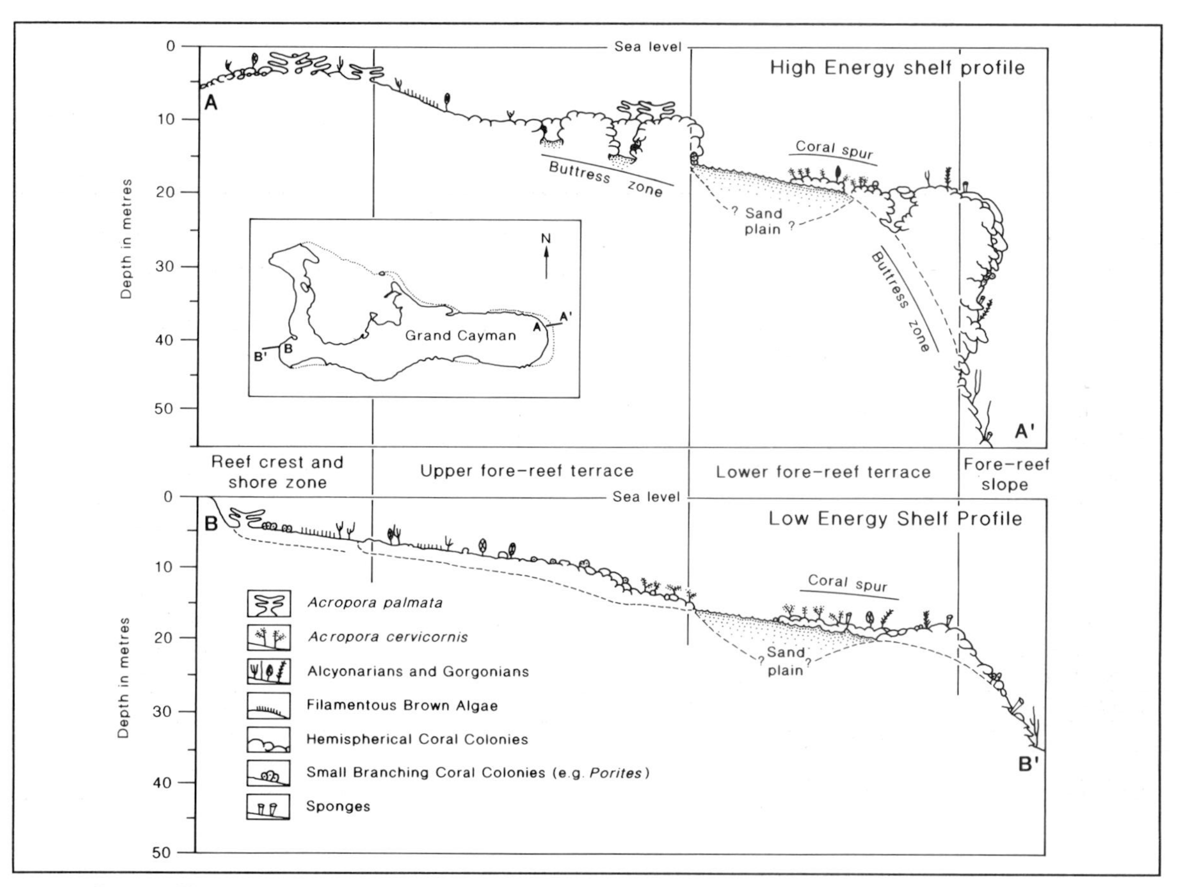

Fig. 6.7b Profiles of reef development from high and low wave energy sectors. After Roberts (1974)

Table 6.3 Form-process relationships on shallow, wave-dominated and deep, current-dominated reefs, Grand Cayman Island, Caribbean Sea

	Major form-process relationships			
	Wave dominated		*Current dominated*	
Parameters	*Near-reef crest*	*Shallow terrace margin*	*Deep terrace*	*Shelf-edge reef*
Zonation	*Acropora palmata* *Millepora alcicornia*	*Acropora palmata* *Agaricia*	*Acropora cerviconis* *Agaricia*	*Montastrea* *Agaricia*
Growth forms	Thickly branched Bladed Encrusting	Branched Bladed	Delicately branched Massive	Massive Platelike
Coral cover	Moderate	Moderate	Abundant	Abundant
Bottom roughness	<2 m	<5 m	<4 m	<30 m
Shelf morphology	Limestone pavement, low-relief spurs and grooves	Moderate-relief spurs and grooves	Moderate–low-relief spurs and grooves	High-relief spurs and grooves
Sediment	Sparse	Thin veneer in grooves	Extensive impounded sediment plains	Off-shelf movement down grooves
Waves	Breaking, high turbulence, high wave force	Moderate wave forces, 20% height reduction from shelf edge	Moderate–low wave force, small-scale turbulence	Low wave force, small-scale turbulence
Currents	Weak, multidirectional flow	Multidirectional flow. 60–70% speed reduction from shelf	Moderate on-shelf rectilinear tidal currents	Strong, on-shelf rectilinear tidal currents (>50 cm s^{-1})

Source: Roberts *et al.* (1977)

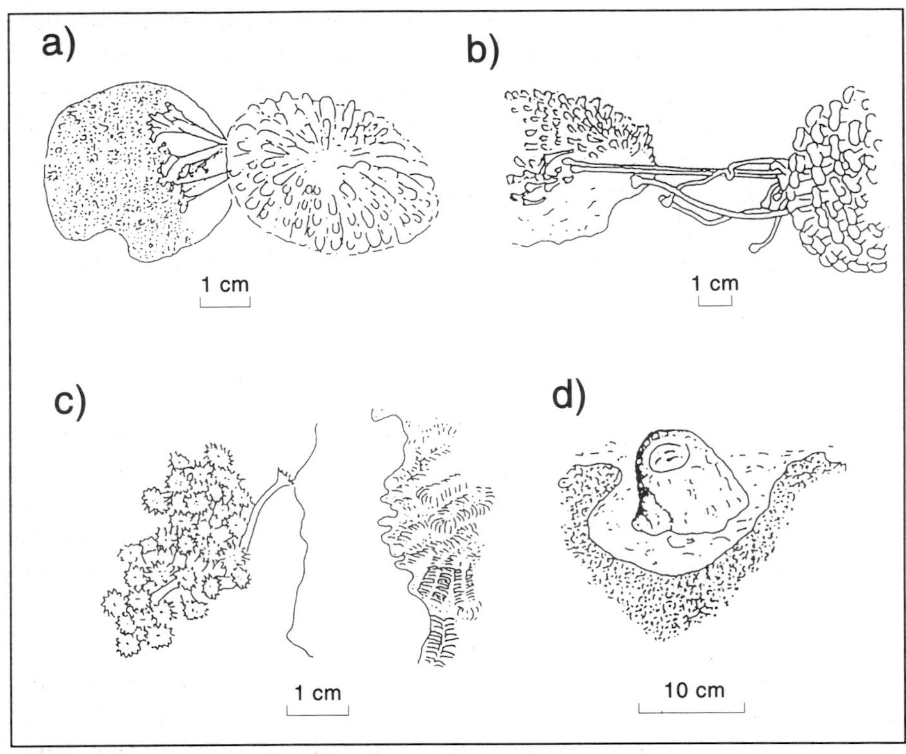

Fig. 6.8 Aggressive mechanisms in coral reef species: a) extracoelenteric digestion, b) sweeper tentacles, c) sweeper polyps and d) probable allochemical aggression by the sponge *Siphanadictyon coralliphagum*. After Hughes (1983)

wound a neighbour by the short-range extrusion of mesenterial filaments only to be attacked subsequently when the initial victim develops and deploys longer range sweeper tentacles. The resulting injuries may produce a sufficient gap between the two species for hostilities to cease; however as soon as the corals grow towards one another the cycle of reciprocal wounding begins again. Another set of mechanisms on the reef is defensive rather than aggressive and takes the form of the inhibition of one species by another through the use of noxious or toxic chemical secretions. Such defence mechanisms are particularly prevalent in the sponges and soft corals in competing with corals.

Another form of interaction is that between reef structure and the role of herbivorous fish. Many damselfish, for example, establish and maintain algal mat gardens, 0.25–0.50 m^2 in area, by excluding other herbivores, by weeding out potentially smothering macroalgae and by nipping back the growth tips of within-garden corals. This latter process may determine reef community structure; on shallow reefs in Panama, reef flat and very shallow (<6 m) environments are dominated by the branching coral *Pocillopora*; damselfish find abundant shelters within the complex topography created by this coral and expand their numbers. This has two effects: first, high damselfish densities aid *Pocillopora* recruitment by excluding organisms which would attack the coral and second, the fish preferentially eat the coral tissue of a competitive species, *Pavona*, which can then be overtopped by neighbouring *Pocillopora* colonies which come to dominate the very shallow reef. On deeper reefs (6–10 m), where there is less topographic complexity, damselfish numbers are lower and *Pavona* dominates over *Pocillopora* (Fig. 6.9).

Redefining spaces: the reef community and storms

Finally, if the reef is seen as a space-filling exercise, then the spaces can be repeatedly redefined and renewed by externally imposed events. Indeed, it has been argued that reefs must be periodically disturbed if they are to maintain their high diversity. These ideas were first put forward by Connell (1978) in his 'intermediate disturbance' hypothesis. Of course, the branching corals are more susceptible to breakage and dislodgement than massive corals and it has been argued that this periodic opening up of the reef community is a mechanism whereby a high diversity of coral species is maintained. On the other hand, alternative arguments suggest that storm fragmentation may help the dispersal of branching species (providing the fragments are large enough; Highsmith *et al.*, 1980) and high densities of rejuvenating coral sticks may smother underlying competitors and promote the development of monospecific stands.

The one event in many parts of the reef seas that impacts coral reefs periodically is the tropical cyclone or hurricane, which is characteristic of zones 7–25° north and south of the Equator. Reef areas which are regularly afflicted by such events include the north-west Pacific Ocean, the south-west Pacific Ocean, the Great Barrier Reef of Australia, parts of the northern and southern Indian Ocean and the Caribbean Sea and Gulf of Mexico. Although many of these areas may suffer perhaps as many as five hurricanes or cyclones in a season, the narrowness of storm tracks means that even in areas of high cyclone incidence individual reefs may only experience four to eight storms per century. Regional impacts very much depend on the interactions of storm track direction and timing and regional bathymetry; thus of two hurricanes of similar intensity, Hurricane Donna (1960) funnelling up Florida Bay and augmenting a high spring tide produced a 4 m storm surge whereas Hurricane Betsy (1965), which approached from the east with no high tide synchroneity, resulted in no appreciable rise over expected water level. Similarly, Hurricane Andrew (1992) produced winds of up to 250 km h^{-1} (among the highest

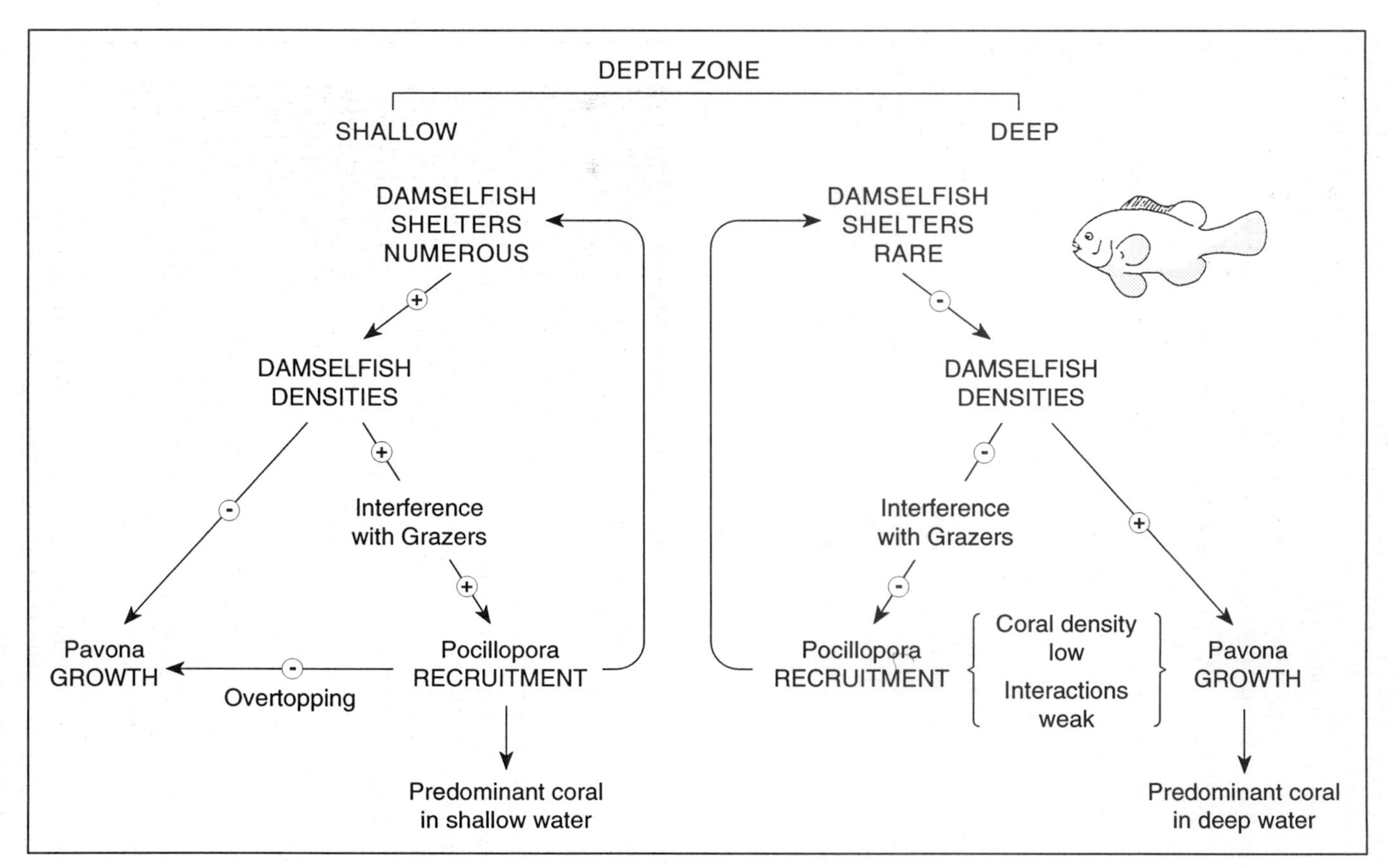

Fig. 6.9 Summary model of the relationships between some physical and biological factors responsible for the maintenance of coral reef zonation, Gulf of Panama. After Wellington (1982)

recorded for any Atlantic hurricane) but did little damage to the reefs on the northern Great Bahama Bank. Here winds were *c.* 180 km h^{-1} and the disturbance passed over the Bank in *c.* 3 hours, limiting the opportunities for the creation of surge conditions (Boss and Neumann, 1993).

At the site scale, impacts may be differentiated by aspect, slope angle, reef topography and even damage patterns around particularly large coral colonies (Woodley *et al.*, 1981). This kind of structural trimming is characteristic of cyclones with typical windspeeds of 120–150 km h^{-1}; in these events damage may be repaired within a decade. Disturbance history may also be an important factor in explaining reef response. Shallow exposed reefs at St John, US Virgin Islands were damaged relatively little by the very severe Hurricane Hugo (1989), having already been damaged by Hurricane Gilbert (1988) and the 1987 coral bleaching episode (Witman, 1992). More severe hurricanes with windspeeds exceeding 200 km h^{-1} may so damage the reef and suspend re-establishment that full recovery may take over 50 years (Stoddart, 1985). The complete flattening of reef structures by very severe disturbances both removes the habitats, and thus biota, associated with topographic complexity and

Plate 6.5 Coral rubble ridge, created by hurricane reef destruction in the 1930s, piled against Tertiary limestones, Great Bluff, Grand Cayman Island, West Indies. Contributing fringing reef offshore right

creates unstable platforms not conducive to the settlement of coral planulae and the re-establishment of reef growth.

If catastrophic storms are destructive of reef growth, they are the only mechanism by which large amounts of coarse sediments can be supplied to reef top environments (Plate 6.5). Corals are typically pruned from water depths of 10 m and moved rapidly onshore to be deposited as rubble 0.6–1.0 m in diameter (Hernandez-Avila *et al.*, 1977). Subsequent more frequent, lower magnitude storms may then serve to transport these accumulations and reconstruct eroded reef islands; a classic example is the addition of 2.8×10^6 tonnes of coral rubble (Stoddart, 1985) to the shoreline of Funafuti Atoll from the shoreward migration of a rubble ridge generated by Cyclone Bebe in 1972 (Baines and McLean, 1976). Long term patterns of motu and sand cay status depend, therefore, upon the relationships between storm frequency and intensity and reef productivity, and hence potential coarse sediment supply to subaerial reef environments, and, to a lesser extent, sediment stabilization by incipient cementation and/or vegetation colonization and binding. Bayliss-Smith (1988) has argued that whereas shingle motu follow this model, sand cays are vulnerable to destruction by storms of even moderate magnitude and thus subject to large and constant variations in size, shape, position and elevation. Sand cay dynamics have been investigated for reef islands on the southern Great Barrier Reef by Flood (1986) who has shown how cays adjust to variations in wind and wave fields (Fig. 6.10a and b). Human interference in these delicate systems, such as the interference with natural processes of sediment re-distribution by boat channel construction at Heron Island, in the southern Great Barrier Reef, Australia, runs the risk of destabilizing the whole system (Fig. 6.10c).

Corals and temperature

Low temperature controls

Lower thermal thresholds to coral function are long established: below 16°C corals lose the ability to capture food, between 14 and 16°C they cease to reproduce sexually and below 14°C many species die after perhaps only a few hours exposure to such temperatures. Close to the 16°C limit, coral growth rates are low and often do not balance bioerosion so that 18°C is a truer threshold between communities of individual corals and structural reefs. Optimal conditions for reef growth are temperatures of 25–29°C, with temperatures in the coldest month not falling below 20°C. The margins of the 'reef seas' are therefore climatically determined. This has been well illustrated by Grigg (1982) who has shown how reef accretion declines with increasing latitude in the Hawaiian islands. Hawaii, at 19° N, has typical reef accretion rates of

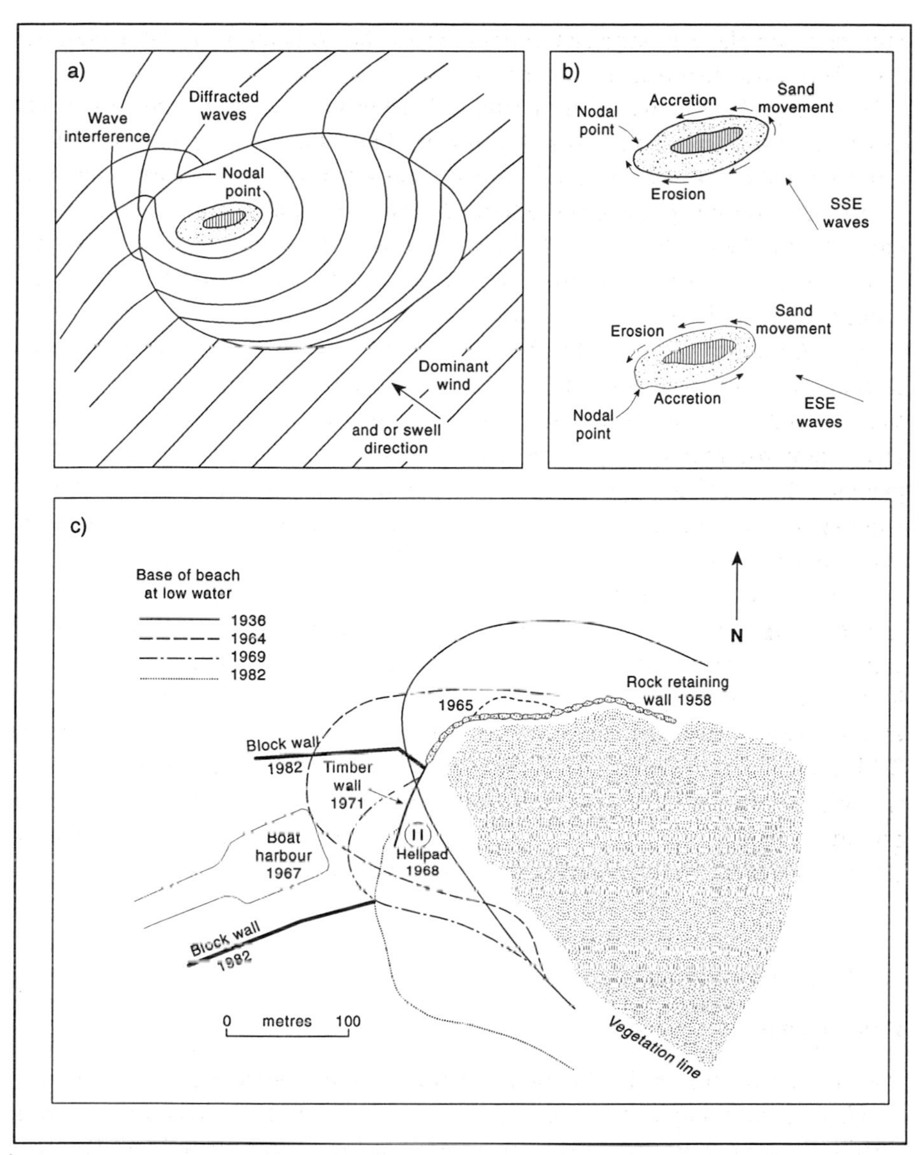

Fig. 6.10 Dynamics of coral reef sand cays. a) Cay formation. Reef top outline shown by black line; cay stippled with stable vegetation portion in black. b) Patterns of erosion and accretion and position of nodal point (minimum wave energy) under SSE and ESE waves. c) Variability of shoreline position, western Heron Island, southern Great Barrier Reef, Australia, 1936–82. After Flood (1986)

11 mm a^{-1}, whereas Kure Atoll, at 29° N the most northerly coral reef in the Pacific Ocean, shows accretion rates of just 0.2 mm a^{-1}.

The exact nature of this control has been well established for the Atlantic reef province by research integrating remote sensing techniques with field monitoring in the region of the Florida Keys and the Bahamas Banks. In winter, outbreaks of cold continental air, known as 'northers', sweep down from the interior of the USA. Thermal infrared and satellite imagery (Walker *et al.*, 1982) show that shallow bank and bay waters have limited heat storage capacity, experience rapid heat loss and can be chilled below the 16°C threshold for several days during these events. One event in January 1977 was accompanied by gale force winds, freezing temperatures and even snowfall on North Andros Island in the Bahamas. Temperatures dropped below 16°C for 7–8 days in Florida Bay and on the northern Bahama Banks (Roberts *et al.*, 1982). In 1981 minimum temperatures of 8.7°C in Florida Bay were recorded (Walker *et al.*, 1982). Coral mortality associated with these events can be high: 80–90 per cent mortality on Hen and Chickens Reef, Florida in 1969–70, 90 per cent mortality of *Acropora cervicornis*, a major framework builder at Dry Tortugas in 1977, and deaths of *Porites asteroides, Montastrea annularis* and *Agaricia* spp. in north Florida in 1981.

High temperature controls

Corals, like many tropical shallow marine organisms, have a narrow thermal tolerance and live near the upper limit of that tolerance, being adapted to local mean maximum summer temperatures (Jokiel and Coles, 1990).

Classic experiments in the 1930s, more recently followed up by studies of corals exposed to warm outflow from power stations, have shown clearly that corals respond to thermal stress by whitening or 'bleaching'. Bleaching is the visible manifestation of the loss of the zooxanthellae from coral tissues and/or the loss of photosynthetic pigmentation from the zooxanthellae. This chapter has shown how the zooxanthellae play a key role in coral metabolism and thus their loss or reduced function is accompanied by reduced carbon fixation and skeletal growth. Furthermore, bleached corals show reduced reproductive ability which may slow coral recruitment after disturbance (Glynn, 1993). Up to several weeks of temperature elevation of +1 to +2°C results in bleaching but limited (< 10 per cent) coral mortality; recovery takes place when temperatures return to normal and full coral pigmentation is achieved after a few months. However, larger temperature excursions of +3 to +4°C, even if only sustained for a few days, can produce mass bleaching of reef communities and coral mortality rates in excess of 90 per cent. Such mass mortality episodes have been readily correlated with temperature and solar irradiance changes associated with the warm phase of ENSO events. During these periods corals in the western Pacific are stressed by high solar irradiance (e.g. Indonesian reef flats, Brown and Suharsono, 1990), whilst eastern Pacific reefs are impacted by high sea surface temperatures (in 1982–3: +3 to +4°C above optimal temperatures of 27–28°C), low nutrient concentrations and lowered salinities (Glynn *et al.*, 1988). Interestingly, the eastern Pacific reefs which are periodically impacted by severe ENSO-related warmings show thin (1–8 m) sequences of Holocene reef growth (compared to typical 20 m + thicknesses elsewhere in the Pacific basin), despite globally average rates of skeletal growth on 'biological' time-scales (Glynn, 1990). ENSO events also show teleconnections to the Caribbean Sea which shows calm and thus clear seas during Pacific ENSO warm phases; it has been argued (Gleason and Wellington, 1993) that under such conditions higher than average intensities of ultraviolet (UV) radiation penetrate the water column and trigger coral bleaching.

Corals, temperature and global warming

It has been argued that bleaching episodes prior to the 1980s were related to small scale, geographically isolated events whereas many of the bleaching events witnessed in the next decade were complex in character, covered large geographical areas and were characteristic of all the world's reef regions, leading to speculation that these episodes represented an early signal of global warming in the oceans (Fig. 6.11; Williams and Bunkley-Williams, 1990;

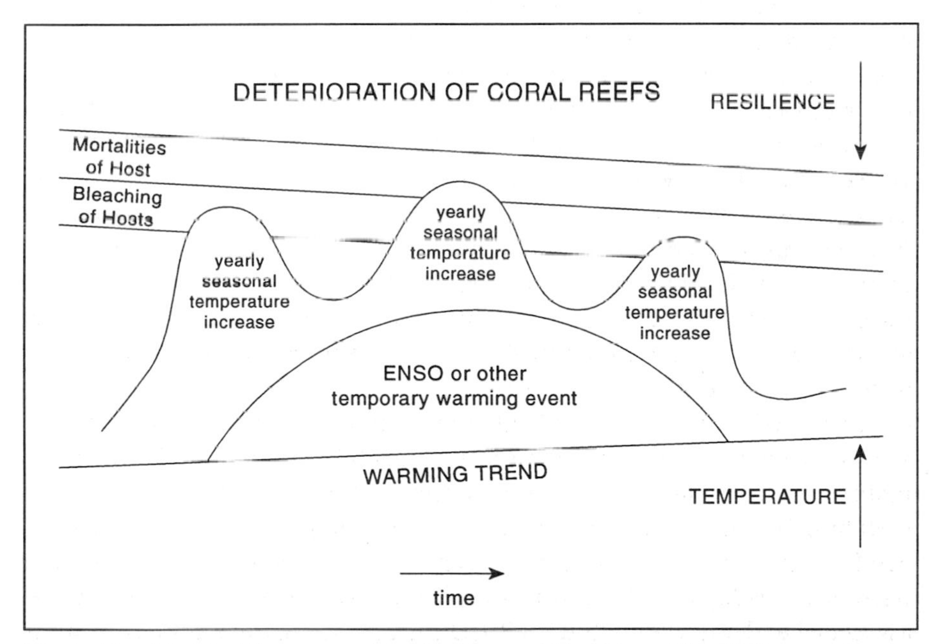

Fig. 6.11 Model of the causes of worldwide coral bleaching. After Williams and Bunkley-Williams (1990)

Glynn, 1993). However, this chapter has shown that coral reef ecosystems are vulnerable to a wide range of environmental stresses and thus not only temperature and temperature-related factors may induce reef bleaching. Although the frequency of coral bleaching episodes appears to have been increasing markedly in recent times (e.g. 1876–1979: three major events, 1979–90: sixty-three events (Glynn, 1993)) this may be due to a more comprehensive and better trained network of reporting scientists rather than a real change in event frequency. Furthermore, attempts to link widespread bleaching episodes to global environmental change indicators such as ocean warming, increased UV radiation flux and ecosystem degradation 'have not been convincing' (Glynn, 1993, p. 1 and see also D'Elia *et al.*, 1991; Grigg, 1992). It is, therefore, difficult at this stage to make predictions about coral reef physiological responses to raised ocean temperatures in the future and doubly difficult to assess the translation of any changes in coral physiology to vertical reef accretion.

'Phase shifts' between hard and soft shallow water communities

The preceding discussion views 'events' such as storms and bleaching as temporarily resetting the reef from some equilibrium state in relation to sea level or as interrupting reef growth towards sea level. This is an over-simplification: these events are not discrete but on-going. Coral diseases are often found in association with bleached (Williams and Bunkley-Williams, 1990) and storm-damaged corals; as live tissue is lost to disease, the remaining coral patches suffer continued mortality (e.g. Jamaica: Knowlton *et al.*, 1990). Furthermore, there is evidence for increased bioerosion on dead and weakened coral frameworks, both from more numerous sea urchin grazers (Glynn, 1990) and boring sponges and mussels (Scott *et al.*, 1988).

Catastrophes may be biological as well as geological. Perhaps the best known biological catastrophes are the repeated outbreaks of the crown-of-thorns starfish, *Acanthaster planci*, in the Pacific, and particularly the Great Barrier Reef (see Case study 6.1). In 1983–4, a mass die-back of the sea urchin *Diadema antillarum* occurred on Caribbean reefs, probably as the result of the diffusion of a waterborne pathogen, and aided by high urchin densities as the result of over-fishing of the urchin's predators. *Diadema* was a major grazer of rock surfaces and as a result stressed or damaged reefs in the Caribbean have become dominated by algal communities.

What is of particular concern, however, in these kinds of situations is the possibility that reef growth will not become re-established but that a threshold

— or 'phase shift' (Done, 1992) — is crossed whereby a 'hard' reef is replaced by a different kind of 'soft' ecosystem, dominated by macroalgae (Fig. 6.12). Coral mortality and reduced reproductive ability and growth in survivors with bleaching and storm impacts may decrease the capacity of corals to compete successfully for space with other reef organisms, particularly coralline algae, macroalgae and algal turf species. Benthic algae rapidly overgrew dead and decaying reefs impacted by the 1982–3 ENSO event in Panama and the Galapagos Islands (Glynn, 1990) and Indonesia (Brown and Suharsono, 1990). Sometimes impacts can be complexly sequential: thus Jamaican reefs initially impacted by Hurricane Allen in 1981 were subsequently subjected to a 1982 mass mortality of the grazing urchin *Diadema antillarum* such that the limited post-hurricane coral settlement was smothered and replaced by a carpet of

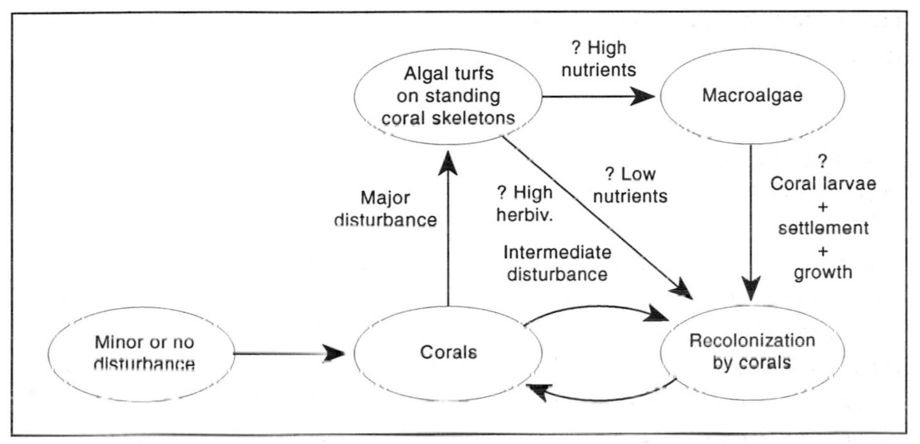

Fig. 6.12 Effects of different levels of disturbance, causative factors and possible 'phase shifts' between hard (coral) and soft (algal) communities. After Done (1992)

fleshy algae (Hughes, 1989). These questions are important in a future context because soft reef structures do not have the ability to track rising sea levels in the way available to accreting hard reef systems.

Coral reefs and environmental degradation

One danger, of course, is that human-induced environmental degradation may force reefs across the hard reef/soft reef threshold. There are large scale interactions between coral reefs and other tropical ecosystems. Reef structures dissipate wave energy and create reef-flat and lagoon environments which permit the opportunistic colonization of seagrasses and mangroves. By stabilizing reef-derived sediments and acting as sinks for terrestrially derived sediments and nutrients, mangrove forests and seagrass beds reduce sediment and nutrient concentrations in the water column and promote reef growth offshore. This delicate balance can be upset by misuse of coastal environments and poorly planned or unplanned coastal development.

Some of the most dramatic and devastating effects on reefs have been caused by war and weapons testing. Guilcher (1988) suggests that dynamiting for fishing began on many south Pacific islands after the Second World War using ammunition left behind by the American troops. Ships were also abandoned or wrecked around many reef coastlines, although the impacts have not been long-lasting owing to the natural regenerative power of reefs. Thus, Japanese vessels sunk in Truk Lagoon, Caroline Islands, are now covered by marine organisms (Guilcher, 1988). The Arabian Gulf War of 1991 was feared at the time to have caused massive oil spill damage to the reefs there, but subsequent surveys have shown that such worries were exaggerated. Nuclear testing has been carried out on Bikini, Enewetak, Christmas, Johnston, Mururoa and Fangataufa Atolls and caused both contamination and destruction of atoll environments. Johnston Atoll now contains a huge store of nerve gas canisters and a facility is being built to incinerate these and other unwanted chemical weapons (Van Dyke, 1991).

Oil pollution poses a real and often long term threat to many coral reefs, especially in those areas close to oil exploration installations in the Red Sea, Arabian Gulf and south-east Asia. A good example is provided by the spill in April 1986 from the oil refinery on Isla Payardi, Panama, when more than 50000 barrels of crude oil were discharged (Garrity and Levings, 1993). The oil came ashore on mangrove coasts at extreme low tide, smothering roots and causing much mortality to bivalves living there. The trapped oil has continued to be reactivated and recycled, and observations after 5 years showed that nearby reef areas were seriously threatened by oil released from the mangroves as well as eroded sediment from the oiled mangroves and seagrass beds. Coral reproduction was being affected, with reduced gonad size in oiled colonies. Mangroves, and therefore the nearby coastal environments as well, may take up to 20 years to recover from such oil spills (Burns *et al.*, 1993).

Blasting, dredging and dispersal of drill muds associated with oil-drilling all create potential sedimentation problems on reefs or alter the transparency of the water column with implications for rates of calcification. There are many documented cases of coral mortality and reef loss with sedimentation impacts, although such events are not necessarily catastrophic. Although corals obviously cannot withstand complete burial, many species effectively shed even high sediment loads by expanding their polyps or by the use of cilia hairs to brush their surfaces clean (albeit through the use of energy that might have been used in growth). Furthermore a curtain of sediment around coral bases may effectively shield corals from the boring and burrowing activities of bio-eroders. One growing problem, however, is the continued sedimentation stress imposed on reef ecosystems by changes in catchment management. Combinations of increased sediment loads in rivers as a result of commercial logging of rain forest catchments, allied to the removal of coastal forests and mangrove swamps for industrial and other development, have increased sedimentation in nearshore areas in the tropics. The state of Madagascan reefs, for example, has not yet been well described, but the acute deforestation problem and massive erosion in recent years on the island is likely to be a major source of problems (Wells, 1988). Evidence from Australia shows that logging along the Queensland coast has increased sedimentation along the Great Barrier Reef during major storms (Hopley, 1988). Kühlmann (1988) illustrates how a range of land-based stresses including intensive constructional activity and agriculture has increased sedimentation and chemical pollution on the reefs of Ishigaki Island, Japan. Mining may well produce similar problems. A study of the Malindi–Watamu reefs on the Kenya coast showed that a combination of three indices of reef damage (injury to living corals, soft coral cover and bare areas) correlated well with the distribution of terrigenous riverborne sediment across the reefs (Van Katwijk *et al.*, 1993). At the resort of Green Island on the Great Barrier Reef sewage has led to an increase in the area of seagrass beds, largely at the expense of hard corals. Over the period 1945–78 seagrass bed area increased from *c.* 900 m^2 to >130 000 m^2 (Hopley, 1988). This has had further repercussions for sediment movements in the area, as the seagrasses trap and bind sediment which normally circulates around the cay.

Increasing development of land immediately adjacent to coral reefs, often promoted by tourism, is also a growing threat to the health of many reef ecosystems. As coral reefs become ever more popular sites for tourist activities, so worries about the damage such tourism can cause grow. The survey of the Malindi–Watamu reefs in Kenya by Van Katwijk *et al.* (1993) showed that here at least tourist impact was negligible, except in the Coral Gardens of Watamu where 35 per cent of coral colonies had been damaged by trampling. Studies at the popular resort of Sharm-el-Sheikh (Egypt) showed that the percentage of bare rock, rubble and damaged corals increased after trampling by SCUBA divers, although the species composition of live corals has not been affected (Hawkins and Roberts, 1993). Some tourist-related effects can be spectacular and abrupt. In the Cayman Islands it is reported that a cruise ship anchoring for one day caused the destruction of 3150 m^2 of previously intact coral reef. Recovery may take more than 50 years (Smith, 1988). Hopley (1988) reports on the environmental considerations faced by those planning a new offshore floating hotel at John Brewer Reef, Great Barrier Reef; effluent from an on-board desalinization plant may affect nearby reef communities. However, as the green tourism movement gains momentum, awareness of such harmful effects is growing and organizations such as Coral Cay Conservation in Belize are helping to provide base level data and education about the dangers of over-use.

Still, however, much stress on the reef comes more directly as a result of utilization (sometimes over-utilization) of reef resources (Plate 6.6). The coral itself is often a valued building and aggregate material, especially for islands where such resources are scarce on land. Lindén (1990) records that in Sri Lanka, *c.* 40 per cent of raw material for the cement industry comes from coral and coral sand. On Madagascar, massive corals, such as *Porites somaliensis*, were used as building materials, and now they are used in cesspools and septic tanks. Reef fish and other key organisms, such as prawns, lobsters and shellfish, are harvested for local consumption and export. Some fishing techniques, such as reef dynamiting and the bashing of coral with heavy weights and sticks, are very damaging to the total reef structure. According to Lindén (1990) the explosion from one stick of dynamite leads to the death of all organisms and the pulverization of corals over an area of 10–100 m^2.

The whole range of human stresses on reefs documented above, and more fully in Guilcher (1988), act with natural stresses, and also with 'quasi-natural' changes linked to global environmental change, to produce a dynamic reef environment. As we have seen, the exact nature of reef response to this complex suite of stresses is unknown, but management and conservation schemes have been designed to ensure that human stresses are kept to a minimum where possible. There are now many conserved and managed reefs worldwide and more planned, although enforcement of use regulations is highly variable. The success of reef management and conservation schemes depends upon their design and how well they prove to be able to accommodate or prevent a range of reef uses. Such schemes must be developed in close co-operation with local people, and can bring in much needed investment. Thus, Meganck (1991) discusses the proposed Pitons National Park on the west coast of St Lucia, which would provide economic help to this depressed area, with the creation of 400 permanent jobs and an investment of around US$ 1 million. Long term data are required before we can assess the success of such schemes, and the schemes need to be adaptable in the face of changing environmental conditions. Barbados, with a population of almost 300 000 is one of the most densely populated islands in the world, with burgeoning tourist developments. Allied nutrient enrichment of coastal waters poses a serious future threat here (Wells, 1988). Yet the Barbados Marine Reserve (Fig. 6.13) is a good example of a small, relatively successful, reserve. On the west coast of the island, it includes two fringing reefs and an offshore bank reef. The reserve extends 1 km offshore, and down to −50 m a.s.l., and covers a total of 250 ha. It was officially established

Plate 6.6 Pearls from the Manihiki Atoll lagoon, Northern Cook Islands, south Pacific Ocean. Successful pearl farming requires a knowledge of atoll hydrology and circulation systems

in 1980 and is managed by the National Conservation Commission of the Ministry of Tourism and the Environment. The reserve is divided into four zones: the Scientific Zone in the north where tourism is limited and scientific projects are permitted, the North and South Watersports zones where watersports are

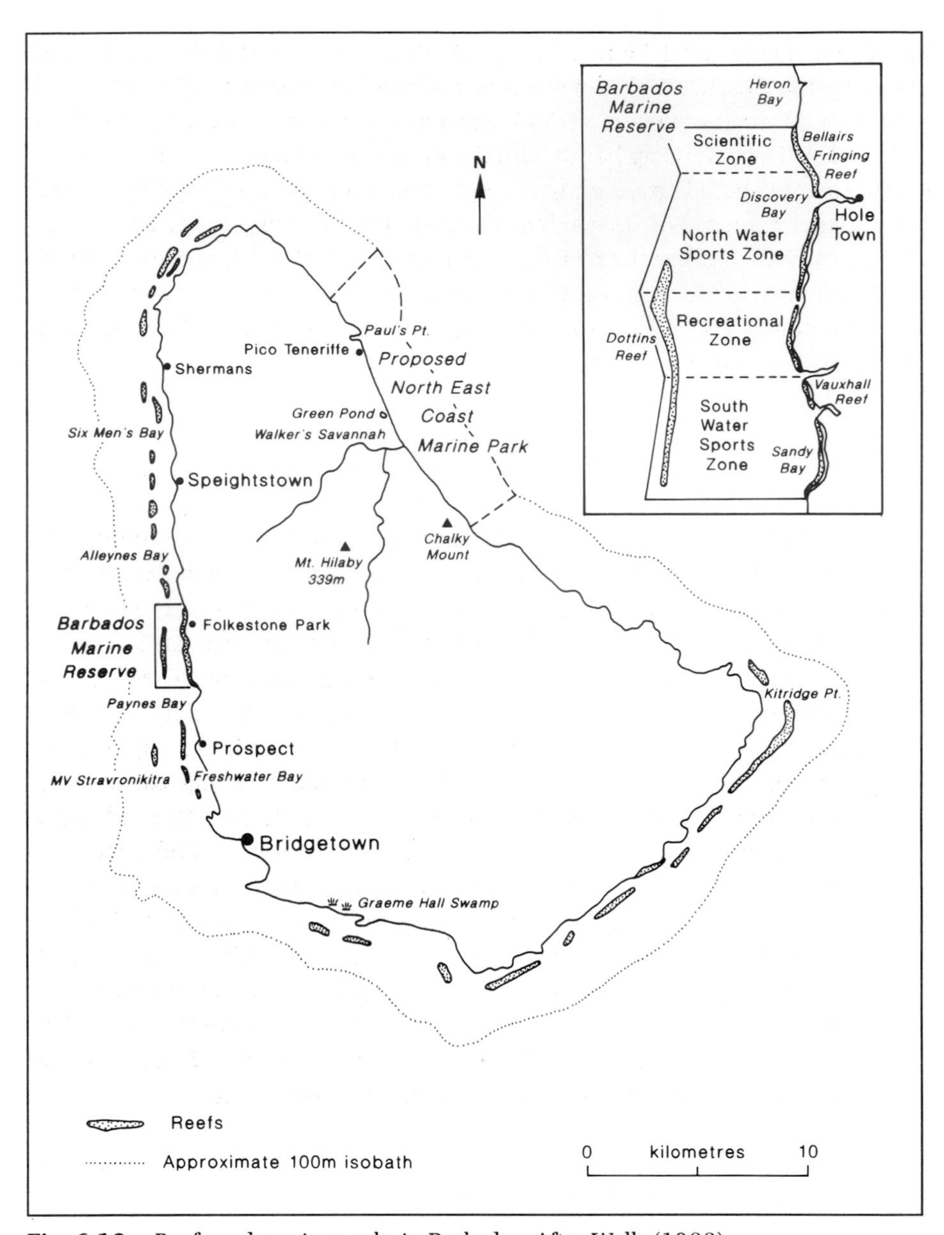

Fig. 6.13 Reefs and marine parks in Barbados. After Wells (1988)

encouraged, and the Recreational Zone in the middle where snorkelling, diving and glass-bottomed boat activities are allowed.

One highly successful management system is that for Australia's Great Barrier Reef, which was recognized as a UNESCO World Heritage Site in 1981.

The Great Barrier Reef Marine Park (GBRMP) was created following an act passed by the Australian Government in 1975; it incorporates 2904 individual reefs with a total reef area of 20 335 km^2 along a 200 km long coastline (Parnell, 1988). The Authority has divided the Park into five sections, within which management is based on a zoning process which aims to partition different reef uses. Such zonation aims to combine considerations of the present state of the reef with what is best for the reef, the reef users and the local economy. In practice this means that in some areas most reef uses are allowed, whereas in others only traditional fishing and scientific research are permitted. Oil drilling and mining are banned throughout.

Reef robustness and fragility: a few concluding remarks

Recently considerable debate has surrounded the fundamental nature of reef systems. Some authors have seen reefs as specialized and fragile ecosystems in a delicate balance with a largely stable tropical environment with the possibility of no recovery after human-induced degradation of the reef environment. Alternatively, reefs have been seen as subject to constant change and dominated by processes of rejuvenation, recovery and species replacement. As Grigg (1994) points out, to some extent these differences are complementary as they refer to different time-scales, the 'fragile' line of argument being particularly appropriate to small scale 'biological' processes over space and time, whereas the 'robust' viewpoint is more applicable to the long time-scales and large spatial scales of 'geological' processes. Nevertheless, this is a real problem which is difficult to resolve: some of the key elements of reef systems — e.g. the life history of massive framework-builders — mean that it is difficult to evaluate reef health on the basis of even a decade's careful monitoring and unfortunately the geological record is too 'smoothed' to extract event-level information easily. These questions need resolution if improved management of increasingly threatened reef ecosystems is to be achieved into the next century.

Case study 6.1: Kanehoe Bay, Hawaii

Kanehoe Bay lies on the windward north-east coast of the island of Oahu in the Hawaiian islands, north central Pacific. The bay is 13 km in length and 4 km in width. It is surrounded by fringing reefs up to 1 km wide and has a 4 km long barrier reef 1–2 km offshore in the centre of the bay (Fig. 6.14a). The bay can be divided into three sections: a deep, well-flushed northern sector, a central section and a more enclosed southern area. The volcanic terrain around the

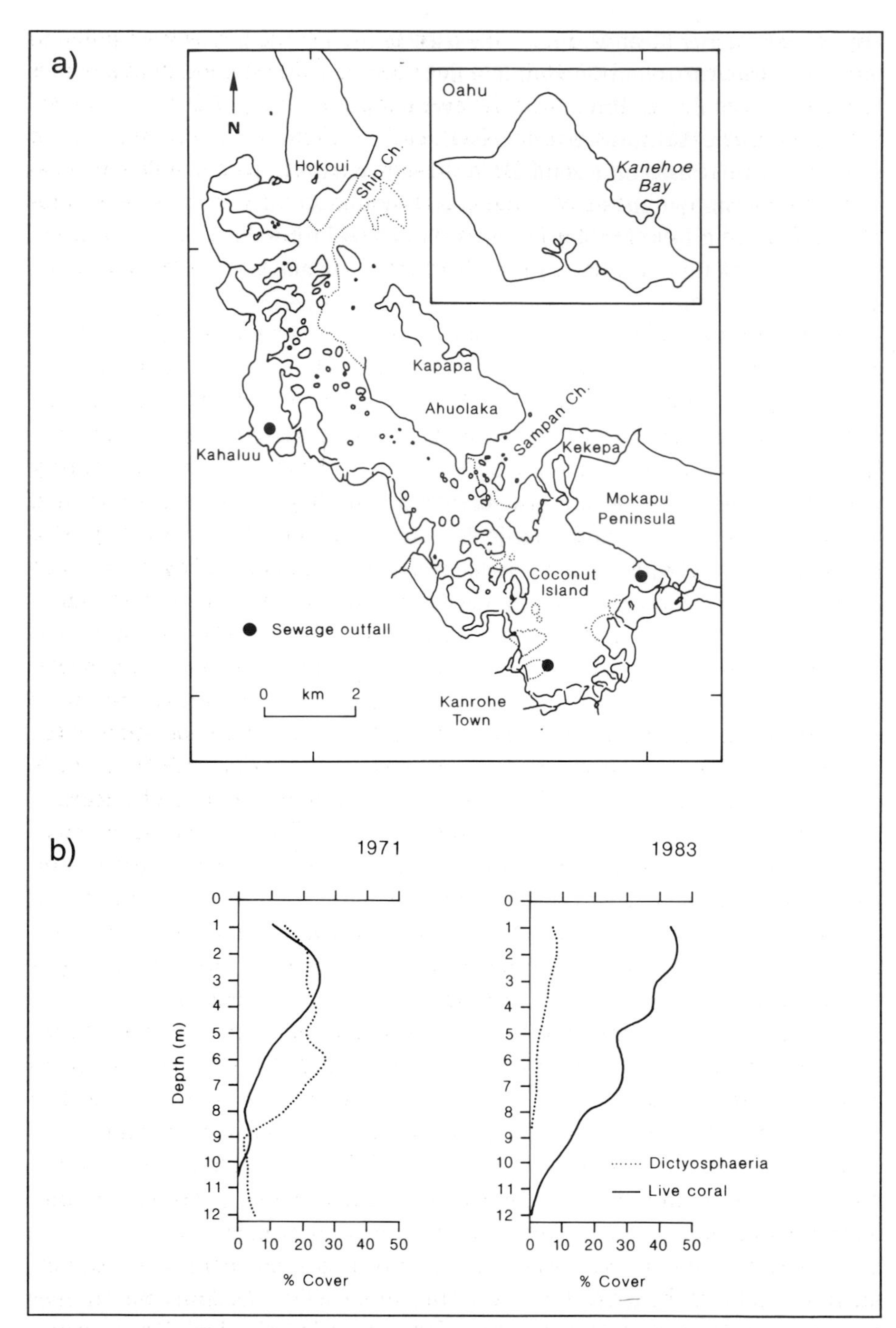

Fig. 6.14 a) Kanehoe Bay, Hawaii with sewage outfalls. After Smith *et al.* (1981). b) Live coverage of coral and algae *Dictyosphaeria* in 1971 and 1983 by depth. After Maragos *et al.* (1985)

bay is steep and with limited lagoon–ocean exchange the bay is susceptible to large freshwater inputs. In 1965, a major flood introduced enough freshwater to form a layer 27 cm thick over the entire lagoon, while in 1987 a greater-than-100 year event rainstorm on New Year's Eve reduced bay salinities from *c.* 30 to *c.* 15 parts per thousand. Both these events were accompanied by massive coral mortality on shallow fringing reefs and lagoon patch reefs, with complete loss of some species (e.g *Pocillopora damicornis* and *Montipora verrucosa*). However, most components of the reef system recovered in 1–2 years after the 1987 event.

Prior to Western contact, Polynesian populations altered bay environments by the construction of elaborate systems of fish ponds, but these activities were in decline by the nineteenth century when ranching and agriculture in the surrounding catchments became the dominant forms of resource use in the area. However, early scientific accounts from the first two decades of this century remark on the profusion of coral growth in the bay. Major environmental impacts began in the 1930s with dredging and spoil dumping in the bay with the construction of military facilities at the southern end of the bay; it has been estimated that 110 ha of reefs were filled and over 11 million m^3 of reef material dredged between 1938 and 1945. Environmental impacts continued into the 1940s with the development of the bay's catchments as residential areas. Urbanization continued in the 1960s with the improvement of road links from Kanehoe to the city of Honolulu so that by 1980 the resident population of the bay area had grown to over 60000 from a base of less than 5000 in 1920. Urbanization led to land disturbance and an increase in the area of impermeable surfaces in the bay's catchments. Increased runoff and increased sediment loads during storms led to terrestrially derived sediment blanketing the lagoon floor. Sewage discharges into the lagoon grew with population growth so that by 1977, 20000 m^3 d^{-1} of effluent was being discharged into the bay, introducing a nutrient 'subsidy' — in the form of high levels of inorganic nitrogen and inorganic phosphoros — into the southern, restricted lagoon.

Scientific concern over deteriorating coral communities and then public concern over the health of the bay led the County and Federal Government to remove sewage discharges from the southern lagoon in the late 1970s and to divert them to deep ocean outfalls outside the bay. Re-surveys in 1983 of coral cover, growth and mortality initially monitored in 1970–1 allow some estimation to be made of the effects of sedimentation and sewage impacts on reef communities and their recovery after environmental improvement.

The environmental deterioration of the bay was accompanied by replacement of coral reefs by oysters, sponges and barnacle filter-feeding communities and in particular by explosive growths of the 'green bubble alga' *Dictyosphaeria cavernosa* which smothered living corals and covered surfaces that might have been colonized by coral planulae. The alga was more common than any single coral species and its abundance equivalent to all coral species combined in 1970–1. By 1983, with the diversion of sewage inputs, nutrient levels diminished, phyto- and zooplankton levels fell and water transparency increased. *Dictyosphaeria* showed a dramatic decline in coverage with levels of abundance being only 25 per cent of previous levels. At the same time, increases in the abundance and distribution of coral species were reported, with species like *Porites compressa* and *Montipora verrucosa* doubling in their abundance across the monitoring stations after a slow start to recovery (Fig. 6.14b).

This example shows the underlying resilience of reef communities and their potential for recovery with an improvement in environmental conditions. It also demonstrates, however, the interaction of natural impacts (in this case freshwater runoff events) with human-induced stresses. Comparison of reef recovery after the two freshwater floods of 1965 and 1987 suggests that coral reefs can recover rapidly from natural disturbances but only under non-polluted conditions.

Selected references

JOKIEL, P.L., HUNTER, C.L., TAGUCHI, S. and WATARAI, L. 1993: Ecological impact of a fresh-water "reef kill" in Kanehoe Bay, Oahu, Hawaii. *Coral Reefs* 12, 177–84.

MARAGOS, J.E., EVANS, C.W. and HOLTHUS, P.F. 1985: Reef corals in Kanehoe Bay six years before and after termination of sewage discharges. *Proceedings of the 5th International Coral Reef Congress* 4, 189–94.

SMITH, S.V., KIMMERER, W.J., LAWS, E.A., BROCK, R.E. and WALSH, T.W. 1981: Kanehoe Bay sewage diversion experiment: perspectives on ecosystem responses to natural perturbation. *Pacific Science* 35, 1–402.

Case study 6.2: The crown-of-thorns starfish: a highly complex biological phenomenon

The large, multi-rayed crown-of-thorns starfish, *Acanthaster planci* (Plate 6.7), preys on coral polyps and is capable of the massive destruction of reef-building corals. *Acanthaster* normally occurs at low population densities (perhaps two to three individuals per square kilometre) on coral reefs but in the 1960s population explosions of tens to hundreds of thousands of the starfish on individual reefs began to be reported from many areas of the Indo-Pacific region, leading one commentator to remark that 'there is a possibility that we are witnessing the initial phases of extinction of madreporian corals in the Pacific' (Chesher,

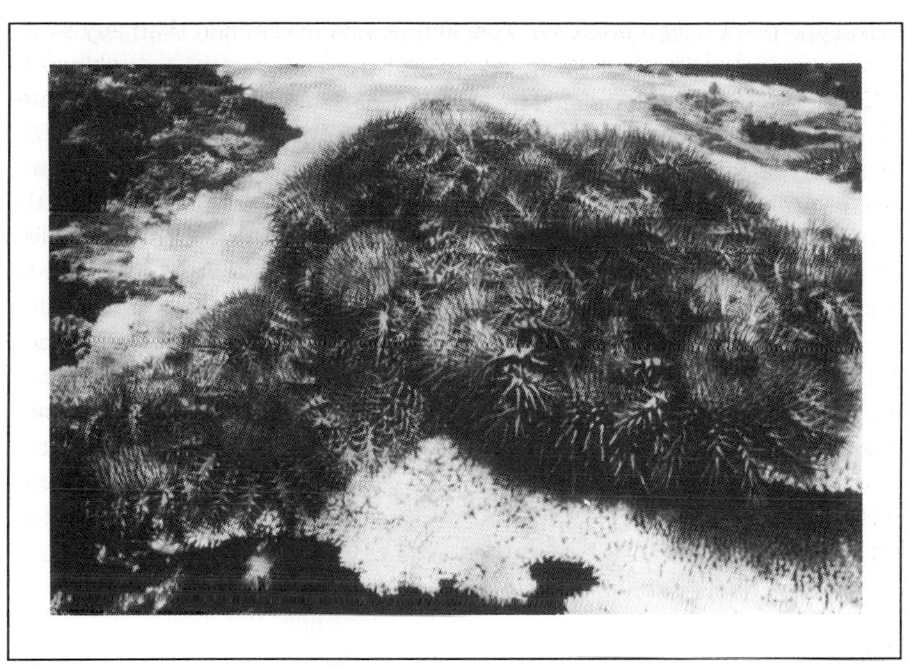

Plate 6.7 *Acanthaster* predation on corals (photograph: C. Wilkinson)

1969). Infestations on individual reefs typically last 3–4 years before the starfish exhaust their food supply, with often dramatic impacts. In some locations, coral mortality may reach 95 per cent, with a typical coral cover of 78 per cent being reduced to 2 per cent in 6 months around entire reef perimeters and being replaced by algal communities. Coral cover may be restored within 10–20 years but it may take much longer for the species diversity and the framework structure of the reef to resemble its pre-infestation character.

The dynamics of *Acanthaster planci* populations are best known for the Great Barrier Reef where the first well documented outbreak occurred at Green Island in 1962. Sixty per cent of the reefs in the central section of the reef (Lizard Island– Townsville) have been affected to varying degrees over the last 10 years: a peak of activity between 1966 and 1975 was followed by a further episode between 1981 and 1989. The outbreaks occur in a series of southward-moving waves travelling at 60–90 km a^{-1} and following the direction, and at the speed, of regional currents; thus it is thought that the infestations result from larval dispersal and then the development of adult starfish.

There have been a wide range of explanations for the population explosions witnessed in recent years. These explanations can be split into two groups: those that infer some aspect of human interference and environmental degradation (and thus an increasing frequency for a previously rare event) and those that argue for natural cycles of starfish abundance and decline. Anthropogenic theories include, for example, the enhanced survival of starfish larvae as a result of feeding on phytoplankton blooms associated with nutrient-enriched terrestrial runoff, and the removal of the natural predators of the starfish — for example, the collection of the giant triton, *Charonia tritonis*, by shell collectors. Alternatively, it has been argued that the population explosions simply represent better reporting of natural variations in *Acanthaster*'s breeding cycle. These latter arguments have been strengthened in recent years by the discovery of *Acanthaster* spines in Great Barrier Reef sediments, suggesting, controversially, that current outbreaks of high starfish densities are comparable to many episodes which have occurred over the last 7000 years. However, these data have not gone unchallenged, not least on the grounds that it is difficult to make direct comparisons between the inevitably smoothed geological record and the contemporary biological record to confirm that the same kinds of events are being compared.

It is still difficult to answer the question as to whether *Acanthaster* outbreaks are 'normal', human-induced or human-increased, although earlier 'Domesday' scenarios for the future of the world's reefs under the *Acanthaster* threat appear to have been misplaced. Despite the enormous research effort that has gone into the '*Acanthaster* problem' the life history and population dynamics of the starfish are still poorly understood. Explanations of *Acanthaster* outbreaks probably relate to a combination of several causes. This case study shows that problems like this phenomenon are complex and difficult to resolve, particularly when the phenomenon under study shows great 'patchiness' in time and space. Much work still needs to be done to understand it.

Finally, there is the question of large scale control. All present methods rely on treatment of individual starfish; the most effective method is injection of copper sulphate. Although it has been estimated that 25 million starfish have been killed in the last 15 years, even the most efficient programmes show that divers can kill less than 150 starfish an hour and at a high cost (A$35 per starfish). Thus impacts on aggregations in excess of 100 000 individuals are minimal.

Selected references

(It is instructive to follow this debate chronologically.)

CHESHER, R.H. 1969: Destruction of Pacific corals by the sea star *Acanthaster planci*. *Science* 165, 280–3.

ENDEAN, R. 1973: Population explosions of *Acanthaster planci* and associated destruction of hermatypic corals in the Indo-Pacific region. In Jones, O.A. and Endean, R. (eds), *Biology and ecology of coral reefs, vol. II: Biology 1*. New York: Academic Press, 389–438.

MORAN, P. 1986: The *Acanthaster* phenomenon. *Annual Reviews in Oceanography and Marine Biology* 24, 398–480.

WALBRAN, P.D., HENDERSON, R.A., FAITHFUL, J.W., POLACH, H.A., SPARKS, R.J., WALLACE, G. and LOWE, D.C. 1989: Crown-of-thorns starfish outbreaks on the Great Barrier Reef: a geological perspective based on the sediment record. *Coral Reefs* 8, 67–78.

Coral Reefs vol. 9 (1990): Special Issue on *Acanthaster planci.*

Coral Reefs vol. 11 (1992): Special Issue on *Acanthaster planci.*

Chapter Seven

COLD COASTS: PERMAFROST, GLACIERS, SEA ICE AND FJORDS

Introduction

Cold coasts may be defined as those 'where there is or has been abundant sea ice, lake ice, water-terminating glaciers or deeply frozen ground' (Taylor and McCann, 1983, after Nichols, 1961). These various types of ice exert a major control on the morphodynamics of cold coasts, which stretch from Antarctica to the fjord coast of New Zealand and Patagonia in the southern hemisphere, and from the fjord coast of Scandinavia, the Maine coast on the eastern side of North America and Vancouver Island on the west, up to the northern shores of Russia, Greenland, Canada and Alaska in the northern hemisphere (Fig. 7.1). Ice is not, of course, ever-present on these shores, as there are temporal variations in sea ice cover, permafrost depths and glacial extent, and furthermore other 'normal' shore processes operate here, but it is ice which makes these coasts unique.

There are key differences between the Arctic and Antarctic areas, in terms of the arrangement of land and sea, which have important ramifications for their coastal environments. The Arctic Ocean is fringed by wide continental shelves, backed by generally low-lying terrain. Conversely, the ice-covered Antarctic continent is fringed by relatively thin continental shelves and surrounded by a large ocean. The Antarctic terrestrial ecosystem has been described as very poor (in terms of species diversity, nutrient levels and primary productivity) whereas the marine system seems very rich. Conversely, in the Arctic the terrestrial ecosystem is rich and the marine system poor (Sugden, 1982). Many workers have started to question such simple statements, however, finding that marine primary productivity is more variable and perhaps overall less high in the Antarctic waters than previously thought (Knox, 1983).

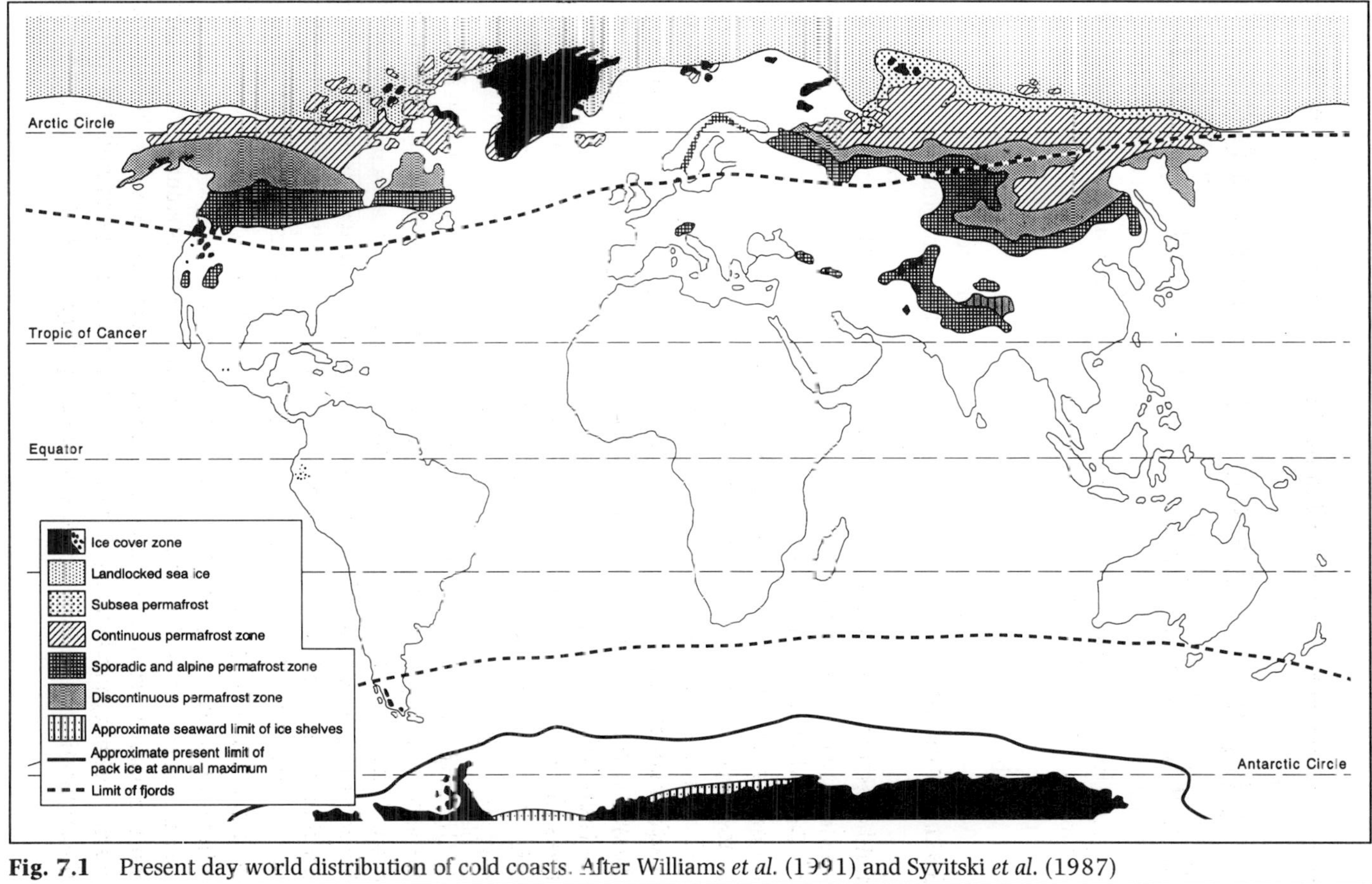

Fig. 7.1 Present day world distribution of cold coasts. After Williams *et al.* (1991) and Syvitski *et al.* (1987)

Ice at the coast

Ice is important to both the geomorphological development of coasts and terrestrial and marine ecology in three main ways: as a moving entity (glaciers, sea ice), as a static phenomenon (permafrost), and as a result of its phase changes between water and ice. Table 7.1 reviews some geomorphological processes associated with different types of coastal ice. Sea ice provides a clear interlinkage between atmospheric and oceanic circulations and energy flows. It forms on the sea surface in thin layers, developing in thickness and relief under suitable temperature conditions (saline water freezes at around −1.9°C). Generally, sea ice forms from a suspension of small ice crystals (called 'frazil ice'), which grow and coalesce to form a thin skin ('nilas'), or a field of 'pancake ice' (where each pancake is around 3 m in diameter and up to 50 cm thick). Sea ice is more widely distributed than one might imagine, as shown in Fig. 7.1. In the Alaskan Beaufort Sea, for example, sea ice covers 100 per cent of the surface for 9–10 months of the year, and freezes up to 2.4 m thick in one season (Norton and Weller, 1984). According to Taylor and McCann (1983), 90 per cent of the Canadian coast is affected by sea ice, mostly only seasonal in nature.

Where sea ice does not melt appreciably in summer, multi-year ice forms, as is the case for over half of Arctic sea ice (Wadhams, 1991). In the Antarctic, however, much of the sea ice melts annually as it drifts generally northwards towards warmer seas, and is usually as a consequence thinner and easier for

Table 7.1 Some different types of coastal ice, with examples of their geomorphological roles

Ice type	*Coastal geomorphological roles*
Glacial ice (including ice shelves)	Tidewater calving and ice-contact deposits Iceberg drifting and grounding pits, wallows and ice-rafted deposits
Permafrost and ground ice	Erosion by mechanical and thaw failure
Snow and ice on beaches	Icefoot development and allied sediment deposition
Surface ice cover on seas, pressure ridges, and ice floes	Ice acts as a protective barrier against surface wave motion Enhances hydrodynamic scour Ice scour Ice ride-up and pile-up producing shore ridges Ice rafting producing boulder-strewn surfaces
Frazil ice, slush ice and anchor ice	Sediment entrainment by ice aids shoreface profile adjustment

Source: modified from Forbes and Taylor (1994).

ships to move through. Multi-layer ice is only found in the Weddell, Ross and Bellingshausen Seas in the Antarctic (Wadhams, 1991). Detailed studies in many sea ice areas have revealed more about regional variations and movements. Thus, in the Weddell Sea, Antarctica, sea ice is often 2–5 m thick in the western part and persists for 2 or more years, whereas it is only *c.* 0.6 m thick in the eastern part. This reflects local and regional differences in circulation (Ackley, 1991). Studies in the Arctic, in the Bering and Chikchi Seas, show how much inter-annual variability there often is in sea ice cover, with sea ice stretching far south to the Alaskan coast in 1988, and being anomalously far north in 1987 (Muench *et al.*, 1991).

Ice can also enter the sea from ice shelf break up (where ice shelves are floating ice sheets attached either to land or to a grounded ice sheet, according to Souchez and Lorrain, 1991), or calving from ice sheets or tidewater glaciers (which enter the sea directly). Distributions of ice shelf break up, and tidewater glacier movements, are highly variable, but show distinct temporal trends in many areas. The Wordie ice sheet, on the west coast of the Antarctic peninsula, has been steadily retreating since the 1960s, and a large phase of ice front calving occurred during 1988 and 1989 (Doake and Vaughan, 1991). In the Arctic, about 90 per cent of glaciers in Spitsbergen are tidewater, and nearly 20 per cent of the coastline here is composed of ice cliffs. Calculations at the Hornsund fjord, where the Hansbreen valley glacier enters the Arctic, show that ice calving along the 1.5 km long ice cliff produces an influx of *c.* 22×10^6 m^3 a^{-1} of water (Glazovsky *et al.*, 1991). A more general study of twenty-two cold and temperate tidewater glaciers shows that, for reasons not yet fully understood, the iceberg calving rates are controlled by water depth at the terminus over periods of a year or more (Pelto and Warren, 1991). However, the authors of this study acknowledge that there are many areas with very limited data, such as Patagonia, Alaska and south Georgia.

Individual iceberg calving events can be spectacular. The 70 m high ice cliffs of the Ventisquerro San Rafael, Chile were monitored in February 1991 when an ice mass *c.* 210 m wide and 50 m deep calved. With some smaller blocks, an estimated 1 560 000 m^3 of ice calved into the Laguna San Rafael (which is connected to the sea by a narrow channel) (Harrison, 1992). Tidewater glacier calving has complex dynamics, with many feedbacks, producing an often tenuous relationship with climatic changes. However, for the San Rafael glacier mentioned above, there seems to be a clear relationship between winter precipitation amounts and calving rate (Warren, 1993).

There are considerable differences between the icebergs produced by such calving events in the Arctic and the Antarctic. In Antarctica, an estimated *c.* 5000 icebergs a year enter the ocean from the Ross, Fichner and Amery ice shelves. Characteristically, they are 250 m in depth. In comparison, *c.* 25 000 icebergs of around 50 m depth are produced each year in the Arctic, coming mainly from glaciers on the east and west coasts of Greenland and Ellesmere Island (Pisarevskaya and Popov, 1991).

Souchez and Lorrain (1991) point to a further, damming, influence of ice shelves on the coast, as with the Ward Hunt ice shelf on the northern coast of Ellesmere Island, 440 km^2 in area, which acts as a floating dam at the entrance of Disraeli fjord. Alley (1991) makes the further point that tidewater glaciers introduce not only water, but also sediment into the marine environment, with for example a 'moraine shoal' being deposited in front, which itself may influence the rate of ice calving. The nature of these and other sedimentary bodies deposited in the glacimarine environment (which is defined by Powell and Elverhøi, 1989, as including 'all sediment deposited below sea level after release from grounded or floating glacial or sea ice') has received much recent attention from marine geologists and Quaternary scientists.

When the protective sea ice is removed in summer, occasional large storm events, with associated high wave energies, can have a large effect on cold coastal landscapes. Along the Alaskan Beaufort coast a 5 m storm surge once destroyed nineteen buildings at Barrow, and caused localized shore retreat of up to 18 m (Norton and Weller, 1984). Most such coastlines are backed by ice-rich terrestrial environments, especially permafrost. Permafrost, defined as ground where temperature is continuously below 0°C summer and winter (Sugden, 1982) occupies much of the land surface of the Arctic. It outcrops on coastal bluffs and is also present under beaches and on the continental shelf (Fig. 7.1). During warm summer conditions the upper few metres or centimetres of ice in the permafrost melts creating an active layer of alternate freezing and melting, which has considerable geomorphic effects.

There has been considerable debate in the scientific literature over the years about whether sea ice is dominantly an agent of coastal erosion, or whether it mainly acts to protect the coast. Opinions vary enormously, largely depending upon the areas studied. As Hansom and Kirk (1989) and Forbes and Taylor (1994) note, the role of ice varies regionally and over a range of time-scales at any one site. The balance between ice processes and the action of waves and tides determines the suite of coastal features which exists at any one site.

Types of cold coast

A variety of factors influence the geomorphology of cold coasts at different scales. At the large scale geological structure, tectonic activity, glacial and post-glacial changes to relative sea level, and glacial and fluvial sediment production are important factors. Large scale geological structure determines whether coasts are high or low relief. Faulting provides the major lineaments of the coastline and determines its trend and variability. The size and supply characteristics of coastal sediments are important determinants of large scale depositional landforms such as deltas. Coarse-grained sediments are common in many cold region coastal areas where there is a ready supply of glacial debris.

The tectonic and glacial histories are key determinants of the form of cold coasts, as they influence present day relative sea level changes. Ice at the glacial maximum covered a much larger area in both the northern and southern hemispheres (Fig. 7.2), creating much lower sea levels. Isostatic rebound is still occurring on areas formerly covered by extensive ice sheets, such as Arctic Canada. There is considerable debate over the exact extent of ice cover in the Arctic during the glacial maximum. One hypothesis suggests that large ice domes covered the Barents and Kara Seas and perhaps parts of eastern Siberia, eventually coalescing with each other and the north-west European ice sheet. A more recent viewpoint is that there were instead a series of smaller ice caps over the Arctic islands, and there are, of course, a whole suite of intermediate viewpoints. Recent work on raised beaches on the island of Alexandra Land in Franz Joseph Land, where there are now two ice cap complexes, has revealed that this area was deglaciated by 6800 ^{14}C years BP. The highest beach is 23.5 m above present sea level, and the average rate of upwards shore displacement since 6800 BP has been calculated as 0.3 m a^{-1} (Glazovsky *et al.*, 1991).

Current tectonic activity is also important in determining coastal geomorphology in areas such as south Alaska, which was hit by an earthquake of *c.* 8.6 on the Richter scale in 1964. This event uplifted some coastal areas and submerged others along the Kenai peninsula, giving a maximum crustal downwarping of 2.3 m (Ward *et al.*, 1987). Areas submerged in that event have steep bedrock slopes, a very narrow intertidal zone and relatively steep coarse-grained beaches. Conversely, the uplifted areas possess a wider intertidal zone and both beaches and shore platforms are more common.

Large scale coastal features in cold regions include fjords and other estuaries, strandflats, deltas, headlands and embayments. Of these, fjords and strandflats are characteristic of glaciated coasts, although several workers have stressed the tectonic control of fjords and relegated the importance of glacial erosion (e.g. Gregory, 1913). Generally, fjords occur in two belts (*see* Fig. 7.1) which roughly coincide with the northern and southern zones of our cold coastal area. According to Syvitski *et al.* (1987) the usual scientific definition of a fjord is 'a deep, high-latitude estuary which has been (or is presently being) excavated or modified by land-based ice'. They are, therefore, predominantly found along high relief coasts, such as the coasts of Norway, southern Alaska, Svalbard, Ice-

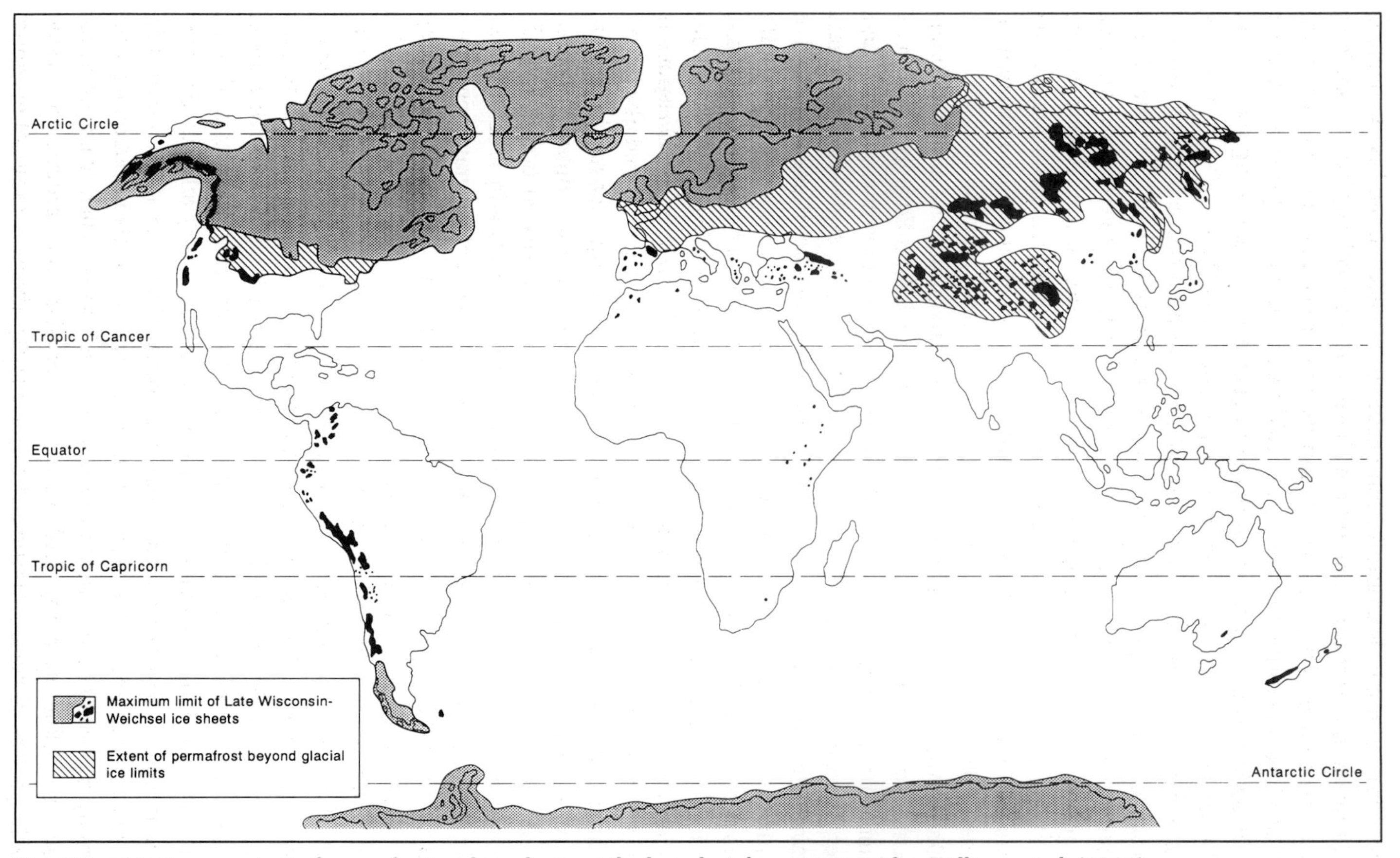

Fig. 7.2 Maximum extent of permafrost and ice sheets at the last glacial maximum. After Williams *et al.* (1991)

land, Greenland and Scotland in the northern hemisphere, and New Zealand and Chile in the southern hemisphere. Fjords characteristically have a shallow entrance (known as the sill), with a narrow, steep-sided and deep basin behind.

Strandflats are also characteristic of cold coasts, being at least partly caused by ice action. The term was introduced to describe the undulating rocky, partially submerged, lowland in west Norway (Sugden in Goudie, 1994) which is backed by a steep coast. Similar strandflats have been described from Greenland, the Antarctic peninsula, Iceland and other areas (e.g. Guilcher *et al.*, 1986). There has been considerable debate in the literature over the role that ice plays in strandflat formation, as eloquently summarized by Washburn (1980, pp. 258–9). A comparative study of strandflats from many high latitude areas by Guilcher *et al.* (1986) reveals that a different combination of formative processes is involved in each area.

On the medium scale cold region coasts are characterized by a suite of landforms, most of which are found in all latitudes. Depositional landforms such as beaches (gravel and sandy), spits, intertidal flats (sandy and muddy), salt marshes and dunes are common. Erosional features present in cold regions include cliffs and shore platforms. Sheer cliffs along the rocky southern coast of Alaska, for example, reach heights of over 300 m on the Kenai peninsula (Ward *et al.*, 1987). Cliffs cut in glacial debris often have very high retreat rates, as found by Hequette and Barnes (1990) on the Beaufort Sea coast of Canada. The only medium scale coastal feature that is clearly unique to cold regions is the ice cliff, i.e. a steep, coastal slope entirely composed of land-based glacier ice. As John and Sugden (1975, p. 78) put it 'Coasts cut along a glacier front have the rare property of being formed on and adjacent to a mass which is in continual movement'. Raised beaches, although found in many parts of the earth, are particularly well developed in some cold region coasts where relative sea level has fallen over a long period. Fairbridge (1983), for example, describes flights of raised beaches on the east coast of Hudson Bay, Canada, extending up to 315 m a.s.l., which have been produced by glacio-isostatic uplift.

It is on the small scale that coastal landforms in cold regions reflect the great importance of ice. Thus Sugden (1982, pp. 120–1) says 'The coastal geomorphology of polar regions is marked by the presence of ice — ice on the sea, ice in the ground and under the beach and ice on the beach'. In the intertidal zone, boulder pavements and barricades are commonly found. According to Dionne (1989a), such boulder barricades on the St Lawrence estuary and gulf coasts of Canada are a result of ice action, pushing sediment seawards and landwards. On beaches ice pushed ridges are common, as are various thermokarst features (mainly hollows and mounds) formed by the melting of beach ice. Similar ice scars are visible on many cold region salt marshes, and produce a 'jigsaw puzzle' pattern of ponds (Martini, 1986). On rocky shores a range of erosional features is produced by sea ice and iceberg grounding, including scratches, scour marks and polishing.

Perhaps the most remarkable small scale coastal landform found in cold regions is the ice foot, which is a multi-layered mass of ice which freezes sequentially onto part or all of a beach during extreme conditions. The ice foot protects the beach from wave action. Recent investigations at Kuujjuarapik Beach on the Canadian Hudson Bay coast show that in this subarctic environment, sand-laden ice from the ice foot is an important source of sediment for incipient foredunes produced by niveo-aeolian sedimentation (Ruz and Allard, 1994). At this site sea ice freezes up in November, and thaws in May, and thus foredune establishment and development proceeds more slowly here than on temperate beaches. Other features owe their origin indirectly to the presence of ice, such as shallow, south-facing pits indenting coastal foredunes on the southwest coasts of Hudson Bay, which are dug by aestivating polar bears for use as 'sunbathing' spots (Martini, 1986).

Geomorphic change and ecology of cold coasts

As we have seen, there is a great variety of cold coast types, many of which are really just like any sandy, rocky or muddy coasts, but with the added influence of ice. As such, the natural processes within these systems are in many respects similar to those covered in earlier chapters. Here it is only necessary to show how ice in its various guises contributes to the process regime in cold region coasts. This section considers the different coastal types found within the cold regions, i.e. permafrost coasts, totally glaciated coasts and those where glacial activity today is limited or non-existent, such as fjords, which are subjected to a number of stresses from human activity in various countries and, because of their unique characteristics, react differently to 'normal' estuaries.

Permafrost coasts

Permafrost coasts are those where the landward side (and often the beach and other coastal sediments themselves) is underlain by frozen ground. Mostly, these coasts are affected by sea ice as well. Ice adds an extra dimension to geomorphological and ecological processes within these areas. On shore platforms, for example, various processes involving ice have been seen to produce recognizable platform morphologies. Guilcher (1981) discusses the importance of 'cryoplanation' to platforms in the St Lawrence Estuary, Canada. Trenhaile (1983) stresses the importance of frost, coastal ice and isostatic changes to platform development in high latitudes. In the South Shetland Islands, Antarctica,

Hansom (1983) finds that ice freezing-on, quarrying by impact and abrasion are the dominant processes producing near-horizontal platforms in sheltered areas. On rocky coasts in general, scour from sea ice and small icebergs creates a whole range of recognizable marks, including grooves, friction cracks and polished surfaces.

On cold coast beaches five types of ice action can be recognized. First, where beach permafrost occurs the effectiveness of erosion by sea ice and waves is limited. Second, sea ice can cause ice-push features such as ridges. Ice rafting of large pieces of material can create boulder ridges in the intertidal zone. As a fourth mechanism, melting of ice within the beach can create a 'thermokarst' topography of pits and mounds. Finally, the ice foot can cover large areas of beach with ice, which releases sediment on ablation.

On cliffed sections of cold coasts, specifically those developed in porous media such as glacial deposits and many sedimentary rocks, ice encourages fast retreat rates. Freeze–thaw activity helps break up cliff material, creating large talus deposits on the beaches in some areas (as observed on Sakhalin, in the east of the former USSR by Guilcher, 1980). Thermal erosion of cliffs occurs where ice within the cliff mass itself melts, causing slumping. Hequette and Barnes (1990) have observed this mechanism on the Beaufort Sea coast of Canada. In this case, studies have shown that the cliff retreat rate of up to 10 m a^{-1} is bound tightly to nearshore gouging by sea ice — illustrating once more the close linkages between coastal and nearshore systems. Along the arctic coast of Alaska, observations of coastal bluff erosion show average rates varying from 0.3 m a^{-1} on the northern Chukchi Sea coast, to 4.5 m a^{-1} on Beaufort Sea coasts (Batten and Murray, 1993). Very high erosion rates were found to be associated with two storms in 1986 when there was long fetch over an ice-free ocean, showing a rather different relationship with sea ice (Walker, 1991). On Baffin Island, the mid-June to mid-July break up of sea ice leads to a 5–7 day dynamic phase of sediment transport and redistribution in the intertidal zone (McCann and Dale, 1986).

On cold coast marshes ice can again be an important influence. Dionne (1989b) shows how ice floes move marsh clumps and make an important contribution to the sediment budget of *Spartina* marshes on the St Lawrence Estuary, Canada. As Taylor and McCann (1983, p. 70) remark about cold coasts in general 'A primary effect of sea ice is the reworking of surficial coastal sediments'. In the Mackenzie Delta, on the Canadian Beaufort Sea coast, there is a general erosional trend related to rising sea level, allied with localized sedimentation. The sedimentation here is governed by the interplay of sea ice-covered and ice-free periods which affect both marine and fluvial sedimentation. Thus, in winter when both the Beaufort Sea and the delta area are ice covered, there is very little fluvial sediment discharge. This increases during spring and peaks in summer, when storms and storm surges are important controls on sediment movements (Jenner and Hill, 1991).

A survey of the entire Canadian Beaufort Sea coast from the Alaskan–Yukon border to Baillie Island showed the overall importance of ice to morphology and process. Six coastal types were identified, five of which (ice-rich cliffs, ice-poor cliffs, low tundra cliffs, inundated tundra and deltas) cover 80 per cent of the coast and are eroding; and the last one being accreting barrier islands and spits which cover the remaining 20 per cent of the coast (Harper, 1990). The mean rate of coastal retreat here is 1 m a^{-1}, although a maximum rate of 18 m a^{-1} has been recorded. Harper (1990) concludes that ice in coastal sediments (both in pores and as massive 'ice beds') appears to be a major cause of the regional coast retreat, allied with sea level rise.

The different forms of ice which so affect the geomorphological processes on cold coasts also affect the ecological processes. Many permafrost coasts are north or south of the treeline and the landward vegetation is therefore tundra of one form or another. According to Sugden (1982) there are three main types of tundra vegetation, low Arctic tundra which occurs in the southernmost parts of the Arctic tundra, high Arctic tundra and polar desert. Low Arctic tundra is characterized by dwarf shrubs, mosses, sedges, grasses and lichens. High Arctic tundra has generally $<$ 80 per cent plant cover of herbs and mosses. Polar desert areas are characterized by $<$ 10 per cent plant cover, and a very limited range of plant types. In Antarctica, for example, there are few flowering plants, and lichens dominate. Higher parts of the food chain are also limited in tundra areas and in the Arctic include musk ox, caribou, hares, lemmings, birds, insects, wolves and foxes. Despite the relative simplicity of the tundra biome, there are many important links between plants and geomorphology (Williams, 1988). Plants influence permafrost levels, for example, and lichens have been shown to be important weathering agents in the Antarctic.

In the low Arctic there are only a few species found on dunes and gravel beaches, all of which show clear adatations to the harsh environment. The bluff vegetation is usually identical to that of inland tundra areas here, as saltspray is limited and thus maritime influences are low. On salt marshes in the low Arctic *Carex ramenskii*, *C. subspathacea*, *Puccinellia phrygonodes* and *Cochlearia officinalis* are all common species, and the total plant cover is often less than 15–25 per cent, with individual plants only reaching *c.* 5 cm high (Bliss, 1993). In the high Arctic coastal vegetation on beaches, dunes and cliffs is even more limited in productivity, cover and diversity. On tidal mudflats in Svalbard, for example, *Puccinellia phrygonodes* dominates on low, frequently inundated surfaces, with *Carex ursina* and *Stellaria humifusa* on higher mudflats. Cyanobacteria may be very important in such marshes as carbon accumulators and nitrogen fixers. Along the Antarctic coast desert and semi-desert polar vegetation exists at only a few locations, specifically along the western side of the Antarctic peninsula and offshore islands. Here, lichens dominate on lower parts of sea cliffs, especially crustose *Verrucaria* species, being replaced by species of *Caloplaca* and *Xanthoria elegans* at higher levels (Fig. 7.3). Sea bird nesting is common here, and avian excreta stimulates a varied lichen community (Smith, 1993). In all these ecosystems the low species diversity, low biomass, low productivity, short foodwebs and high degrees of specialization make them especially vulnerable to human-induced and natural disturbances.

The marine part of the coastal ecosystem is, as in other parts of the world, made up of phytoplankton, zooplankton, birds, fish, molluscs and marine mammals (such as polar bears in the Arctic and a range of whales). In the Antarctic there is a unique community of algae (mainly diatoms) which live under the sea ice in what is called the epontic habitat. Algae also live within and on sea ice (Knox, 1983). The whole structure and function of marine and intertidal life in cold areas is affected by the presence of ice. Most creatures cannot live under ice, and ice moving over rocky and sediment substrates can kill organisms living in those habitats. Intertidal and subtidal life in many areas is quite modest in species numbers, growth rates etc.

A detailed study of the 35 km long Simpson Lagoon, between Prudhoe Bay and the Colville River delta on the Alaskan Beaufort Sea coast, provides some useful data on marine ecosystems (Craig *et al.*, 1984). There is a short ice-free period here lasting from early July to early October, whereas in winter the surface ice in the lagoon reaches 2 m in thickness. Food chains are very short, with birds and fishes eating epibenthic (i.e. living on bottom sediments) mysids and amphipods which in turn eat plankton. There are relatively few species involved, with birds such as oldsquaw duck (*Clangula hyernalis*) and two species of phalarope (*Phalaropus lobatus* and *P. fulicarius*). Fish include two species of cisco (*Coregonus autumnalis* and *C. sardinella*), arctic char and arctic cod. Observations of consumer diets here showed that epibenthic crustacea are the most popular food for both birds and fish, and that there is *c.* 50 times as much food available as is needed. In summer, there appears to be considerable overlap between the dietary preferences of different species. Net primary productivity in the lagoon is estimated at $1.2–1.7 \times 10^6$ kg C a^{-1} (or 5–7 g C m^{-2} a^{-1}) which, coupled with material advected in from the marine environment, drives the lagoon ecosystem. The lagoon appears to be an ecologically disturbed environment, kept in an early successional stage by physical stresses such as high flushing rate, seasonal freezing, variable salinities and pronounced turbidity. The successful epibenthic organisms are very versatile, tolerant colonizers which can cope with such a harsh environment.

Often these organisms are closely linked to geomorphological processes. Boring cyanobacteria and green algae have been found to be important bio-

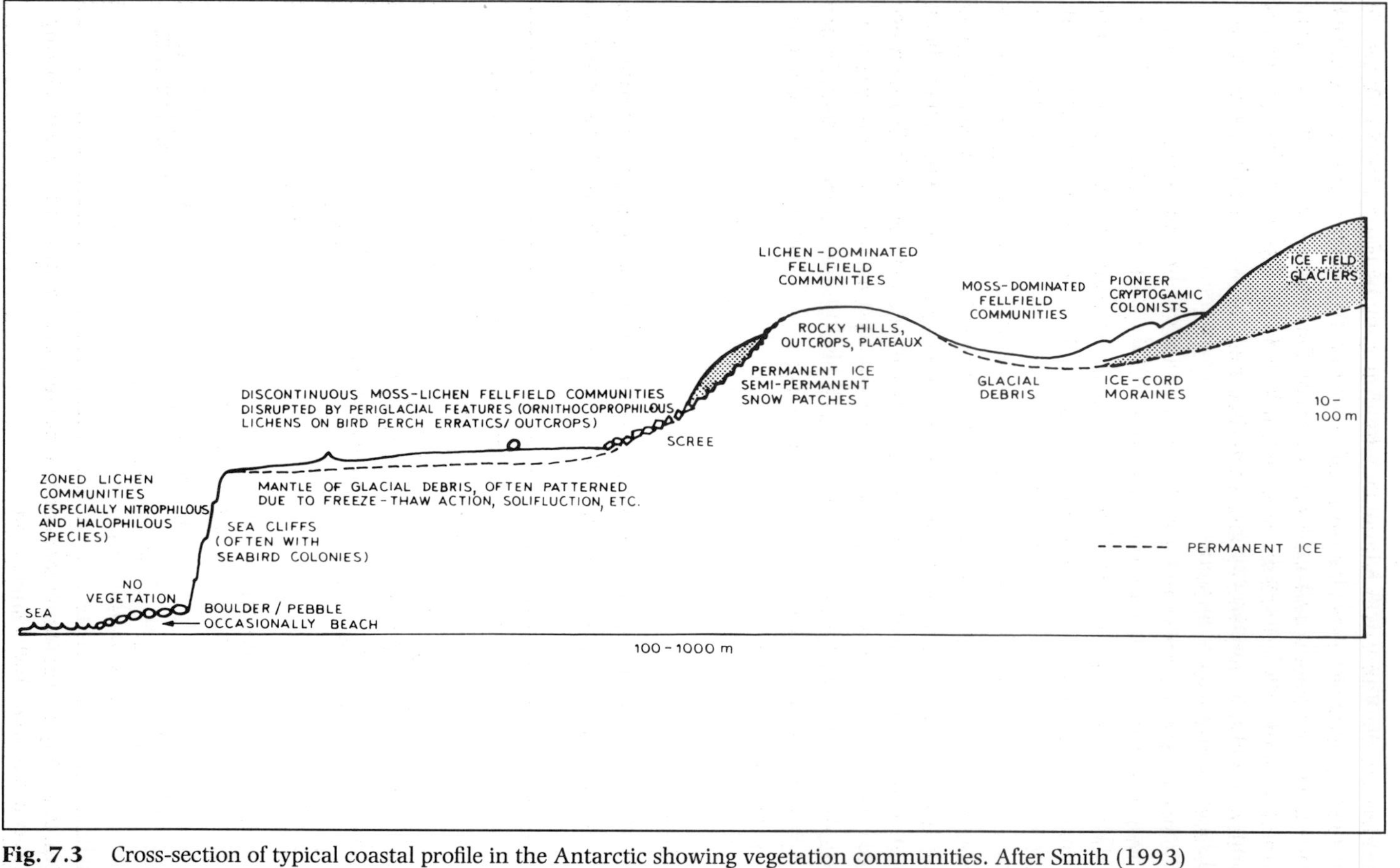

Fig. 7.3 Cross-section of typical coastal profile in the Antarctic showing vegetation communities. After Smith (1993)

eroders on the Vancouver Island shelf by Young and Nelson (1988). Aitken *et al.* (1988) describe the bioturbating effects of intertidal communities of bivalves, sedentary polychaetes and other organisms on Baffin Island which work in conjunction with drift ice action. Several workers have suggested that penguins, which can obtain very high densities in their nesting sites, can abrade rock surfaces (Splettstoesser, 1985).

Once again, ice can have a major impact on littoral and sublittoral organisms. Along the northern Bering Sea coast of Alaska, the zonation of kelp and other sublittoral species has been studied at St Matthew and St Lawrence Islands. Observations by divers showed how effective ice scour, associated with the break up of sea ice in May, is in influencing sublittoral zonations down to a depth of 12 m (Fig. 7.4). Flat topped rocks were denuded of species altogether, even crustose coralline algae were removed by the ice. More protected, sloping surfaces were less severely affected, and mature algae were able to survive here. A diversity gradient was found regionally, with the most northerly St Lawrence Island facing a more severe ice scour regime and possessing a less diverse community of algae and sessile invertebrates. Under an intermediate disturbance regime in the more southerly St Matthew Island a higher diversity of species was found (Heine, 1989).

Glacial coasts

All those coastlines whose landward portion is composed mainly or totally of ice are included in this category. Almost invariably, such coastlines are also

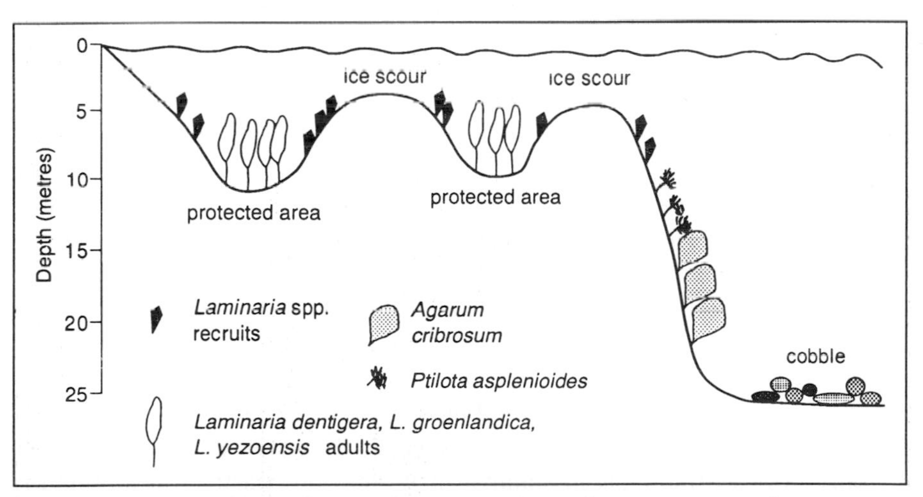

Fig. 7.4 Typical sublittoral zonation in areas subjected to ice scour, northern Bering Sea coast, Alaska. After Heine (1989)

characterized by seasonal sea ice on their seaward margin. Some fjords fulfil these conditions, including many in Antarctica and some in Greenland and Alaska. Many other fjords have some portion covered by terrestrial glaciers, including many in Greenland, Svalbard, Baffin Island and Ellesmere Island, and some in Alaska and Chile (Syvitski *et al.*, 1987).

In general terms there are two main types of glacial coast, depending on ice-flow dynamics and water depth (Pfirman and Solheim, 1989), i.e. grounded tidewater glacier or floating ice shelf coasts. In Antarctica, ice shelves occupy *c.* 47 per cent of permanent ice coasts (Alley *et al.*, 1989) with the largest examples being the Ross and Filchner–Ronne ice shelves. There are few examples elsewhere in the world today, although ice shelf coasts were undoubtedly more common in glacial periods. Ice shelf coasts are commonly bordered on the seaward side by a steep ice cliff. The major processes involved in ice shelf coasts are ice erosion at the seaward side, and the deposition of glacial sediment in the form of grounding-line till deltas and moraine shoals (Alley *et al.*, 1989).

Processes occurring on tidewater glacial coasts are influenced by the temperature and dynamics of the glacier, the substrate relief and the inland relief and lithology. Ice-front melting, glaciofluvial discharge and iceberg calving are all important mechanisms involved in sediment deposition at the ice front (Syvitski, 1989). Some types of tidewater glacial sedimentation are shown in Fig. 7.5. Once sediment is released in the proglacial region, currents are a dominant influence on its fate and any resultant landforms. In many cases surging glaciers are involved (e.g. around the northern Barents Sea coast according to Elverhøi *et al.*, 1989) which experience periodical advances and subsequent influences on the coastline. These glacial surges themselves can cause coastal problems, especially dramatic, large scale events which involve the displacement of a large ice mass. Pfirman and Solheim (1989) have studied hydrological and sedimentological processes along a tidewater glacial coast in Svalbard. Here the ice front is *c.* 200 km long with terminal ice cliffs extending 20–35 m above sea level. Two large and stable meltwater outflows were found in the ice front. Coarse sediments were being deposited as ridge deposits, whereas fine sediments formed a plume extending 15 km away from the ice front. Subglacial meltwater seems to be the major source of freshwater and sediment at the coast in this case. Griffith and Anderson (1989) show how climatic variations are reflected in changing coastal sediment types along glacial coasts in Antarctica.

Glacial coasts have more terrestrial biota than one might expect, although moving glaciers are hostile environments for life. However, cyanobacteria and diatoms form benthic mats in summer melt pools on coastal and marine ice shelves in the Antarctic (Young, 1991). On the seaward side, despite the frequent presence of sea ice, there is often a rich biota, especially in the Antarctic.

Fig. 7.5 Sediment inputs at the front of a tidewater glacier. 1 = supraglacial material, 2 = englacial material, 3 = basal material, 4 = iceberg-rafted, 5 = aeolian sediments, 6 = lateral (kame) deltas. After Syvitski *et al.* (1987)

The Southern Ocean, which surrounds Antarctica and nearby islands, contains eleven species of krill which act as the dominant herbivore, thought to account for up to 50 per cent of the zooplankton standing crop. Krill act as a 'keystone predator' in the Antarctic (Knox, 1983). The dominance of a few species in terms of biomass makes the Antarctic marine environment potentially vulnerable to future changes, such as over-fishing. Over 100 species of fish have been recorded in the Southern Ocean, and 38 species of bird breed or feed there. On pack ice, the most important birds by far are the penguins, especially Adélie, Emperor and Chinstrap species.

Several studies have attempted to measure ecosystem processes in cold seas. Dunbar *et al.* (1989) describe the very high rate of organic carbon accumulation in McMurdo Sound, coming partly from primary productivity within and below sea ice. They estimate annual primary productivity within sea ice here to be 40–80 g C m^{-2} a^{-1}, with open water areas producing 16–100 g C m^{-2} a^{-1}. In the Ross Sea, the ice algal community is thought to provide *c.* 20 per cent of the total productivity, being utilized by ice-dwelling invertebrates and fish as well as being melted out in spring (Knox, 1983). Other workers have suggested an often complex relationship between coastal ice and coastal organisms (usually subtidal as the intertidal zone is dominantly ice). Dayton *et al.* (1969), for example, discuss the effects of anchor ice (i.e. ice platelets frozen on to the bottom substrate at depths down to 33 m) in killing off and moving organisms as it forms and rises to the surface.

Fjords

Fjords are characterized by diversity. Generalizations are hard to make because of the wide range of conditions experienced in different fjords (Plate 7.1). It is clear, however, that fjords are often sinks for sediment and carbon, and can also be sinks for pollutants. They have many similarities with lower latitude estuaries, but are characterized by high seasonality in biological production, sediment and water inputs, especially fjords at very high latitudes (polar or arctic fjords). Seventy per cent of fjords are influenced by glacial and/or sea ice. Fjords often have low flushing rates coupled with high sedimentation rates: this is around 0.01 cm a^{-1} in European fjords; but in some glacial fjords may reach concentrations in the summer months of 10 m a^{-1}. Seven end members in a continuum of fjord types have been recognized by Syvitski *et al.* (1987) in their monumental synthesis of knowledge on many aspects of fjords: high sedimentation rate fjords low sedimentation rate fjords, well-mixed fjords with circulations dominated by wave and tidal action, polar fjords with permafrost on land and sea ice, glacial fjords where glacial ice tongues control circulations and sediment dynamics, fjords subjected to semi-continuous subaqueous slope

Plate 7.1 Milford Sound: a relatively pristine fjord on the south-west coast of South Island, New Zealand (photograph: H.A. Viles)

failures and fjords containing anoxic basin waters, with marked temporal and spatial biogeochemical gradients. The regional characteristics of fjords in different parts of the world are shown in Table 7.2.

Major processes operating within fjords are hydrological, geomorphological and biogeochemical in nature. The hydrology of fjords is highly dependent on the input characteristics, i.e. volume and seasonality of freshwater input. The 'classic' fjord circulation is known as the two-layer flow and is described thus by Syvitski *et al.* (1987, p. 32): 'an outward flowing surface layer and an inward moving compensating current, compensating for the loss of salt entrained into the surface zone' (*see* Fig. 7.6). In many deep fjords other circulation cells at depth may also be present. The circulation system is an important influence on the transport and deposition of sediment, the biota within the fjord and the nature of geochemical transformations, and thus will influence the response of fjords to pollution.

The hydrology of fjords, which is itself controlled by fjord morphology and regional climate, has a major influence on energy flow, biogeochemistry and ecology within the fjord. The sill height is a key control on hydrology, with a complex water flow layer above sill level, and a more stable basin environment below (*see* Fig. 7.6). Where sills are shallow, intermittent stagnation of the bot-

Table 7.2 Typical characteristics of the world's major fjord coastlines

Fjord district	*Fjord stage*[a]	*Tidal range*[b]	*River discharge*[c]	*Climate*	*Sedimentation rate*[d]	*Human influence*
Greenland	1,2	Low	Medium to high	Subarctic to arctic maritime	Medium to high	Low
Alaska	1,2,3,4	High	Low to high	Subarctic maritime	Medium to high	Low to moderate
British Columbia	3,4	High	Medium to high	Temperate maritime	Medium to high	Low to high
Canadian maritime (N.S., Nfld., Quebec, Labrador)	4,5	Low to medium	Low to high	Subarctic to temperate maritime	Low	Low to moderate
Canadian Arctic archipelago	1,2,3,4	Low to high	Low to medium	Arctic desert to maritime	Low to medium	Low
Norwegian mainland	3,4	Low	Low to medium	Subarctic to temperate maritime	Low	Moderate to high
Svalbard	2,3	Low	Low	Arctic island	Medium	Low
New Zealand	4,5	Medium	Low to medium	Temperate maritime	Low to medium	Low
Chile	2,3,4	Low	Low to high	Temperate to subarctic maritime	Medium to high	Low
Scotland	4,5	Low to high	Low	Temperate maritime	Low	Moderate

[a] Stage 1: glacier filled; 2: retreating tidewater glaciers; 3: hinterland glaciers; 4: completely deglaciated; 5: fjords infilled
[b] Low: < 2 m mean range; medium: 2–4 m; high: > 4 m mean range
[c] Low: < 50 $m^3 s^{-1}$ mean annual discharge; medium: 50–200 $m^3 s^{-1}$; high: > 200 $m^3 s^{-1}$
[d] Low: < 1 mm a^{-1} averaged over the entire fjord basin; medium: 1–10 mm a^{-1}; high: > 10 mm a^{-1}
Source: Syvitski *et al.* (1987)

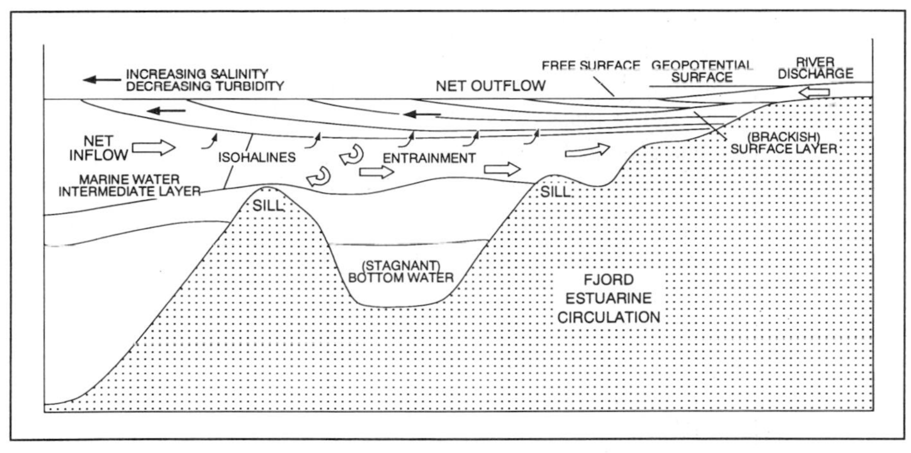

Fig. 7.6 Typical two-layered fjord circulation. After Syvitski *et al.* (1987)

tom waters is encouraged. Pearson (1988) provides a useful review of studies of carbon and nutrient budgets from a wide range of fjords, and makes a relatively simple classifiation of fjords according to mixing characteristics and latitude (Table 7.3). Thus, well-mixed fjords are net exporters of nutrients for much of the year, whereas stagnant fjords at all latitudes seem to act as carbon and nutrient sinks. However, Pearson concludes that there is a dearth of information from high latitude fjords on energy flows and budgets. Seasonality is very important, as many fjords have lower basin waters which are isolated in winter and flushed out in spring and early summer.

Fjords generally have steep subaerial slopes, often mantled with talus, sometimes vegetated and affected by waterfalls. Intertidal zones are often narrow (because of the steep slopes). Beneath the water surface there is a highly vari-

Table 7.3 Generalized characteristics of fjord ecosystems as a function of their overall physical conditions

Hydrodynamic energy input	*Low latitude*	*High latitude*
High	High biomass High diversity High productivity Nutrient limited most of the year	High biomass Medium diversity Low productivity Carbon limited most of the year
Low	Low biomass Low diversity Intermittently high productivity Carbon and nutrient sink	Low biomass Low diversity Low productivity Carbon and nutrient sink

Source: modified from Pearson (1988)

able cover of sediment. The fjord-head delta is a characteristic feature, whose nature is highly influenced by the presence or absence of vegetation (which depends on climate). Delta progradation is an important depositional process within fjords. Subaqueous mass movements are important in many fjords, producing dramatic changes in bottom topography, with consequent impact on benthic fauna.

The geochemistry of fjords is of great significance to studies of pollution, as when freshwater meets saltwater, aggregation and changes in water chemistry occur. The oxygen regime of fjords (itself influenced by circulation and the supply of organic detritus) is an important control on the nature of the chemical reactions taking place in the decomposition of organic detritus which involves microorganisms. In fjords where organic detritus is supplied at a fast rate and oxygen resupply rates are low, anoxic conditions in the bottom sediments (and sometimes in the bottom water also) result. Anaerobic respiration is less efficient than aerobic respiration, so much of the organic detritus in anoxic fjords remains in the sediments as a large sink of carbon. Under both anoxic and oxic conditions the degradation of organic matter is associated with the release of manganese and iron.

The ecology of fjords is characterized by small intertidal and shallow subtidal communities, often dominated by kelp and seagrass. Dale *et al.* (1989) describe the macrofaunal communities of Canadian arctic fjords, finding that the intertidal zone biota is dominated by bivalves, polychaetes and fucoid algae. Species composition and species diversity differs between subarctic and true arctic fjords, with the latter having fewer species. A survey of twenty transects over eight fjords in subarctic eastern Iceland showed that the species richness of rocky eulittoral communities (dominated by fucoids) increased from head to mouth (Hansen and Ingólfsson, 1993). This may be because of increased temperature fluctuations towards the head of fjords. As a general statement Syvitski *et al.* (1987) assert that open marine productivity in fjords is higher than that in the intertidal and subtidal zones, contrary to estuaries in general.

Many studies have been made of the benthic fauna living on and in sediments and rocky outcrops on the bottom of fjords. There appears to be a complex spatial patterning of organisms, rather than a homogeneous cover. Important findings are that bioturbation is common (although less important than in lower latitude estuaries) and helps aerate bottom sediments, and that bottom communities are stressed by high magnitude events such as mass movements, by oxygen depletion and by high sedimentation rates. These stresses may occur relatively frequently, and their effects may be confused with human-induced stresses such as pollution (Table 7.4). Characteristically, disturbed benthic environments in fjords are characterized by a low diversity, resilient community of '*r*' selected species (pioneer species) which do not pro-

Table 7.4 Changes in fjord soft-bottom fauna over time after disturbance

Severely polluted/disturbed	*Normal*
'*r*' strategists	'*k*' strategists
Little impact of organisms on substrate	Intimate contact with substrate
Shallow redox front, anaerobic sediments	Deeper redox front, aerobic sediments
Low diversity, resilient community	Relatively stable, patchy community

duce much bioturbation: as the environment recovers a higher diversity, patchy community dominated by '*k*' selected, bioturbating species develops. Hein and Syvitski (1989) suggest that large sea mammals may also play a role in bioturbation in Baffin Island fjords, by creating large gouges and pits on the sea floor. As a further sediment–organism interaction, Sancetta (1989) produces evidence that faecal pellets help sediment transport in fjords in British Columbia, as has been found elsewhere.

Human impacts on cold coasts

Historical background

Cold coasts, especially those above 60° N and S, are commonly sparsely populated. The environments are usually hostile, thereby making settlement difficult, and these areas are also a long way from major concentrations of population, making them more isolated. Much of the spectacular, high relief coastline of cold areas (such as fjord coasts) is extremely difficult to settle, even below the tree line in relatively warm climates. Population densities are usually very low, although many areas have a long history of settlement. Thus, the Alaskan Beaufort Sea coast has been inhabited by the Tareumiut people for at least 2000 years (Norton and Weller, 1984). The vast expanse of Arctic Canada had a population of *c.* 43000 in 1981 (Sugden, 1982) with only 33 per cent of these living north of the tree line. Many areas are, to all intents and purposes, wilderness and untouched by human artefacts such as roads, towns and mines. Perhaps the greatest total population occupying cold coastal areas is found in the former USSR. There has been a long history of settlement in the far north of the USSR although up until the late 1950s numbers were probably low, possibly in the region of 30000 (Sugden, 1982).

Over the past 100 years or so, the impact of man on cold coasts has increased enormously, for four main reasons: military development, resource extraction, tourism and scientific research. These have brought with them allied problems of urban and industrial development. Along the Hudson Bay shore, for example, the estuaries of Eastmain River and La Grande Rivière have been altered by a large scale hydro-electric power project with a 10000 MW capacity (Messier *et al.*, 1986). The Arctic coasts assumed a special strategic importance after the Second World War (1939–45) and during the Cold War period when the proximity of Western zone (Europe and America) and Eastern zone (especially former USSR) territory was of concern. In the Arctic USA, Canada and Greenland, for example, the Distant Early Warning (DEW) line of forty radar stations was built between 1952 and 1957 along the 70° N line of latitude. The situation is rather different in the Antarctic, although the Falklands War in the early 1980s between Great Britain and Argentina, and political arguments over the 'ownership' of the continent of Antarctica show that these areas are also of great strategic importance. Political strategies were also responsible for populating areas in the far north of the former USSR with enforced labour camps in the Stalinist era. Now the indigenous peoples in many of the Arctic areas are pressing for better economic prospects in areas which, relative to the more southerly parts of the countries, are economically depressed.

Despite the cold and ice, there are some important resources being exploited in the cold regions, many of which affect coastal areas. Important oil and gas reserves are being tapped in Alaska, at Prudhoe Bay on the north coast where reserves of 10 billion barrels of oil were discovered in 1968, and on the Kenai Peninsula in the south. In the north of Russia, gas is produced around the Obskaya Guba inlet off the Kara Sea. Gas is also produced on Melville Island and Mackenzie Delta, Northwest Territories, Canada. In other cold coast areas many minerals are mined, including gold, coal and aluminium. Both Norway and Russia have coal mining settlements on Svalbard, for example, which would otherwise be unpopulated. These mining activities are often combined with mineral processing facilities. South of the tree line, several fjords have paper mills and chemical plants located on their sides causing pollution. A total of twenty-two fjords in seven countries are recognized by Syvitski *et al.* (1987) as having pollution problems of one sort or another. A further problem associated with resource development is the transport of oil, gas etc. to the major markets. Accidental spillage or dumping of cargoes at sea along the major shipping routes can lead to coastal pollution, as seen for example with the *Exxon Valdez* spill of 1989 (*see* Case study 7.1).

Tourism and scientific research may seem to have little obvious environmental impact, but in cold areas where terrestrial decomposition rates are low, they can cause many problems. Antarctica, despite being 98 per cent ice-covered, has a population of *c.* 800 in winter and 2000 in summer all of which is connected with scientific research and based in over forty, largely coastal, bases. In the past, whaling was an important exploitative activity in the

Antarctic, with many seasonal whaling camps established. Abandoned whaling camps now provide environmental pollution problems. Now tourism is providing a similar transient population, with up to 2000 visitors per year coming to the southern continent. Finally, however, we can see that there is an increasing desire for conservation and protection of both Arctic and Antarctic environments. Sanguin (1989), for example, discusses the development of a National Park along 112 km of the Pacific coast of Vancouver Island, British Columbia, Canada which was established in 1970 and tries to reconcile recreation and conservation activities.

Environmental hazards

Despite low total populations in cold coastal areas there is increasing evidence that such areas are under threat from human impacts. In turn, the very hostile conditions can pose acute problems to the inhabitants. Looking first at natural hazards on cold coasts, ice must pose the greatest threat. Sea ice limits the possibility of navigation and can damage submarines, ships and oil rigs. Coastal permafrost poses a problem for any engineering and building work. If buildings and roads warm the subsurface, permafrost collapse and thermokarst topography (usually hollows and mounds) result, causing damage to the buildings and roads themselves. Several cold coastal areas are prone to earthquake activity. The southern coast of Alaska, for example, was badly affected by the Good Friday 1964 earthquake, causing damage to buildings in the area (Ward *et al.*, 1987).

Two examples illustrate the highly dynamic nature of some cold coasts. Firstly, the Gulf of Alaska coast between Cape Suckling and Cape Spencer is one of the most dynamic coastlines on earth, being influenced by more than one hundred glaciers and by earthquakes (with over eight in the last 80 years reaching over 7.3 on the Richter scale). Winter storms, glacial meltwater sediments and tectonic activity produce a highly active coast. Individual events can be spectacular. In the east of the area, an earthquake in 1888 at Lituya Bay produced a rockslide containing 30 million m^3 of rock, which generated a tsunamic wave in the bay and led to a huge water surge of over 520 m in height which sank boats and destroyed vegetation over a 10 km^2 area (Molnia, 1985). Secondly, the south coast of Iceland has one of the highest wave energy regimes in the world (Nummedal *et al.*, 1987) and is frequently affected by extratropical cyclones. This coastline is also affected by another form of high magnitude–low frequency natural event, in the form of glacier bursts (jokulhlaups) which are generally caused by subglacial volcanic activity. The resultant floods have produced vast changes in the progradation of parts of the coast, as shown in Fig. 7.7.

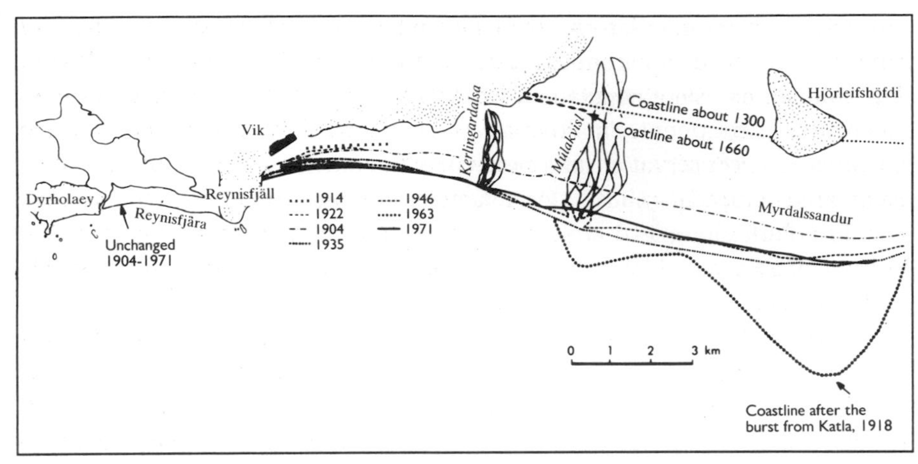

Fig. 7.7 Changes in the shoreline near Vik in Myrdal, Iceland, since 1300 AD. After Nummedal *et al.* (1987)

Permafrost coast erosion

As we have seen above many permafrost coasts are prone to erosion because of a combination of rising sea levels and ground ice. There is some evidence to suggest that settlements exacerbate the erosion problem. Thus, during storm events in 1986 on the Arctic Alaskan coast, localized bluff erosion was much more serious in front of the settlements of Barrow and Wainwright than in the intervening coastal section. In both places gravel beach material has been removed for airport runway construction, producing a narrowing of the beach. Limited bluff protection measures had been implemented in 1985, involving fifty-six asphalt tanks placed at the bluff tops, but these were damaged and proved to be totally inadequte in the 1986 storms. The North Slope Borough Public Works Department is now planning beach nourishment, with material dredged from the offshore zone, to prevent further accelerated erosion (Walker, 1991). At the nearby town of Nome there is a small port which is also often affected by severe storms. Local property owners built their own impromptu structures to prevent storm damage during the early years of the twentieth century. These were often ineffective. In 1957 a 1 km long seawall was built by the US Army Corps of Engineers to protect the shore, using granitic rock from Cape Nome (Walker, 1988b).

Pollution of fjords

Pollution of many forms is a major problem for many cold coast areas. The climatic conditions produce slow growth rates and low productivity of much terrestrial and intertidal biota especially where ice causes seasonal lack of growth, or damage to organisms. Pollution may therefore be a great blow to plant and animal communities here, and they may take a long time to recover. Geochemical and circulation characteristics of many fjords tend to encourage the concentration of pollutants in sediments and organisms. Essentially, cold coasts are stressed by many natural factors anyway, and an additional stress such as pollution may prove to be the final straw. Conversely, however, the presence of natural stresses makes the severity of the additional human impacts hard to judge.

As suggested above many, but not all, fjords can clearly be seen to be sensitive to pollution, possessing low flushing rates and therefore acting to concentrate the pollutants in waters, sediments and organisms within the fjord. However, looking at the issue from the other perspective, fjords could be regarded as useful stores of pollution, as they contain its impacts, and prevent much wider areas of coastline from becoming contaminated. Pollutants enter fjords through direct industrial effluent (both liquid and solid wastes), atmospheric inputs (often from industrial chimneys), land use changes (agriculture and forestry), direct seepage from reclaimed and polluted landfill sites, and urban sewage outfalls. In many cases, the source of pollution is easy to identify. In Sørfjord, an extension of Hardangerfjord in western Norway, which has been suggested as being the most metal-polluted fjord in the world, heavy metals, which seriously pollute sediments, water and organisms in the fjord, come primarily from a zinc plant at the head of the fjord in the town of Odda (Molvaer and Skei, 1988).

Two case studies show the severity and nature of the problems. The interested reader should pursue the many case studies presented in Syvitski *et al.* (1987) to gain a wider appreciation of the issues. On the Arctic west coast of Greenland, Agfardlikavsâ fjord has been contaminated since 1973 by a lead/zinc flotation mill which has pumped its waste out via a submarine outfall. An estimated 500 000 t a^{-1} of metal-rich waste has been discharged into the fjord, where natural sedimentation rates of up to *c.* 2 cm a^{-1} are found. Studies have shown high, but seasonally variable, levels of zinc, copper and lead in water, suspended sediments, seaweeds and blue mussels (Loring and Asmund, 1989). Curiously, the contamination of the sediments increases away from the source, apparently because the fine-grained sediments have the highest metal content, and are entrained more easily and carried further down the fjord (Syvitski *et al.*, 1987).

In contrast, Kristiansfjord in the temperate far south of the eastern Skagerrak coast of Norway, receives pollution from a nickel plant (metals and organic micropollutants) as well as untreated wastewater from a population of 80 000. Particular pollutants of concern are hexachlorobenzene (HCB), nickel, copper, arsenic and cobalt. A study of the sediments in the area around the nickel plant outfall showed very high levels of copper, nickel, arsenic and cobalt (Fig. 7.8). The levels were also 3–5 times background levels in surface sediments 6 km away from the plant. HCB and EPOCl (extractable persistent organic chlorine) levels were up to 10000 times background levels in bottom sediments in the harbour. These sediment concentrations are matched by pollution levels in flora and fauna. In the harbour, much of the bottom flora and fauna has been totally wiped out over an area of 3–5 km^2. Cod, flounder and mussels even in the outer part of the fjord showed high levels of HCB and EPOCl (Molvaer and Skei, 1988). As a result, warnings are given against eating fish or shellfish from the fjord. Municipal sewage outputs produce limited eutrophication on top of these other pollution problems.

Cold coasts in the global greenhouse

The future of cold coasts under conditions of global warming will depend largely on the response of the various forms of ice to climatic forcing. As we have seen in Chapter 2, there are many controversies remaining over the nature of future climatic changes, and the melting of sea ice, glaciers and ice sheets will play an important feedback role in these changes. Many studies are being carried out at the moment to try and clarify the relationship of cold coast ice changes to warming, in order to elucidate the exact importance of such feedbacks.

Permafrost coasts will probably face accelerated rates of coastal erosion as permafrost melts in coastal bluffs and beaches, rendering them more prone to collapse, erosion and retreat. Such effects will be compounded if thaw settlement occurs leading to inundation. However, there have been few detailed studies of how permafrost melting will influence coastal retreat. Basal erosion is also required, as well as the production of erodible sediments and slumped profiles. It is likely that the reduction in sea ice (over time and space) will increase the vulnerability of many permafrost coasts to storm wave attack, thereby further encouraging retreat, whilst also leading to an increase in precipitation as snow (a negative feedback). A rise in sea level, also linked to global warming, will be an additional factor favouring erosion. These problems are likely to be particularly severe along currently eroding coasts, such as the Alaskan Beaufort Sea coast. Woo *et al.* (1992) give a balanced account of the likely impacts of global warming on the permafrost coast of Canada, and suggest that some effects (such as thermokarst produced by permafrost melting) will be only transient features. Coastal retrogressive thaw slumping, a major erosional process on tundra coasts, will increase and erosion will be enhanced by melting of subsea permafrost which will lead to thaw settling on the sea bottom.

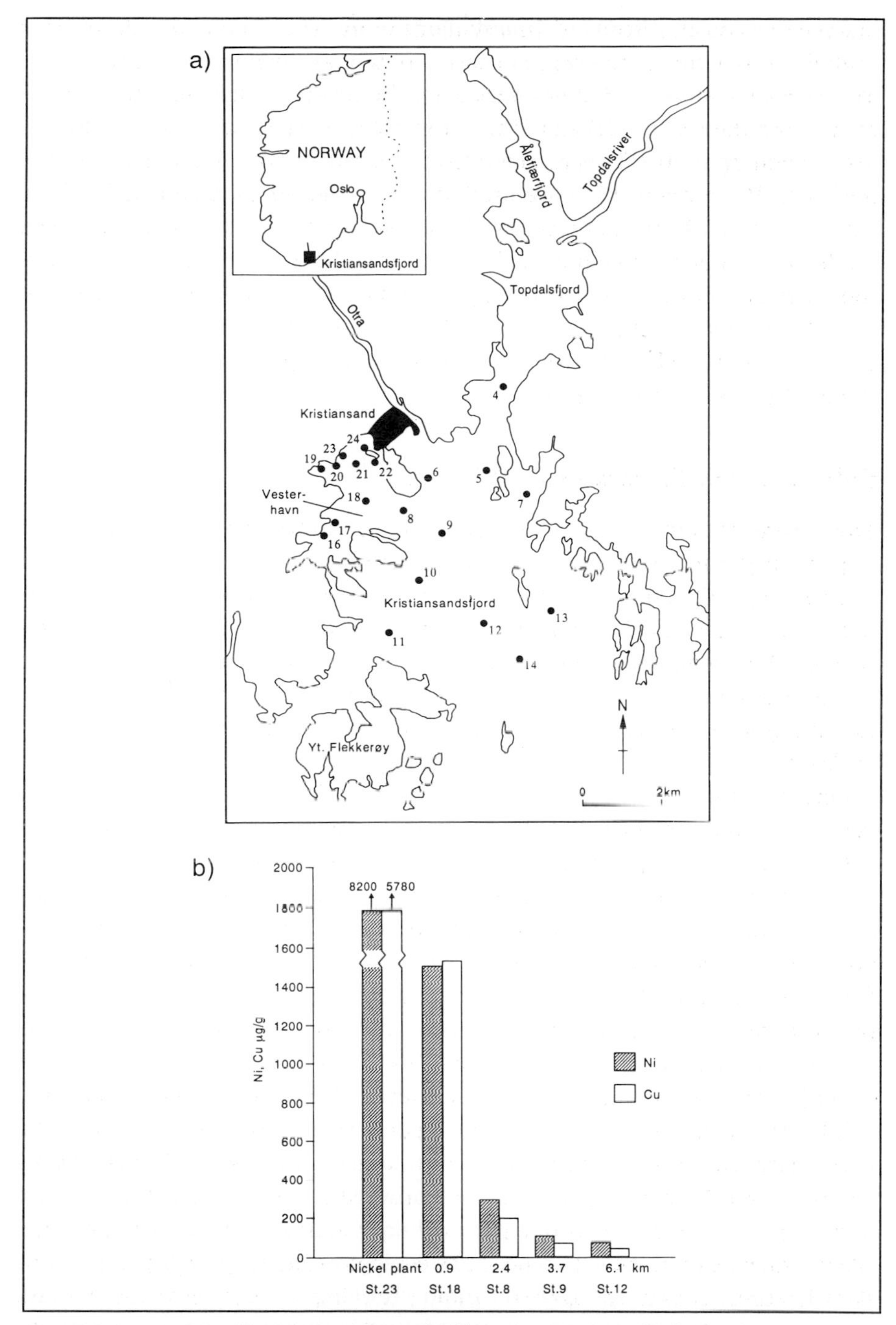

Fig. 7.8 Sources and amounts of metal pollution in Kristiansfjord, Norway. a) Location map with sampling locations, b) levels of Ni (shaded) and Cu at sampled locations in surface sediments. After Molvaer and Skei (1988)

Glaciated coasts, including many fjords, are likely to face considerable change as melting and calving increases (not necessarily in a predictable, linear way) with warming. However, it is clear from many studies that parts of the Antarctic ice sheets (such as the West Antarctic ice sheet) may increase in volume in a warmer world (as precipitation increases, leading to a net accumulation of ice). Drewry (1991) suggests that such a short term response will be coupled with a longer term response (*c.* 500 years or so) of increased discharge of ice to sea. Sugden (1991) shows how modelling indicates that topography influences calving, as well as climatic forcing, producing a stepped form of growth and decay. One good outcome of all this research effort on the response of ice sheets and tidewater glaciers is that there has been an increasing focus of scientific interest in these areas. It is clear that, in most areas, the local human impacts of such changes will be quite small (as populations are usually tiny in such hostile environments), in comparison with their potential global impact on future climate and sea level.

Sea ice may also not behave predictably in a warmer world. Several models suggest that even the thin first-year Antarctic sea ice will be quite resilient and resistant to the impact of warming (Wadhams, 1991). A debate in the literature over the melting of sea ice in the Arctic has led to estimates of the necessary warming varying between 2 and 10°C to produce catastrophic melting (Stocker, 1994). Certainly, coastlines currently affected by sea ice may suffer radical changes to the ice regime, thereby altering the seasonal pattern of sedimentation and erosion, as well as influencing floral and faunal distributions. Fjord circulations will also be affected by changes in freshwater inputs as glaciers melt, as well as by global sea level rise and regional oceanographic changes which will alter flushing regimes and sedimentation processes. Thus, the fate of pollutants entering fjords in the future may be very different to today. The diversity of fjord environments makes any more concrete predictions impossible.

Overview

Cold coasts display huge variety in both geomorphology and ecology, and there are still many areas requiring more in-depth study. Clearly, many cold coasts are only suffering from localized human-induced coastal problems, such as pollution. Natural hazards affect large areas, however, and future greenhouse conditions are likely to affect vast swathes of cold coasts through changes to ice cover on land and sea and associated ecological changes (Plate 7.2). Low growth rates, low species diversities and short food chains make cold coast ecosystems vulnerable; at least in terms of their resilience, which is low because of the long time necessary for regrowth to occur. The following two case stud-

Plate 7.2 The Laguna San Rafael, southern Chile receives the San Rafael glacier — the nearest tidewater glacier to the equator. The glacier front (middle distance) is 3 km wide and 100 m high; the larger icebergs in this plate, calved from the glacier, are 100 m long and 50 m high (photograph by D.J. Reed, December 1987)

ies examine how pollution and direct environmental modification can affect Arctic and Antarctic environments.

Case study 7.1: The *Exxon Valdez* oil spill, Alaska: catastrophe or not?

In the early hours of the morning of 24 March 1989 the tanker *Exxon Valdez*, which contained 1.2 million barrels of North Slope crude oil, ran aground on Bligh Reef in Prince William Sound, Alaska. In total 240 000 barrels of oil were spilled, and only 4 per cent of this amount was recovered (Fig. 7.9). Within a week the oil had spread over 900 square miles of water, and when it came ashore it affected 1100 miles of coast. Eventually, the oil covered 25 000 km^2 of coastal and offshore waters, inhabited at the time by 600 000 marine birds. The oil spill occurred just when migratory birds were beginning to enter the area, and its effects on wildlife would have been even more serious if it had occurred in summer or autumn. Several studies were made immediately after the spill, and months or years later to investigate the impacts of this, the most devastating oil spill to affect the US shoreline. However, legal debates between Exxon

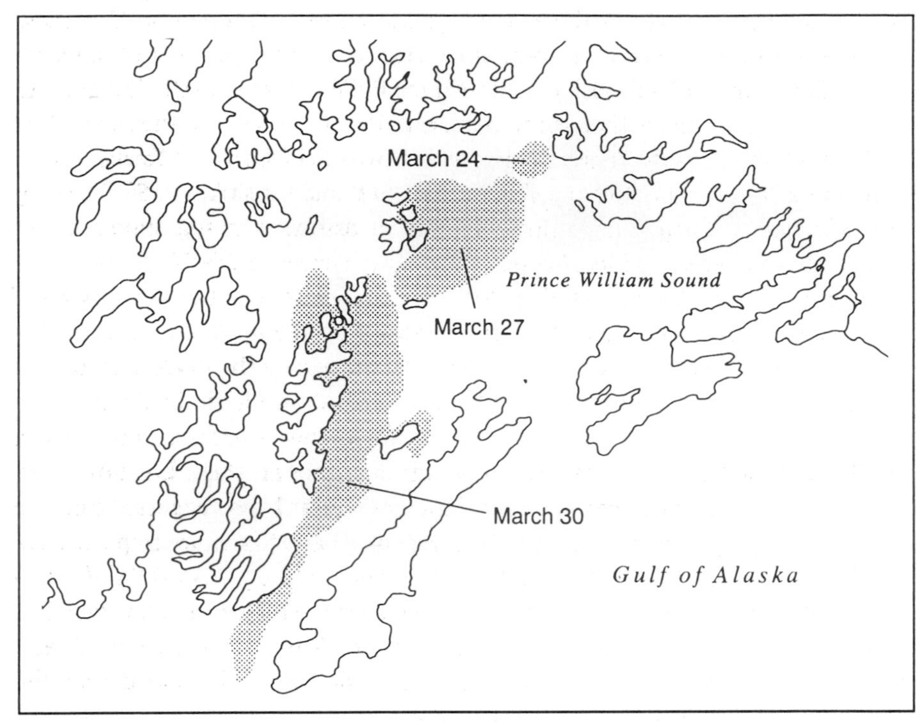

Fig. 7.9 Floating oil distribution in Prince William Sound, Alaska following the *Exxon Valdez* oil spill on 24 March 1989. After Galt *et al.* (1991)

and the US Government have made it difficult to get hold of information, and there is still much controversy over the severity of the impacts in the long and short term, as well as the success of the cleaning operation. Furthermore, the event was treated to intense international media coverage, which often gave an unrepresentative view of what was really going on, and has perpetuated a number of myths about the event.

The spill can, in fact, be viewed as two events, one inside Prince William Sound (which is a complex fjord-type estuary) caused by rapid release of oil from the tanker, and a second in the Gulf of Alaska caused by a slow leakage of oil out of Montague Strait. The oil moved in a generally south-western direction across Prince William Sound. On 26 March there was a large storm in the area, which changed the direction of movement of the oil slick and aided mixing, emulsification and promoted weathering of the oil. Two weeks after the spill, approximately 20 to 25 per cent of the spilled oil had moved out into the Gulf of Alaska (Fig. 7.9). Approximately 35 per cent of the oil is thought to have evaporated or dispersed into the water column in the Sound, mainly in the first 2 weeks. An additional 40 per cent affected the coastline inside the Sound (Galt *et al.*, 1991).

The major immediate effects of the oil were on bird populations. In all 35 000 dead birds were retrieved by the end of September 1989, although 5000 of these are thought to have died from natural causes. Around 90 per cent of the birds were killed in the Gulf of Alaska, not in Prince William Sound itself. The number of dead birds collected is thought to represent 10–30 per cent of the total kill of 100 000–300 000 birds. This compares with *c.* 30 000 killed by the *Torrey Canyon* spill, and 20 000 by the *Amoco Cadiz* disaster (Piatt and Lensink, 1989). The long term effect on bird population numbers is hard to assess, as immigration may act to balance the losses, and local populations may recover in 20–70 years.

There are only a few accessible reports on the impacts of oil on other parts of the coastal environment here. The initial clean-up used applications of fertilizers on beaches and other coastal substrates to encourage the biodegradation of oil residues by microorganisms. In Arctic areas, although biodegradation occurs, it works at a much slower rate than in the temperate zone (Atlas, 1985). Spectacular results seemed to be found, although there has now been much debate over the efficacy of the techniques. It has been suggested that the force of the spray used to apply the fertilizer may have had more impact. Over the three summers 1989, 1990 and 1991, Exxon applied 104 000 lbs of nitrogen fertilizer, which was claimed to have increased the biodegradation rate 3–5 times. Stone (1992) is particularly critical of the clean-up methods used, and suggests that observations 3 years after the spill occurred showed that the clean-up had not worked as well as first thought.

Interestingly, studies of the oil residues on beaches in the area found isotopically different residues as well as those from the *Exxon Valdez*. These are thought to be from the 1964 earthquake, which obliterated the old Valdez townsite, spilling asphalt into the fjord of Port Valdez, which then is assumed to have flowed down into Prince William Sound (Kvenvolden *et al.*, 1993). This suggests that at least some parts of the oil may remain in the sediments in and around the Sound for a few decades at least. The impacts on slow-growing, species-poor tundra vegetation in the area have been much debated, and no consensus has emerged. Atlas (1985) writing about oil pollution in the Arctic before the spill occurred said that 'the dangers of pollution are of the same sort and magnitude as elsewhere', thus implying that the fragility of Arctic ecosystems is more imaginary than real. The *Exxon Valdez* spill provides an invaluable opportunity to test such ideas, and it is to be hoped that information collected about the spill and its aftermath will be made widely available in the near future.

Selected references

ATLAS, R.M. 1985: Effects of hydrocarbons on microorganisms and petroleum biodegradation in Arctic ecosystems. In Engelhardt, F.R. (ed.), *Petroleum effects in the Arctic environment*. London: Elsevier Applied Science, 63–99.

DUNBAR, R.B., LEVENTER, A.R. and STOCKTON, W.L.K. 1989: Biogenic sedimentation in McMurdo Sound, Antarctica. *Marine Geology* 85, 155–79.
GALT, J.A., LEHR, W.J. and PAYTON, D.C. 1991: Fate and transport of the Exxon Valdez oil spill. *Environmental Science and Technology* 25, 202–9.
KVENVOLDEN, K.A., COWLSON, P.A., THREKELD, C.N. and WARDEN, A. 1993: Possible connections between two Alaskan catastrophes occurring 25 years apart (1964 and 1989). Geology 21, 813-16.
PIATT, J.F. and LENSINK, C.J. 1989: Exxon Valdez bird toll. *Nature* 342, 865–6.
STONE, R. 1992: Environmental research: oil clean-up method questioned. *Science* 257, 320–1.

Case study 7.2: The Antarctic coast: exploitation or conservation?

The Antarctic has been described as the last remaining wilderness on earth, and although undoubtedly the environment is no longer completely pristine, human impacts have been relatively limited in their history and spatial extent. It is, of course, the very harshness of the Antarctic environment, with 98 per cent of the area covered by permanent ice, which has precluded exploitation in the past. Now, however, there are several scientific bases in the area (Fig. 7.10), and plans afoot to exploit both the marine environment (especially krill and whale), and the terrestrial environment (oil and platinum extraction). Eager to avoid the environmental degradation associated with much resource exploitation in other areas in the past, there have been several important moves, including establishing the Convention for the Conservation of Antarctic Marine Living Resources (CCAMLR) and the Scientific Committee on Antarctic Research (SCAR). Some examples of recent events in the Antarctic area indicate the kinds of problems likely to be encountered by further resource exploitation.

In Winter Quarters Bay, McMurdo Sound there have been found to be locally high levels of pollution in subsurface sediments. Hydrocarbons have been found in concentrations as high as 4500 ppm, and metal concentrations are also high, and comparable with levels in many temperate zone polluted coastal areas. The pollution has not, as yet, been transported outside the bay, and is clearly related to settlements within the area (Lenihan *et al.*, 1990). The bay will probably take many decades to recover from the pollution. Near the Australian station Casey (Fig. 7.10) a petroleum spill of around 90 000 litres occurred in June 1990, of which an estimated 40 000 litres is still unrecovered. A further 19 000 litres was spilled in January 1991 in the same area. Studies have shown that biodegradation of the petroleum is hampered in the Antarctic environment by severe cold and lack of nutrients. It is likely that the petroleum

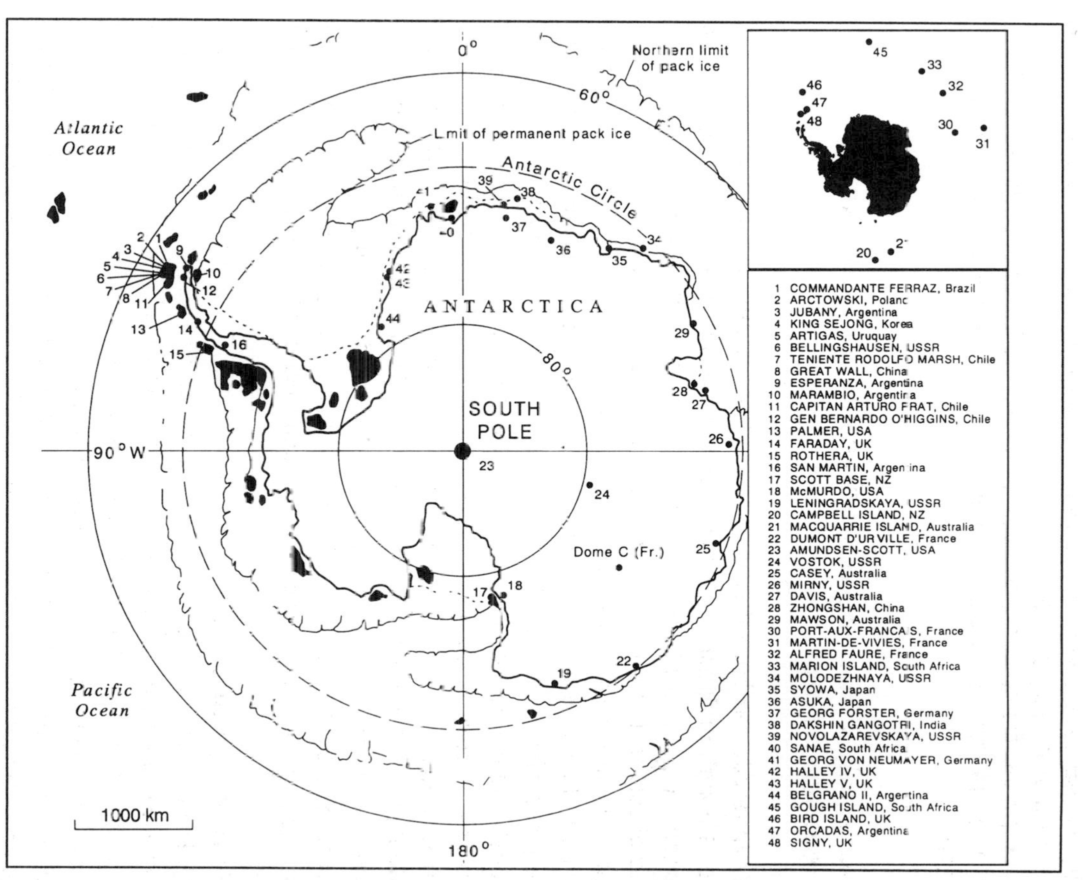

Fig. 7.10 Antarctic research stations active in winter 1990. After Anon (1990)

will take up to 600 years to be degraded (compared with 40 years in the Arctic), but experiments suggest that fertilizer applications will speed things up, although debate over the *Exxon Valdez* clean-up operation implies that they should be done with extreme care.

A dramatic diesel spill occurred in January 1989, when the Argentinian supply ship *Bahia Paraiso*, carrying 250000 gallons of diesel, ran aground on the Antarctic Peninsula near the US Palmer Station (Fig. 7.10). The toxic diesel killed off krill and limpets very quickly, and also affected a nearby Adélie penguin rookery. Penguins became covered with diesel, and also ate contaminated krill. Diesel is more toxic, but less persistent, than crude oil, so the long term effects are likely to be negligible. Palmer Station staff were able to help with the clean-up, which was made easier because the spill occurred in the Antarctic summer when there was no problem with ice floes restricting access to the area (Barinaga and Lindley, 1989).

Pollution is not the only likely impact of human interaction in the Antarctic. At the French Dumont d'Urville base engineers have created a new airstrip. Unable to use the ice sheet surface, they have had to blow up six offshore islands. They have used the rubble to form a dyke. There were concerns that the ecology of the area would be affected, specifically that Adélie penguin nesting on the islands would die out and Emperor penguin migration routes would be affected. Surveys so far have suggested that both species have adapted well.

It is clear from these examples that the Antarctic coast has suffered relatively little from human activity to date, but that localized problems have occurred which may be of a long term nature because of the very slow recovery (low resilience) of the Antarctic environment. Scientific research bases clearly do have an environmental impact through pollution and environmental modification which should not be underestimated. Any future influx of scientists to Antarctica (related to global climatic and environmental change monitoring) must be carefully managed to ensure that this almost-pristine coastal environment does not become damaged unnecessarily.

Selected references

ANON 1990: Research in Antarctica. *Nature* 350, 280.

BARINAGA, M. and LINDLEY, D. 1989: Wrecked ship causes damage to Antarctic ecosystem. *Nature* 337, 495.

LENIHAN, H.S., OLIVER, J.S., OAKDEN, J.M. and STEPHENSON, M.D. 1990: Intense and localised benthic marine pollution around McMurdo Sound, Antarctica. *Marine Pollution Bulletin* 21, 422–30.

Chapter Eight

MANAGING THE COAST: COPING WITH COASTAL PROBLEMS

Managing the coastal zone — regaining a holistic perspective on coastal problems

So far in this book we have dealt with a range of different coastal problems in specific environments, illustrating the natural processes and disturbances involved. It has become apparent that each coastal type faces a huge range of problems now, and that these are likely to increase in the future as a result of intensification of coastal zone use as coastal populations grow, and also because of likely future climatic alterations. Each coastal environment faces a different range of problems, although many of the basic problems, such as erosion, pollution and vegetation disturbance are common to all coasts. Furthermore, many of the problems faced by one area of the coast (such as wetlands) are linked to those experienced in other, nearby coastal areas, such as adjoining beach and dune systems.

In order to be able to cope with these problems, we need not only to understand how the coastal environments function naturally, and how disturbance regimes operate upon these environments, but also what the human dimensions of the problems are. It is this human side of the issue which we want to focus on now, before bringing all the factors together in considering two case study areas in some depth, the Bay of Bengal and the Mediterranean Sea.

Human influences on the coast

As demonstrated in Chapter 1, the coastal zone is of ever-increasing importance to societies worldwide. Currently, about 60 per cent of the world's population (or nearly 3 billion people) live in the coastal zone, and two-thirds of the world's cities with populations of over 2.5 million people are near estuaries (Tolba and

El-Kholy, 1992). Within the USA alone, population is growing faster by the coast than elsewhere (Fig. 8.1). Human activity is at least partly responsible for the following list of impacts seen to be affecting the American coast:

- loss of habitat
- interception of water and sediment
- invasions by exotic species
- increased rate of sea level rise
- increased pollution of nearshore environments
- increased nutrients in nearshore environments.

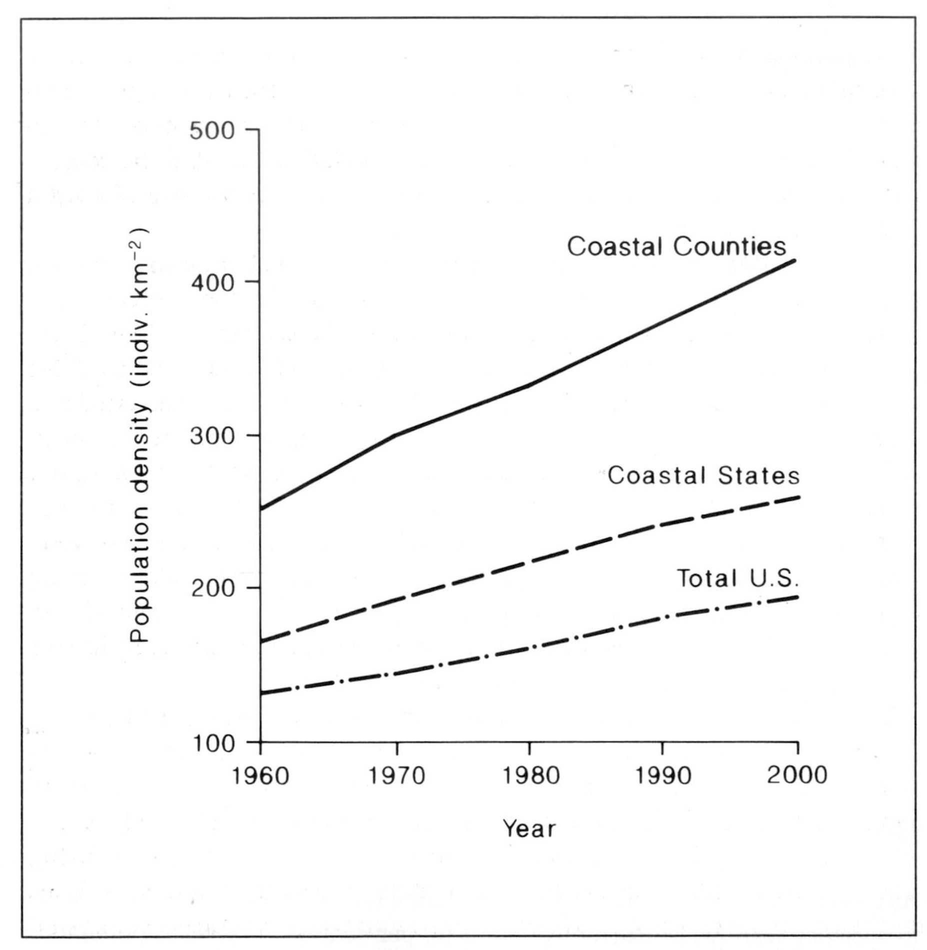

Fig. 8.1 Growth in population 1960–2000 in coastal counties and coastal states in comparison with the whole of the USA. After LMER Co-ordinating Committee (1992)

Of these, the LMER Co-ordinating Committee (1992) identify the last as the most serious one. Clearly, the scale of the nutrient loading problem has increased rapidly in recent years. Land use changes in the Mississippi River basin, for example, have led to a doubling of dissolved nitrogen contents in the river over the last 100 years. Recent evidence has shown that this nutrient increase has caused coastal eutrophication of the continental shelf near the river delta outflow. Changes in the deposition of diatom skeletons (derived from phytoplankton) on the continental shelf are seen to reflect eutrophication, and there is a pronounced increase in the deposition rate after 1980 (Turner and Rabalais, 1994). Studies of macroinvertebrate assemblages on the San Diego shelf, on the other side of the USA, indicate that here a wastewater outfall is an important control on benthic communities, but that El Niño events contribute to changes over time as well (Zmarzly *et al.*, 1994).

Human activities offshore, within the coastal zone and in the hinterland all contribute towards these and other more localized problems. The range of activities contributing to coastal problems is huge and thus the challenge for successful integrated coastal zone management (especially given that the coast is naturally dynamic) is enormous. Globally, we can identify a range of coastal areas suffering particularly acute problems (Fig. 8.2).

The Philippines provides a good example of the range of problems involved. Approximately 65 per cent of the Filipino population live in coastal areas, within an archipelago of 7100 islands having a total coastline of *c.* 18 100 km. The ten largest cities are all coastal. Particularly stressed areas of the coastline are the mangroves and coral reefs. Mangroves have been widely converted into brackish water fishponds for aquaculture. 106000 ha of mangroves remain, but a further 176000 ha have been converted to fishponds. Some mangrove swamps have been reclaimed for mine tailings disposal. Coral reefs have been damaged by siltation, pollution and destructive fishing techniques (Talavera, 1985). Manila Bay is now highly polluted, with the pollution mainly coming from domestic sources (which account for 75 per cent of the biochemical oxygen demand (BOD) totalling 130000 t a^{-1} in the bay) and airborne pollutants from four nearby power stations (Nakamura, 1985).

The coastline of the Philippines is also being affected by onshore activities, as deforestation inland increases erosion and allied sedimentation at the coast. In Larap Bay, on Luzon Island, sediment from iron ore quarrying has been carried down rivers and has aided coastal progradation. The coast itself is highly active, with volcanic eruptions (such as the recent Mount Pinatubo event) producing new pyroclastic sediments, earthquakes leading to local uplift and subsidence and helping trigger coastal landslides, and tsunamis which produce severe coastal flooding episodes (Bird, 1985a; Bird and Schwartz, 1985). Tropical cyclones also provide a major coastal flood hazard, and affect the Philippines coast frequently.

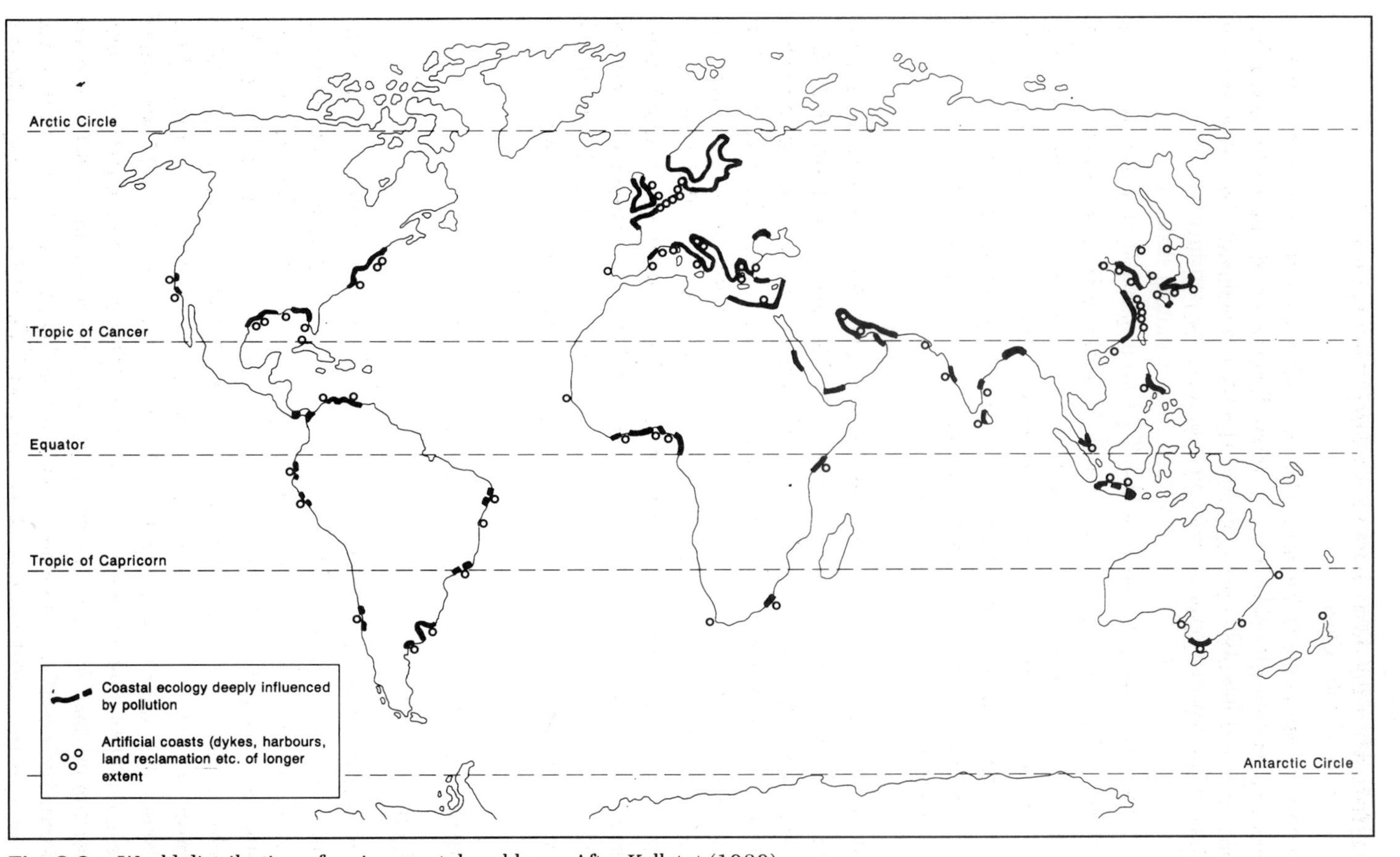

Fig. 8.2 World distribution of major coastal problems. After Kelletat (1989)

Sustainable coastal zone use and management

Much has been made of the phrase 'ecologically sustainable use of resources' in recent years, and it is now a key (if often ill-defined and not easily attainable) concept in coastal zone management. The UK Department of the Environment report 'Managing the Coast' (1993) for example states that:

> The Government's strategic aim is to promote the sustainable use of the coast. Its objective is to encourage the management of all aspects of the human use of the coast, including estuaries, to yield the greatest benefit to the present population, while maintaining the potential of coastal systems to meet the aims and aspirations of future generations. Management of the coast needs to reflect its human uses, both social and economic, as well as its nature conservation value.

This is a considerable task, and involves integrating successfully a range of activities reviewed in Chapter 1 (Table 1.1) whilst acting to reduce their deleterious effects (individually and often in combination) without, presumably, transferring problems elsewhere (e.g. further out to sea). Clearly, it will be impossible to 'solve' all coastal problems, and part of any 'sustainable use' plan must be to recognize that the environment cannot be controlled as such, and that environmental management must be flexible, and involve as much management of human, as well as physical, aspects. Thus, Goldberg (1994) sees the control of coastal zone population as the most important goal, without which any attempts to manage will, as in the past, prove largely inadequate.

The coast of Bangladesh, Bay of Bengal

The physical setting

Bangladesh is situated at the top of the Bay of Bengal where the Indo-Gangetic plain meets the Indian Ocean. Tectonically, the area is highly active, being at the meeting point of three major tectonic plates, and earthquakes are common. Between 1833 and 1971 over 200 earthquakes of between 5.0 and 8.5 on the Richter scale affected Bangladesh (Mollah, 1993). The climate of Bangladesh is monsoonal with a warm season from March to May, a rainy season from June to September and a cool season from October to February. Up to 95 per cent of the annual preciptation falls between April and October. The predominant wind direction reverses from south-west in the rainy season to north-east in the dry season. Each year the monsoon season brings floods to *c.* 18 per cent of the country, with allied deposition of sediment (Mahtab, 1992). The Bay of Bengal coast has a mean tidal range of 2.84 m at Chittagong in the east, rising to 3.64 m at springs. In the west along the Sundarbans coast, the tidal range is higher, approaching 7 m during monsoon storms (Snead, 1985).

Bangladesh occupies the largest delta in the world, where the Ganges–Meghna and Brahmaputra Rivers and their tributaries meet the sea in the Bay of Bengal. About 80 per cent of Bangladesh is on river floodplain and the three major rivers carry around 2 billion tons a^{-1} of sediment out to the delta front. According to Broadus (1993) this sedimentation is just about able to maintain the delta given the current rate of sea level rise in the area, although Brammer (1993) suggests that there should be some accretion at the moment because of increased erosion upstream. Current rates of subsidence of 0.6–5.5 mm a^{-1} have been calculated, although locally rates may be more than 20 mm a^{-1} (Mahtab, 1992). A relative sea level rise of *c.* 1 mm a^{-1} is being experienced along the delta coast at the moment. The delta itself is characterized by diversity, with five different types of floodplain recognized by Brammer (1990a).

The coast of Bangladesh has been classified into four zones (Snead, 1985) which are shown in Fig. 8.3. In the west, the heavily wooded Sundarbans area (which is the older deltaic plain of the Ganges), is dominated by mangroves and nipa palm swamps. To the east of this region a 68 km strip of coastal forest has been cleared for paddy farming with scattered remaining mangrove areas. The central part of the coast is occupied by the main Meghna channel delta, comprising mainly unvegetated mud and sand shoals (the Meghna flats). Finally, the eastern part of the coast is the Chittagong coast which varies greatly in coastal plain width. To the east of Bangladesh, the Burmese coast contains another major delta, the Irrawaddy, with extensive mangrove development. This delta appears to be prograding by up to 10 m a^{-1} (Bird 1985c). Subsidence is also occurring here, as well as along the eastern Indian coast on the other side of the Bay of Bengal. To the west, in Indian West Bengal, the delta environment is no longer active, and silting of rivers has become a common problem (Farmer, 1983).

The Bay of Bengal acts as a giant funnel for storm events, creating the most severe storm surge problem in the world (Murty *et al.*, 1986). Storm surges, as described in Chapter 2, cause an elevation of sea level (because of meteorological factors) above the tidal height, which if they occur at spring maxima may have devastating effects on a low-lying coast such as that of Bangladesh. Eight severe cyclones were recorded during 1960–70, including a devastating one in 1970 which killed an estimated 300 000 people. In 1985 a cyclone surge event caused *c.* 11 000 deaths in the Meghna estuary, and in 1991 another cyclone caused deaths estimated at well over 100 000.

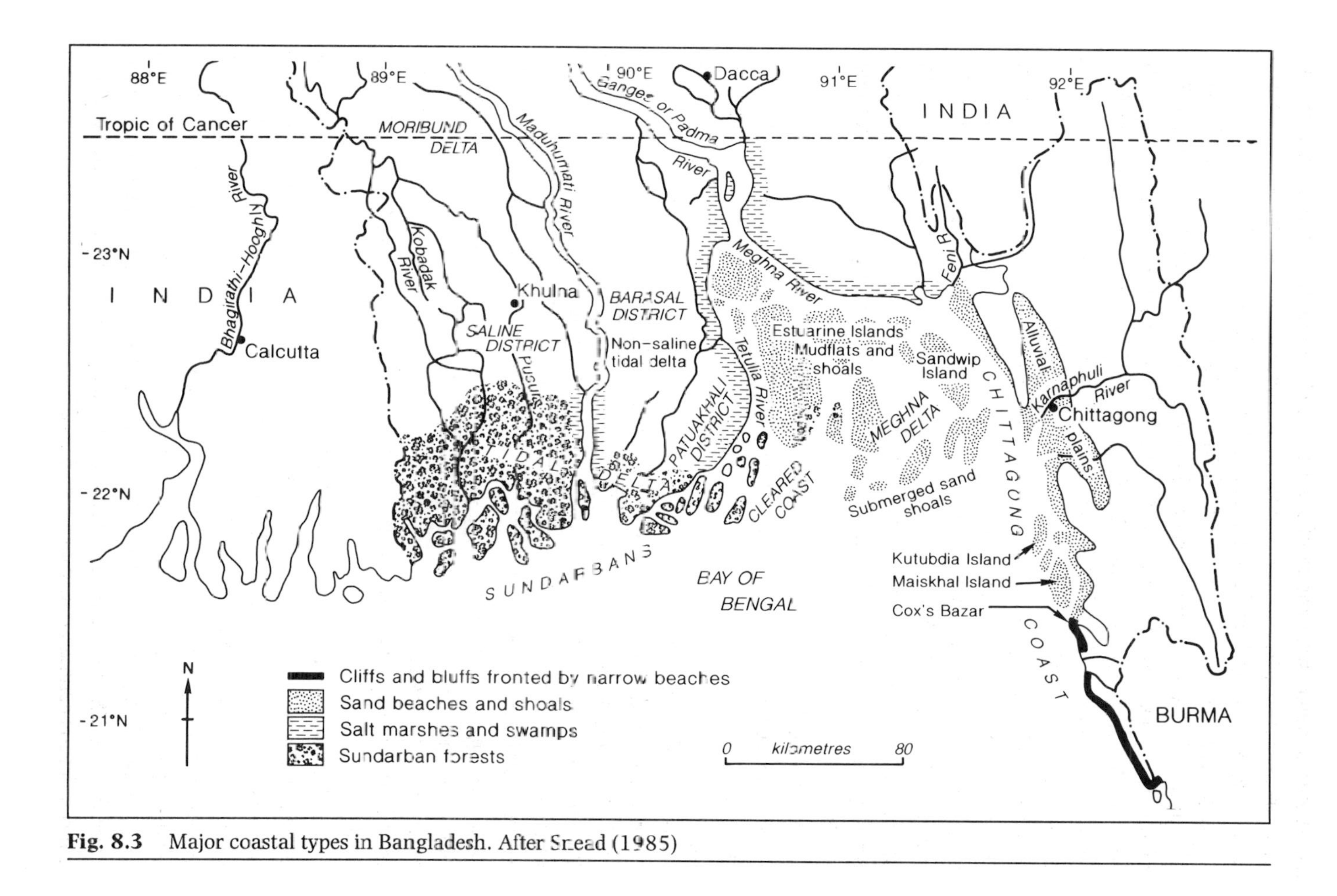

Fig. 8.3 Major coastal types in Bangladesh. After Snead (1985)

Population, society and economy

The above descriptions provide a stark physical background to the coastal problems of Bangladesh and adjacent areas — a tectonically active, dynamic, shifting and low-lying floodplain coast, prone to river flooding (with serious floods in 1987 and 1988) and also affected by catastrophic storm surges. The human dimensions of the problems also demand further attention. Bangladesh became an independent state in 1971. Once, the whole Bengal area (including Bangladesh and West Bengal in India) was known as the rice bowl of India and was very prosperous; now, however, Bangladesh is extremely poor. The population of Bangladesh is over 115 million, and has been estimated to rise by the years 2000 and 2030 to 145 and 232 million, respectively. Per capita income is US$170 per year, making Bangladesh one of the poorest countries in the world. Agriculture employs around 60 per cent of the civilian workforce, and around 80 per cent of the population is rural, mainly inhabiting the floodplain areas. Life in the floodplain is adapted to the environmental stresses: earth mounds are used as settlement sites above normal flood levels, communication lines are built on embankments, and rice cultivation accounts for around 80 per cent of the agricultural area. Most farms are small on the densely populated parts of the Meghna delta, where densities of 1000 km^{-2} are found, and settlements occupy 20–40 per cent of the land area. The Sundarbans are the only part of the Bangladesh coast to remain largely natural, with 401 600 ha of mangroves remaining which act as a reserve for many important species of wildlife, e.g. cobras, crocodiles, Bengal tigers and leopards, as well as protecting the rich agricultural land behind. As Brammer (1993) points out, these natural areas are not static. Over the past 30 years the health of the main brackish water forest trees in the northern Sundarbans (the Sundri or *Hevetiera minor*) has declined, probably as a result of saltwater intrusion, linked to the decline of freshwater supply caused by the diversion of much Ganges water by the Farraka Barrages (Mahtab, 1992). This scheme was designed to increase the flow of the Bhagirathi and Hogli Rivers to help flush silt out of the port of Calcutta in West Bengal (Farmer, 1983).

Thus, the Bangladesh coast can be seen to be a vulnerable environment inhabited by an equally vulnerable population. However, such blanket statements are dangerous, given the human and natural diversity of the country. As we will see below, there are clearly serious problems faced by the coastal population, which may become exacerbated in the future.

The coastal problems

The most immediately serious, life-threatening problem is the vulnerability to storm surge flood events. As we have seen, this is a direct consequence of the natural environment, i.e. a large, active and subsiding delta in a monsoon climate and a tectonically active setting. Many workers have also argued that the severity of the problems has increased as settlement has made the area more vulnerable. Deforestation has produced erosion and loss of vital mangrove buffer zones. As we have also seen, wetland changes may also be indirectly due to human activity, through changes to the freshwater–saline water balance caused by upstream water extraction.

Other, smaller scale problems can also be seen as part of this same scenario. Mollah (1993) points to the geotechnical problems of many soils within the delta area, which in turn create problems for buildings, roads and bridges. Thus subsidence is a problem faced by many embankment roads, where 1.5–3 m high embankments may subside by up to 40 cm, causing cracking and severe surface distortions. Allied to this are problems of lack of suitable aggregates to make good concrete, and a highly saline and corrosive environment within the delta plain which leads to accelerated weathering of concrete structures and foundations.

Pollution is a growing problem for much of the delta area, although quantitative data are scarce. In the western part of the Bay of Bengal, red tides off the east coast of India were first observed in 1981, causing several cases of shellfish poisoning. Red tides are produced by algal blooms which occur as a result of high nutrient loadings from river inputs, producing eutrophic conditions. Some of these algal blooms contain species with toxic effects on shellfish, which can be transferred to humans who eat them resulting in paralytic shellfish poisoning (PSP) and other illnesses (UNEP, Environmental Data Report, 1993–4). Within the Bangladesh delta environment, pollution from newly established industries, as well as from agricultural fertilizers and pesticides, is likely to be affecting coastal waters (Mollah, 1993).

It is, of course, the threat of an acceleration and growth in these linked problems from greenhouse gas-induced global warming that has proved to be of particular concern in recent years. Early predictions suggested that large areas of the country would be inundated, with for example a 3 m rise in sea level by 2100 (caused by a 2 m global rise in sea level coupled with a 1 m rise caused by local subsidence) leading to a loss of 26 per cent of habitable land (Broadus, 1993). However, it is likely that a whole suite of other changes would accompany such a situation, such as increasing probability of storm surges, loss of valuable buffer mangrove swamps and increased vulnerability to flooding, reduction in agricultural productivity and associated economic problems, and increased saltwater intrusion into aquifers and irrigation waters with concomitant effects on vegetation (Broadus, 1993).

Simple future predictions are hard to sustain, given the diversity and dynamism that characterizes the Bangladeshi environment. As Brammer

(1993) points out, there will be vast changes to the delta anyway over the next 50–100 years, regardless of whether accelerated sea level rise occurs or not, as the river courses have fluctuated wildly in the past and will continue to do so in the future. Increased sedimentation, consequent upon increased erosion in the Himalayas and Assam Hills will also influence the delta environment. If sea level rise does occur, Brammer suggests that its impacts will mainly be felt inland and not at the coast, where sedimentation of levées will compensate for rising water levels. Hardoy *et al.* (1992) echo this concern, indicating that salt intrusion into groundwater could be a major future problem for Dhaka's public drinking water supply. Flather and Khandker (1993) have used a numerical model to estimate the effects of a 2 m increase in mean sea level on storm surges and tides in the Bay of Bengal. They predict increased tidal amplitudes in the north-eastern corner of the bay and decreased ones in the north-west (of the order of *c.* 10 cm). They suggest that the highest flood level of a storm surge event would be raised by about 1.8 m, although a reduction in the main surge generating force (wind stress/water depth) means that the effects are not perhaps as dramatic as one might fear. Flather and Khandker (1993) also point out that changes in cyclone tracks and intensities, as well as in river flows, which may also be a result of global warming, will affect these storm surge scenarios. However, it is very difficult to predict changes to cyclone activity, as the scale of general circulation models cannot resolve such smaller scale atmospheric processes.

Solving the coastal problems of Bangladesh?

Is there anything that can be done about such serious problems, now and in the future? The responses so far indicate the options and problems. Several projects have already been carried out, with variable success, each tackling a portion of the overall range of problems. One of the most innovative and interesting has been the Mangrove Afforestation Programme. Recognition of the natural protection provided by the Sundarbans mangroves against cyclone surges led the Forest Department to start planting trials in 1966. These proved very successful, and led to a long term (World Bank-funded) project.

The newly deposited deltaic land, locally called 'char' lands, is not suitable for agriculture until it has been drained and salts removed, and it is vulnerable to submergence during high monsoon seas. Planting of mangroves can aid accretion and protect these lands from erosion. The highly dynamic nature of the Bangladeshi coastal environment means that plantings often fail, and follow-up plantings are often necessary. Saenger and Siddiqi (1993) report however that *c.* 120000 ha of mangroves had been successfully planted at the time of their study. Problems have been encountered, especially relating to the biological vulnerability of mono-specific, mono-aged stands. The severe 1991 cyclone, for example, damaged plantations along the Chittagong coast, but regrowth occurred quickly. Saenger and Siddiqi (1993) describe the mangrove plantations as providing a 'self-repairing system' which facilitates the 'stabilisation and protection of the coast during intense cyclonic conditions'. Furthermore, they are a source of wood and reclaimed land. Thus, approximately 60 000 ha of planted mangroves have now accreted to above the level tolerated by mangrove trees and are now used for forestry and agriculture.

Another project is the Coastal Embankment Project which is a flood control and drainage project comprising of 58 polders and affects most coastal areas except the Sundarbans. Embankments and sluices have been constructed as part of the project, to protect against tidal and river flooding and saline water intrusion (Mahtab, 1992). They have, however, had the result of 'decoupling' areas of land from the natural siltation cycle of the delta, thereby possibly threatening their long term stability. These measures do not help protect against the major storm surge events, and there have been many problems with embankment failures. After the 1987 and 1988 flood events a major international agency-coordinated Flood Action Plan was instigated to strengthen the embankments, as well as other linked projects such as improved flood hazard warning (Brammer, 1990b). The estimated cost is US$146.3 million.

Unfortunately, the natural dynamism, which is often erratic and sporadic, makes it diffcult to assess the success or otherwise of such schemes. The problem of cyclone protection may be better addressed by building emergency shelters on raised ground for temporary use while floods subside, coupled with an effective early warning system and means of getting the population to the shelters in time. Sea level rise may be tackled in three main ways:

1. build new, or enlarge existing, embankments
2. allow low-lying land to be inundated and intervene to aid the natural accretion process
3. a combination of 1 and 2 (Mahtab, 1992).

Rural settlement mounds would need to be raised, but this could easily be done by the inhabitants themselves; other changes would need to be co-ordinated and should bear in mind the total delta setting, as well as the fate of small sections of it. As Brammer (1993) sums it up:

> [E]mbankments and drainage do not provide a simple solution to the problems created by increased flood levels that might follow a rise in sea level. The hydrological problems are complex and will require much study. The solutions proposed may be expensive to provide and to operate.

However, the huge population of the area, based closely on the delta lands for its livelihood, make some planning essential. Letting nature take its course is probably impossible here. Controlling nature is also an impossibility. The key must be to understand the processes involved at all scales, to adopt a wide range of mutually compatible solutions involving local people, regional and national governments and international aid agencies, and to accept that any plans will require constant monitoring and updating. Of course, economic constraints are enormous.

The Mediterranean Sea

The physical setting

The Mediterranean Sea is a relatively shallow, almost enclosed and virtually tideless sea covering *c.* 2.9 million km^2 in a tectonically active area. It straddles the collision zone between Africa and Europe, which has resulted in the formation of a ring of mountains around the sea. Today, the plate boundaries run through Morocco, Algeria, Tunisia, southern Italy, Greece and Turkey creating an earthquake hazard within the region. The tectonic framework has determined the gross coastal morphologies of the Mediterranean, such as the rocky, sediment-starved coasts around the former Yugoslavia, Greece and parts of North Africa. In other places river basins have been created which have fed large river-dominated deltas, such as the Ebro, Po, Rhone and Nile deltas.

The climatic setting is one of a winter rainfall regime, with minimal summer rainfall and high temperatures. Precipitation comes dominantly from cyclonic disturbances originating in the Mediterranean basin (Wigley, 1992). There is a clear north–south annual rainfall and temperature gradient in the Mediterranean, with much of the south-eastern part of the coast of Egypt, Libya and Tunisia being hot and receiving less than 200 mm of rain per year, whereas some parts of the temperate northern Adriatic coast receive over 1000 mm of rain per year (Milliman *et al.*, 1992). The climate encourages high evaporation, and this, coupled with seasonal rainfall, creates seasonal water shortages which are particularly severe in the southern Mediterranean countries.

The Mediterranean Sea has a microtidal regime, with tidal ranges of 20–30 cm, and is a low wave energy environment. This means that coastal processes are restricted to a very narrow intertidal zone, encouraging development of hotels very near sea level. Because of its enclosed nature, and the climatic encouragement of evaporation, plus the reduction in river discharges by dams (e.g. the Nile), the Mediterranean has a negative water balance. This means that water entering from the Atlantic via the Straits of Gibraltar becomes more saline before finally being flushed out. This high salinity and low flushing rate limits biological productivity and leads to problems of pollution dispersal.

Population and tourism

The Mediterranean Basin has been an important seat of civilization for thousands of years, and there has been a long interaction of society with the coastal environment, as illustrated by the many archaeological sites around the Greek and Turkish coasts. Recent archaeological investigations at the harbour of Cosa, north of Rome in Italy, for example, have shown that it was a key port in the second and first centuries BC with concrete piers built which have lasted for almost 2000 years (McCann, 1994) and an elaborate commercial complex.

Now, the Mediterranean is home to an ever-growing population. In 1985, the eighteen countries (now twenty with the division of Yugoslavia) had a combined population of 352 million, of which 37 per cent (133 million) lived in the coastal zone (Milliman *et al.*, 1992). UNEP estimates that the population will rise to 430 million by 2000 AD and 545 million by 2025. The distribution of population will also change, as growth rates increase in the southern countries and stable populations result in the northern ones. The population of Egypt, for example, is predicted to burgeon from 58 million today to over 100 million by 2050 (Stanley and Warne, 1993). Concomitantly, the population will become increasingly urbanized. In 1980 it was 57 per cent urban in these countries, and this is predicted to rise to 75 per cent by 2025. Many of the cities are located directly on the coast, and most of the rest are within easy reach of it. As well as this resident population, the Mediterranean attracts a large number of seasonal tourists. 100 million people visited in 1984, and 170–340 million are predicted to visit per year by 2025. Most of these head for the coast, with 90 per cent of international visitors to Tunisia and the former Yugoslavia visiting the Mediterranean shores (Goldberg, 1994). Tourism growth also seems highest in the southern and eastern parts of the Mediterranean, although within the south both Algeria and Libya have little tourist development.

The vast settled and visiting populations around the Mediterranean have brought with them large urban developments, roads, marinas and resort complexes. Water consumption, in an area already plagued by water shortages, has shot up and placed an additional stress on the environment. Waste disposal to a shallow, slowly flushing sea has also expanded. This is the human and natural setting for a whole host of coastal problems, many of which show extreme patchiness over space and time. As with Bangladesh, there are many worries about the future in a 'greenhouse' world, but there are also many more pressing problems today.

The coastal problems of the Mediterranean

The combination of fragile soils and vegetation cover, flash floods and catchment degradation over many centuries resulted in the development of alluvial fans and coastal plains in many parts of the Mediterranean and in delta progradation. However, channelization of flows and protection against flooding has reduced sediment supply to the coastal areas in more recent times.

In some areas, such as the Nile delta, the problems of erosion are serious and intractable, relating to the natural subsidence of the delta environment, coupled with a depletion of sediment coming down the river Nile as a result of the closure of the High Aswan Dam in 1964. Before the dam was built the Nile carried some 124×10^6 t. of sediment to the delta each year (Stanley and Warne, 1993). Now only minor amounts of sediment arrive, either blown in, or carried by longshore drift. Some parts of the Nile delta coast have spectacular rates of erosion, such as at Ras El-Bar where 1800 m of erosion occurred between 1902 and 1960 (Bird, 1985a). Since the Aswan Dam was built, the problem has become worse, especially at the Rosetta, Damietta and Burullus promontories (Sestini, 1992; Fig. 8.4). Other delta coasts are experiencing similar problems. The Ebro River, for example, now has 96 per cent of its sediment load trapped by structures before it reaches the delta, and in the Rhone delta sediment delivery has reduced from *c.* 40 million t a^{-1} at the end of the nineteenth century to 4 million t. a^{-1} in 1970. Thus, many of these sedimentary coastal areas are now almost fossil features, where erosion has begun to dominate and a general natural progradation seems unlikely. As Stanley and Warne (1993) put it 'Human intervention . . . has caused Northern Egypt to cease as a balanced delta system'.

Much of the tourism around the Mediterranean is based on beaches, and there have been huge efforts made by many countries to preserve or re-create beaches in order to boost tourism. Such projects, coupled with other developments in the coastal zone such as harbours and marinas, have interrupted natural sediment movements, and led to a patchy erosion–sedimentation problem. Localized quarrying of sand, such as in Israel, has also narrowed beaches.

Accompanying such erosion problems, there is now a severe pollution problem around many parts of the Mediterranean coast, affecting the sea itself. Clark (1989) provides a good overview of the problems showing how there is now chronic oil pollution over most of the Mediterranean, mainly caused by discharge of bilge and ballast water from tankers. Some 250 million t. of oil are transported each year through the Mediterranean, but since 1976 there have been tighter controls over discharges from ships. Oil pollution has affected the fish and sea fish populations, especially in areas such as the Gulf of Naples and the Venice lagoon. Domestic waste, mainly sewage, is also a severe problem in

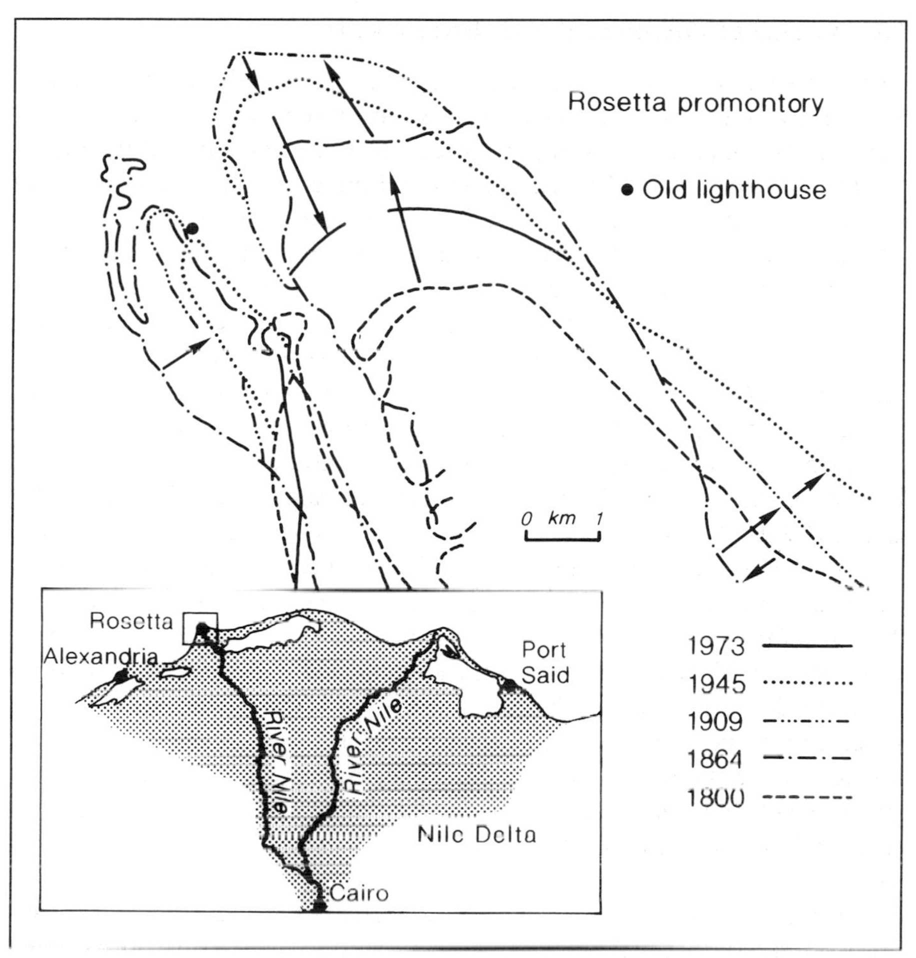

Fig. 8.4 Coastal changes at the Rosetta Nile mouth, 1800–1945. After Sestini (1992)

the Mediterranean (Fig. 8.5), especially around the Spanish, French and Italian coasts. Industrial and urban wastewater, as well as agricultural runoff, is polluting coastal lagoons on the Nile delta, upsetting their ecology and thereby affecting fisheries and migratory birds. Other pollutants, such as organochlorine pesticides (which amount to 90 t. a^{-1} discharged into the Mediterranean from rivers), heavy metals and PCBs (polychlorinated biphenyls) are also proving serious. Imperfectly flushed lagoons, such as the Venice lagoon, are particularly prone to the build-up of pollution in bottom sediments.

Coastal eutrophication is another problem affecting parts of the Mediterranean, caused by increased nutrient inputs from rivers. Massive algal blooms

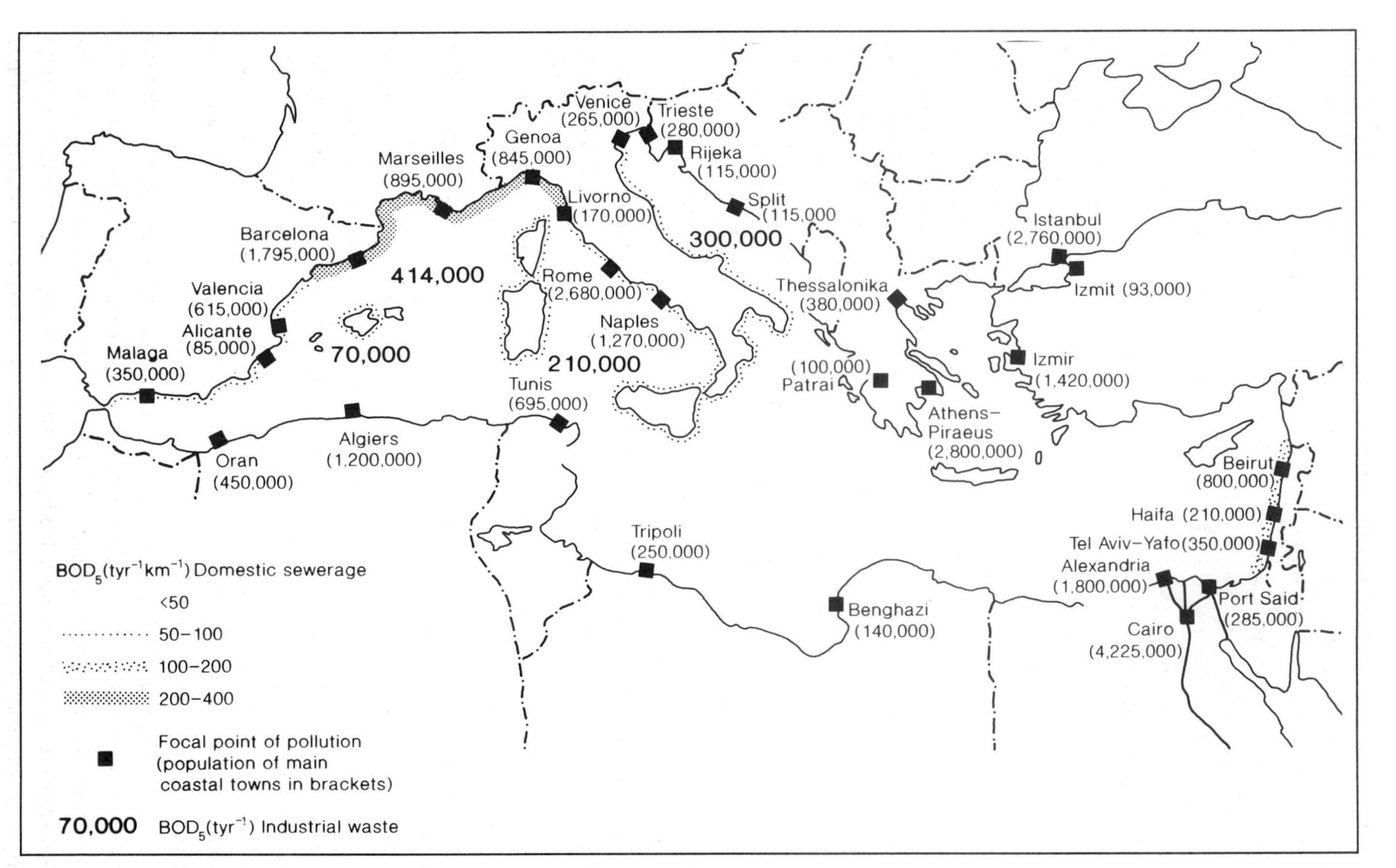

Fig. 8.5 Sewage and industrial waste discharges into the Mediterranean Sea. After Clark (1989)

often result, and have been a problem since first recorded here in 1972 in the Gulf of Venice. A serious problem of eutrophication occurs in the northern Adriatic Sea, where rivers draining Italy bring down a total of *c.* 29×10^3 m^3 a^{-1} of phosphates and over 120×10^3 m^3 a^{-1} of nitrates (Goldberg, 1994). A transnational problem results, in that the beaches of the former Yugoslavia are as seriously affected by the algal blooms as the Italian beaches.

Some examples illustrate the patchy nature of many pollution problems within the Mediterranean. A study of tributyltin (TBT) and its derivatives, undertaken in 1988 along the French, Turkish and Egyptian coasts found that concentrations generally exceeded the no observed effect level (NOEL) of 20 ng l^{-1}. Very high levels (up to 12000 ng l^{-1}) were found around marinas (Gabrielides *et al.*, 1990). TBT is a component of anti-fouling paint used to protect boats from colonization by barnacles and other organisms. It has been found to have major consequences for marine life, producing a condition known as 'imposex' in whelks (the females grow penises). Industrial discharges of pollutants can also be problematical. A tannery in the Gulf of Geras on the Isle of Lesbos discharged tannery wastes into the gulf, which was contaminated with chromium. Other discharges into the bay of olive-press effluents and sewage also contributed to a serious pollution problem here. In 1983 a tannery effluent treatment plant was brought onstream, and subsequent studies of benthic communities in the gulf found that there was evidence of recovery over the 1983–88 period (Papathanassiou and Zenelos, 1993).

Many of the natural coastal wetlands around the Mediterranean Sea have been reclaimed, or converted to other uses, leading to an important habitat loss and potential loss of biodiversity within the area (Plate 8.1). Thus, for example, large areas of coastal wetlands in the Nile delta were drained and cultivated in the nineteenth century and the Abu Qir lagoon was drained and turned into farmland (Stanley and Warne, 1993). Concurrent with these changes, reduction in riverborne sediment loads has also led to increased erosion and instability of these habitats. Along the Po delta, for example, there have been extensive reclamations of coastal wetlands for agricultural land (Sestini, 1992). Land reclamation has also affected wildlife habitats in the Inner Thermaikos gulf in Greece (Georgos and Perissorakis, 1992). There are, however, valuable wildlife habitats left within the Mediterranean, such as the wetlands of the Ebro delta, which attract a large number of migratory birds, and there is considerable enthusiasm for conservation.

Several low-lying Mediterranean coastal areas are suffering from accelerated subsidence and incursion of saline groundwater. As Stanley and Warne (1993) have shown from a study of nearly 100 radiocarbon-dated cores from the Nile delta, subsidence and erosion is leading to increased groundwater salinity here. Venice, to the north of the Po delta, is subsiding naturally at around 1.3 mm

a^{-1}, and reclaimed land nearby has shown additional subsidence as a result of compaction and drying out of sediments. Additionally, a large area near Venice has experienced accelerated subsidence since the 1950s because of over-extraction of groundwater (Sestini, 1992). Although this problem has now been solved, it is estimated that 12–14 cm of extra subsidence occurred until groundwater extraction was regulated, and Venice is now highly vulnerable to flooding. As well as this flood hazard, the encroachment of saline water on land and in aquifers has led to ecological changes and the accelerated weathering of buildings and monuments. The cultural heritage of Venice is now severely threatened.

There have been many recent studies concerning the likely impacts of future climatic changes and allied sea level rise on the Mediterranean, some of which are collected in Jeftic *et al.* (1992). It is likely that, given the relatively modest global sea level rise predicted recently (as discussed in Chapter 2), the impacts of global climatic change will be equalled in severity by the other human impacts on the Mediterranean coast which will increase as population increases. Jeftic *et al.* (1992) suggest that attention should be focused on identifying particularly sensitive areas and devising multiple strategies with which to react.

Plate 8.1 Development and environmental degradation in the Mediterranean: the Rhone delta, southern France

Venice provides a good example of a sensitive area. Local subsidence here will compound the impact of future sea level rise, with a 44 cm global rise by 2070 (based on Warrick and Oerlemans, 1990) translating into a 54 cm rise in Venice itself, and an 88 cm rise along the Po delta. Development of the Venice lagoon area means that a natural response to sea level rise (involving the landward migration of barriers) will be impossible. The immense value of Venice's cultural heritage means that 'managed retreat' will not be a feasible option here, and a long term programme of increasing coastal defences (such as offshore breakwaters) and beach nourishment will be necessary.

Solutions to coastal problems in the Mediterranean

The range of local, regional and basin-wide coastal problems that affect the Mediterranean environment clearly require a number of different management strategies. Some pollution problems, where a source can easily be identified and remedial action taken in time, can be fairly easily solved, as in the case of the tannery on the Isle of Lesbos, discussed above. Localized erosion problems may prove more intractable, as attempts to solve them may only provide short term solutions. So, for example, the many engineering structures built around the Venice and Po delta coasts have only been partially successful in preventing beach erosion as the whole area is now out of equilibrium (because of human activity) with its natural conditions. Thus, for any successful management of beach erosion in this area, the whole regional situation must be taken into account. Localized pollution may also prove a difficult problem to solve where, as in the case of the Venice lagoon, it comes from a whole range of sources. Data from the 1980s indicate that the Venice lagoon is polluted beyond acceptable limits, and is becoming increasingly prone to eutrophication events, as a result of urban, industrial and agricultural wastes (Sestini, 1992). Legislation to control discharges to the lagoon is urgently required.

Complex regional coastal problems are difficult to solve, as exemplified by the case of the Nile delta in the poorer southern Mediterranean. Here, natural subsidence and human activities combine to produce an eroding and polluted coastline, where many wetlands have been reclaimed. Stanley and Warne (1993) suggest that only a large scale project along the lines of the Netherland's coastal protection scheme will be able to control the problems; however, the huge population increases forecast for this country will make even such a large scheme as this unable to cope. Even within the richer northern Mediterranean countries it can be very difficult to solve regional coastal problems whilst reconciling a range of activities. The Camargue, on the Rhone delta in France, is an area of low marshes, ponds and dunes which has had a long history of attempts at environmental control, with dams and seawalls emplaced to try and prevent flooding. During the nineteenth century there was considerable conflict between farmers in the area (who wanted increased area for cultivation) and salters (who wanted increased natural areas for salt production). This was resolved in 1929 when most of the Basse Camargue (the lowest area) was made a reserve (Corré, 1992) of wetlands and lakes. There are many different wetland habitats here which merit conservation, but this semi-managed landscape may be very vulnerable to future accelerated sea level rise. Corré (1992) suggests that 'green tourism' may be a solution; whereby development of the area focuses on its natural attractions without the need for any huge infrastructural developments.

Other regional problems cross national boundaries and require co-operation between nations, as of course do the basin-wide pollution problems faced by the Mediterranean. Thus, pollution in the northern Adriatic, which comes largely from polluted rivers in Italy, affects the beaches of other countries. Considerable progress towards international co-operation has been made since the initiation in 1975, under the auspices of UNEP, of the Action Plan for the Protection of the Mediterranean Sea against Pollution (MAP). This resulted in the Barcelona Convention in 1976, and the development in 1979 of the 'Blue Plan', intended to integrate development plans with environmental protection measures for the Mediterranean Basin (Tolba and El-Kholy, 1992). Since then, MAP has approved protocols to limit coastal pollution from land-based sources, and protect endangered species and critical habitats. However, there are many problems which hamper the translation of these agreements into practice, notably economic and political difficulties. As tourism is such an important part of the Mediterranean coastal environment, and is both affected by, and contributes to, many of the region's coastal problems, it seems that this should be one key industry to involve in attempts to solve these problems.

Conclusions

Detailed consideration of the problems facing the coastal environments of Bangladesh and the Mediterranean shows the contrasts between the situation in the developed world with that in developing nations. Bangladesh, like many developing countries, faces a plethora of natural hazards at the coast, made particularly serious where they interact with intense human pressure on land. By comparison, the human-induced problems of pollution and ecological management are less severe, although growing all the time. Many developed nations, on the other hand, are beset by a nexus of human-induced coastal problems. Both will therefore be affected in different ways by future environmental changes, and will be able to react in quite different ways. The Mediterranean provides an interesting case study, with stark contrasts between the developed northern seaboard and the poorer southern states. Throughout the Mediterranean, however, there is an increasing problem of pollution and human-induced degradation of the environment, as management scheme after management scheme fails to stop the problems.

Along problem coastlines around the world, management options are being refined, as the true costs of natural hazards coupled with human-induced degradation become apparent (Plate 8.2). Tables 8.1 and 8.2 illustrate some of the options open to coastal lowland management in the context of adapting to future sea level rise. These strategies include preventing any further development by legislation, allowing development in the short term, and encouraging people to move in the future. Each policy relies on a combination of legislation and economic forces to encourage people to act as required. Framed in the context of the USA, the assessment in Table 8.2 suggests that policy 7, which uses changes in property ownership rights to manage a threatened lowland coast, would be the most cost-effective, and overall the most successful strategy. However, such a policy is not easily transferable from the developed to the developing world, where coastal populations are often much less mobile, and where low-lying coastal land can be a vital part of the nation's economy. Assessments like those presented in Tables 8.1 and 8.2 must form a vital part of future

Plate 8.2 'Hard' engineering for coastal problems: flood gates on the banks of the Mississippi River, New Orleans, USA

coastal management, as they identify the direct and indirect costs of different strategies in both the long and the short term, but the exact solutions must be tailor-made to fit the social, economic and political realities of the individual country, or group of countries. Very different judgements might be made over the best range of solutions to other coastal problems, such as pollution, and, of course, where a whole family of problems affects one coastal area, decisions have to be made over how best to tackle them together. Where coastal problems transcend national boundaries (as seen in the Mediterranean) then international agreements have to be sought, argued over and policed. Whatever policy is adopted, it is essential that it is grounded firmly in scientific understanding of the nature of the coastal problems and their possible solutions; without this, any hope of maximizing the human use and enjoyment of the world's wondrously diverse and dynamic coastlines will be lost.

Table 8.1 Coastal lowlands and sea level rise: policy options

Policy	*Description*
Prevent areas from being developed	
1. Prohibit development	Statutes — regulatory control
2. Buy coastal land	Agencies, various levels of government purchase land onto which coastal landforms might migrate
Allow development	
(a) Defer action	
3. Order people out later	Ignore sea level rise; government will
4. Buy people out later	require structures to be removed when they threaten migration of coastal landforms
5. Rely on economics	End subsidies to coastal development but otherwise ignore sea level rise; government action will not be necessary as people voluntarily abandon coastal properties
(b) Presumed mobility	
6. Prohibit bulkheads	Do not interfere with private activities today but notify property owners that they will not be able to build protective bulkheads to protect property
7. Leases	Do not interfere now but convert property rights (with compensation if necessary) of current owners to long-term leases which expire after 99 years or conditional leases (e.g. expire after property is inundated). Underlying ownership transferred to private or public conservation group

Source: modified from Titus (1991)

Table 8.2 Social, economic, political and legal aspects of different policy responses (see Table 8.1) to future sea level rise

Policy	*Cost to public taxpayers*	*Social cost (versus no sea level rise)*[b]		*Economic efficiency*	*Performance under uncertainty*		*Constitutional*	*Equitable*	*Political feasibility*	*Risk of back-sliding*	*New institutional requirements*	*Chance of success*
		Present	*Cumulative*		*Sea level*	*Economics*						
1.	None	Speculative premium + <1% of base value[c]	Land	Poor	No	Yes	No	No	None	Possible	Regulatory	Almost certain at first, little in long run
2.	Expensive	Speculative premium + <1% of base value	Land	Poor	No	Yes	Yes	Yes	None	Possible	Acquisition	Almost certain at first, little in long run
3.	None	1% of land + structures	Land + structures	Fair	Yes	Perhaps	Maybe	Doubtful	Low	Very likely	Police	Unlikely
4.	Land and structures	1% of land + structures	Land + structures	Fair	Yes	No	Yes	Yes	Low	Very likely	Acquisition	Unlikely
5.	None[a]	1% of land + structures	Land + structures	Fair (if it works)	Yes	Useless	Yes	Yes	Good	Low	Hazard mitigation	Unlikely
6.	None	<1% land value	Land + residual value of structures	Optional	Yes	Yes	Probably	Usually	Good	Likely	Regulatory	Very likely
7.	<1% land + residual value of structures	<1% land value	Land + residual value of structures	Optional	Yes	Yes	Yes	Yes	Fair	Very unlikely	Change in title of property	Almost certain

[a] Would save taxpayers money as subsidies for flood insurance, coastal protection eliminated
[b] Social cost — includes the cost to people who must yield to the sea
[c] Speculative premium = potential for future development <1% of base value — as happening far into the future ('discounting' current value over 100 years to abandonment)
Source: modified from Titus (1991)

REFERENCES

ABARNOU, A., AVOINE, J., DUPONT, J.P., LAFITTE, K. and SIMON, S. 1987: Role of suspended sediments on the distribution of PCB in the Seine Estuary (France). *Continental Shelf Research* 7, 1345–50.

ABBAS, J.A. and EL-OQLAH, A.A. 1992: Distribution and communities of halophytic plants in Bahrain. *Journal of Arid Environments* 22, 205–18.

ACKLEY, S.F. 1991: The growth, structure and properties of Antarctic sea ice. In Kotlyakov, V.M., Ushakov, A. and Glazovsky, A. (eds), *Glaciers–ocean–atmosphere interactions*. IAHS Publication No. 208, 105–17.

ADAM, P. 1990: *Salt marsh ecology*. Cambridge: Cambridge University Press.

ADEY, W.H. 1975: The algal ridges and coral reefs of St Croix, their structure and Holocene development. *Atoll Research Bulletin* 187, 1–67.

ADEY, W.H. and BURKE, R. 1977: Holocene bioherms of Lesser Antilles — geologic control of development. *American Association of Petroleum Geologists Studies in Geology* 4, 67–81.

AITKEN, A.E., RISK, M.J. and HOWARD, J.D. 1988: Animal–sediment relationships on a subarctic tidal flat, Pangnirtung Fiord, Baffin Island, Canada. *Journal of Sedimentary Petrology* 58, 969–78.

AKILI, W. and TORRANCE, J.R. 1981: The development and geotechnical problems of sabkha, with preliminary experiments on the static penetration resistance of cemented sands. *Quarterly Journal of Engineering Geology* 14, 59–73.

ALBERTS, J.J., PRICE, M.T. and KANIA, M. 1990: Metal concentrations in tissues of *Spartina alterniflora* (Loisel.) and sediments of Georgia salt marshes. *Estuarine, Coastal and Shelf Science* 30, 47–58.

ALEXANDER, D. 1992: *Natural disasters*. London: University College London Press.

ALIZAI, S.A.K. and McMANUS, J. 1980: The significance of reed beds on siltation in the Tay estuary. *Proceedings, Royal Society of Edinburgh* 78B, s1–s13.

ALLEN, J.R.L. 1989: Evolution of salt marsh cliffs in muddy and sandy systems: A qualitative comparison of British west coast estuaries. *Earth Surface Processes and Landforms* 14, 85–92.

ALLEN, J.R.L. 1990a: The post glacial geology and geoarchaeology of the wetlands. *Proceedings, Bristol Naturalists' Society* 50, 28–46.

ALLEN, J.R.L. 1990b: Saltmarsh growth and stratification: a numerical model with special reference to the Severn Estuary, S W Britain. *Marine Geology* 95, 77–96.

ALLEN, J.R.L. 1990c: The formation of coastal peat marshes under an upward tendency of relative sea level. *Journal of Geological Society of London* 147, 743–5.

ALLEN, J.R.L. and PYE, K. 1992: Coastal saltmarshes: their nature and importance. In Allen, J.R.L. and Pye, K. (eds), *Saltmarshes: Morphodynamics, conservation and engineering significance*. Cambridge: Cambridge University Press, 1–18.

ALLEY, R.B. 1991: Sediment processes may cause fluctuations of tidewater glaciers. *Annals of Glaciology* 15, 119–24.

ALLEY, R.B., BLANKENSHIP, D.D., RORNEY, S.T. and BENTLEY, C.R. 1989: Sedimentation beneath ice shelves — the view from ice stream B. *Marine Geology* 85, 101–20.

ALLISON, R.J. 1989: Rates and mechanisms of change on hard rock coastal cliffs. *Zeitschrift für Geomorphologie Supplementband* 73, 125–38.

ALLISON, R.J. (ed.) 1990: *Landslides of the Dorset coast*, B.G.R.G. Field Guide.

ALLISON, R.J. and BRUNSDEN, D. 1990: Some mudslide movement patterns. *Earth Surface Processes and Landforms* 15, 297–312.

ALVEIRINHO DIAS, J.M. and NEAL, W.J. 1992: Sea cliff retreat in southern Portugal: Profiles, processes and problems. *Journal of Coastal Research* 8, 641–54.

ALWELAIE, A.N., CHAUDARY, S.A. and ALEWAID, Y. 1993: Vegetation of some Red Sea islands of the kingdom of Saudi Arabia. *Journal of Arid Environments* 24, 287–96.

ARAYA-VERGARA, J.F. 1985: Sediment supply and morphogenetic response on a high wave energy west coast. *Zeitschrift für Geomorphologie Supplementband* 57, 67–79.

AUGER, F. 1988: Simulation accélérée de la dégradation des matériaux de construction en ambiance aérienne saline. In Marinos, G. and Koukis, G. (eds), *Engineering geology of ancient works, monuments and historic sites*. Rotterdam: Balkema, 792–804.

BAGNOLD, R.A. 1941: *The physics of blown sand and desert dunes*. London: Methuen.

BAINES, G.B.K. and McLEAN, R.F. 1976: Sequential studies of hurricane deposits evolution at Funafuti Atoll. *Marine Geology* 21, M1–M8.

BARD, E., HAMELIN, B. and FAIRBANKS, R.G. 1990: U-Th ages obtained by mass spectrometry in corals from Barbados: sea level during the past 130000 years. *Nature* 346, 456–58.

BARNES, D.J. and CHALKER, B.E. 1990: Calcification and photosynthesis in reef-building corals and algae. In Dubinsky, Z. (ed.), *Coral reefs*. Amsterdam: Elsevier, 109–31.

BARNES, R.S.K. and HUGHES, R.N. 1988: *An introduction to marine ecology*. Oxford: Blackwell Scientific.

BASCOM, W.H. 1951: The relationship between sand size and beach face slope. *Transactions, American Geophysical Union* 32, 866–74.

BATTEN, A.R. and MURRAY, D.F. 1993: Dry coastal ecosystems of Alaska. In Maarel, E. van der (ed.), *Dry coastal ecosystems*. Amsterdam: Elsevier, 23–37.

BATTISTINI, R. 1981: La morphogénèse des plateformes de corrosion littoral dans les grés calcaire (plateforme superieure et plateforme a vasques) et le problème des vasques, d'apres des observations faite a Madagascar. *Revue de Geomorphologie dynamique* 30, 81–94.

BAYLISS-SMITH, T.P. 1988: The role of hurricanes in the development of reef islands, Ontong Java Atoll, Solomon Islands. *Geographical Journal* 154, 377–91.

BIRD, E.C.F. 1985a: *Coastal changes: A global review*. Chichester: Wiley.

BIRD, E.C.F. 1985b: Recent changes in the Somers-Sandy Point coastline, Westernport Bay, Victoria. *Proceedings, Royal Society of Victoria* 97, 115–28.

BIRD, E.C.F. 1985c: Burma. In Bird, E.C.F. and Schwartz, M.L. (eds), *The world's coastline*. New York: Van Nostrand Reinhold, 767–69.

BIRD, E.C.F. 1993: *Submerging coasts*. Chichester: John Wiley.

BIRD, E.C.F. and SCHWARTZ, M.L. (eds) 1985: *The world's coastline*. New York: Van Nostrand Reinhold.

BLASCO, 1977: Outlines of ecology, botany and forestry of the mangals of the Indian subcontinent. In Chapman, V.J. (ed.), *Wet coastal ecosystems*. Amsterdam: Elsevier, 241–60.

BLISS, L.C. 1993: Arctic coastal ecosystems. In Maarel, E. van der (ed.), *Dry coastal ecosystems*. Amsterdam: Elsevier, 15–22.

BODIN, P. 1988: Results of ecological monitoring of three beaches polluted in the Amoco Cadiz oil spill — development of meiofauna from 1978 to 1984. *Marine Ecology Progress Series* 42, 105–23.

BOORMAN, L.A. 1993: Dry coastal ecosystems of Britain: Dunes and shingle beaches. In Maarel, E. van der (ed.), *Dry coastal ecosystems*. Amsterdam: Elsevier, 197–228.

BOORMAN, L.A., GOSS-CUSTARD, J.D. and McGRORTY, S. 1989: *Climatic change, rising sea level and the British Coast*. London: HMSO.

BOSS, S.K. and NEUMANN, A.C. 1993: Impacts of Hurricane Andrew on carbonate platform environments, northern Great Bahama Bank. *Geology* 21, 897–900.

BOSSCHER, P.J., EDIL, T.B. and MICKELSON, D.M. 1988: Evaluation of risks of slope instability along a coastal reach. In Bonnard, C. (ed.), *Landslides*, Vol. 2. Rotterdam: Balkema, 1119–25.

BOURMAN, R.P. 1990: Artificial beach progradation by quarry waste disposal at Rapid Bay, South Australia. *Journal of Coastal Research, Special Issue* 7, 69–76.

BOURMAN, R.P. and MAY, R.I. 1984: Coastal rotational landslump. *Australian Geographer* 16, 144–6.

BOWDLER, S. 1988: Tasmanian aborigines in the Hunter Islands in the Holocene: island resource use and seasonality. In Bailey, G. and Partington, J. (eds), *The archaeology of prehistoric coastlines*. Cambridge: Cambridge University Press, 42–52.

BOWEN, A.J. and INMAN, D.L. 1966: Budget of littoral sands in the vicinity of Point Arguello, California. *Coastal Engineering Research Center Technical Memo* 19, 1–41.

BRAMMER, H. 1990a: Floods in Bangladesh Part I: Geographical background to the 1987 and 1988 floods. *Geographical Journal* 156, 12–22.

BRAMMER, H. 1990b: Floods in Bangladesh Part II: Flood mitigation and environmental aspects. *Geographical Journal* 156, 158–65.

BRAMMER, H. 1993: Geographical complexities of detailed impact assessments for the Ganges–Brahmaputra–Meghna delta of Bangladesh. In Warrick, R.A., Barrow, E.M. and Wigley, T.M.L. (eds), *Climate and sea level change*. Cambridge: Cambridge University Press, 246–62.

BRAY, M.J. 1992: Coastal sediment supply and transport. In Allison, R.J. (ed.), The coastal landforms of West Dorset. *Geologists' Association Guide* 47, 50–61.

BRAY, M.J., CARTER, D.J. and HOOKE, J.M. 1992: *Sea level rise and global warming: Scenarios, physical impacts and policies*. University of Portsmouth, Report to SCOPAC.

BRITSCH, L.D. and KEMP, E.B. 1990: *Land loss rates: Mississippi deltaic plain.* US Army Corps of Engineers Technical Report GL/90/2.

BROADUS, J.M. 1993: Possible impacts of, and adjustments to, sea level rise: the cases of Bangladesh and Egypt. In Warrick, R.A., Barrow, E.M. and Wigley, T.M.L. (eds), *Climate and sea level change.* Cambridge: Cambridge University Press, 263–75.

BROWN, A.C. and McLACHLAN, A. 1990: *Ecology of sandy shores.* Amsterdam: Elsevier.

BROWN, B.E. and SUHARSONO, T. 1990: Damage and recovery of coral reefs affected by El Niño related seawater warming in the Thousand Islands, Indonesia. *Coral Reefs* 8, 163–70.

BRUNSDEN, D. and JONES, D.K.C. 1976: The evolution of landslide slopes in Dorset. *Philosophical Transactions, Royal Society of London A* 283, 605–31.

BRUUN, P. 1962: Sea-level rise as a cause of shore erosion. *Proceedings, American Society of Civil Engineers. Journal of Waterways and Harbor Division* 88, 117–30.

BRUUN, P. 1983: Review of conditions for use of the Bruun rule of erosion. *Coastal Engineering* 7, 77–89.

BRYANT, E. 1985: Rainfall and beach erosion relationships, Stanwell Park, Australia, 1895–1980: worldwide implications for coastal erosion. *Zeitschrift für Geomorphologie Supplementband* 57, 51–65.

BRYANT, E. 1987: CO_2 warming, rising sea level and retreating coasts: review and critique. *Australian Geographer* 18, 101–13.

BRYCESON, I., DE SOUZA, T.F., JEHANGEER, I., NGOILE, M.A.K. and WYNTER, P. 1990: *State of the marine environment in the East African region.* UNEP Regional Seas Reports and Studies No. 113.

BUDDEMEIER, R.W. and KINZIE, R.A. 1976: Coral growth. *Annual Reviews in Oceanography and Marine Biology* 14, 138–225.

BURNS, K.A., GARRITY, S.J. and LEVINGS, S.C. 1993: How many years until mangrove ecosystems recover from catastrophic oil spills? *Marine Pollution Bulletin* 26, 239–48.

BUSH, S. 1992: Andrew shortens lifetime of Louisiana barrier islands. *EOS* 73 (47), 505.

BYRNE, J.V. 1964: An erosional classification for the northern Oregon coast. *Annals, Association of American Geographers* 54, 329–35.

CAMBERS, G. 1976: Temporal scales in coastal systems. *Transactions, Institute of British Geographers, NS* 1, 246–56.

CAPUTO, C., ALESSANDRO, L., LA MONICA, G.B., LANDINI, B. and LUPIA PALIERI, E. 1991: Present erosion and dynamics of Italian beaches. *Zeitschrift für Geomorphologie Supplementband* 81, 31–9.

CAREY, A.E. and OLIVER, F.W. 1918: *Tidal lands: A study of shore problems.* London: Blackie and Son.

CARTER, R.W.G. 1988: *Coastal environments: An introduction to the physical, ecological and cultural systems of coastlines.* London: Academic Press.

CARTER, C. and GUY, D.E. 1988: Coastal erosion: processes, timing and magnitudes at the bluff toe. *Marine Geology* 84, 1–17.

CARTER, R.W.G., and STONE, G.W. 1989: Mechanisms associated with the erosion of sand dune cliffs, Magilligan, Northern Ireland. *Earth Surface Processes and Landforms* 14, 1–10.

CARTER, R.W.G., HESP, P.A. and NORDSTROM, K.F. 1990: Erosional landforms in coastal dunes. In Nordstrom, K.F., Psuty, N.P. and Carter, R.W.G. (eds), *Coastal dunes.* Chichester: John Wiley, 217–50.

CHAPMAN, V.J. 1977a: Introduction. In Chapman, V.J. (ed.), *Wet coastal ecosystems.* Amsterdam: Elsevier.

CHAPMAN, V.J. 1977b: Wet coastal formations of Indo-Malesia and Papua New Guinea. In Chapman, V.J. (ed.), *Wet coastal ecosystems.* Amsterdam: Elsevier, 261–70.

CHENHALL, B.E., YASSINI, I. and JONES, B.G. 1992: Heavy metal concentrations in lagoonal salt marsh species, Illawarra region, south eastern Australia. *Science of the Total Environment* 125, 203–25.

CHRISTIANSON, C. and BOWMAN, D. 1981: Sea level changes, coastal dune building and sand drift, north western Jutland, Denmark. *Geographisches Tidsskrift* 86, 28–31.

CHUNG, C.-H. 1985: The effects of introduced spartina grass on coastal morphology in China. *Zeitschrift für Geomorphologie Supplementband* 57, 169–74.

CLARK, A.R. 1988: The use of Portland Stone armour in coastal protection and sea defence works. *Quarterly Journal of Engineering Geology* 21, 113–36.

CLARK, R.B. 1989: *Marine pollution.* 2nd edition. Oxford: Clarendon Press.

CLARK, J.A., FARRELL, W.E. and PELTIER, W.R. 1978: Global changes in postglacial sea level: a numerical calculation. *Quaternary Research* 9, 265–87.

COASTAL ENGINEERING RESEARCH CENTER 1984: *Shore protection manual.* 4th edition. Washington D.C.: US Government Printing Office.

CODIGNOTTO, J.O. and AGUIRRE, M.C. 1993: Coastal evolution, changes in sea level and molluscan fauna in north eastern Argentina during the late Quaternary. *Marine Geology* 110, 163–75.

COFFROTH, M.A., LASKER, H.R. and OLIVER, J.K. 1990: Coral mortality outside of the eastern Pacific during 1982–3: relationship to El Niño. In Glynn, P.W. (ed.), *Global ecological consequences of the 1982–3 El Niño Southern Oscillation.* Amsterdam: Elsevier, 141–82.

COLES, S.M. 1979: Benthic microalgal populations on intertidal sediments and their role as precursors to salt marsh development. In Jefferies, R.L. and Davy, A.J. (eds), Ecological processes in coastal environments. Oxford: Blackwell Scientific, 25–42.

COLGAN, M.W. 1990: El Niño and the history of eastern Pacific reef building. In Glynn, P.W. (ed.), *Global ecological consequences of the 1982–3 El Niño Southern Oscillation.* Amsterdam: Elsevier, 183–232.

COLINVAUX, P. 1986: *Ecology.* New York: John Wiley.

CONNELL, J.H. 1978: Diversity in tropical rain forests and coral reefs. *Science* 199, 1302–10.

COOPER, W.S. 1958: Coastal sand dunes of Oregon and Washington. *Geological Society of America, Memoirs* 72, 69pp.

CORRÉ, J.-J. 1992: Implications des changements climatiques. Etude de cas: Le Golfe du Lion (France). In Jeftic, L., Milliman, J.D. and Sestini, G. (eds), *Climatic change and the Mediterranean.* London: Edward Arnold, 328–427.

COTTON, C.A. 1951: Sea cliffs of Banks Peninsula and Wellington: some criteria for coastal classification. *New Zealand Geographer* 7, 103–20.

CRAFT, C.B., SENECA, E.D. and BROOME, S.W. 1993: Vertical accretion in microtidal regularly and irregularly flooded estuarine marshes. *Estuarine and Coastal Shelf Science* 37, 371–86.

CRAIG, P.C., GRIFFITHS, W.B. and JOHNSON, S.R. 1984: Trophic dynamics in an Arctic lagoon. In Barnes, P.W., Schell, D.M. and Reimnitz, E. (eds), *The Alaskan Beaufort Sea*. Orlando: Academic Press, 347–81.

CROTHERS, J.H. and HAYNES, S. 1994: Rocky shore distribution patterns along the Somerset coast. *Biological Journal of the Linnean Society* 51, 115–21.

DAHL, A.L. 1984: Oceania's most pressing environmental concerns. *Ambio* 13, 296–301.

DALE, J.E., AITKEN, A.E., GILBERT, R. and RISK, M.J. 1989: Macrofauna of Canadian arctic fiords. *Marine Geology* 85, 331–58.

DARBY, D.A. 1990: Evidence for the Hudson River as the dominant source of sand on the US continental shelf. *Nature* 346, 828–31.

DAVEY, P. 1993: *Spartina*: eliminating the root of the problem. *Enact* 1, 3, 7.

DAVIES, J.L. 1964: A morphogenetic approach to world shorelines. *Zeitschrift für Geomorphologie* 8, 127–42.

DAVIES, J.L. 1972: *Geographical variation in coastline development*. Edinburgh: Oliver and Boyd.

DAVIES, J.L. 1980: *Geographical variation in coastline development*. 2nd edition. Edinburgh: Oliver and Boyd.

DAVIES, P.J. and MONTAGGIONI, L.F. 1985: Reef growth and sea level change: the environmental signature. *Proceedings, 5th International Coral Reef Congress, Tahiti* 3, 477–515.

DAVIS, R.A. 1985: Beach and nearshore zone. In Davis, R.A. (ed.), *Coastal sedimentary environments*. 2nd edition. New York: Springer-Verlag, 379–444.

DAY, J.W., BUTLER, T.J. and CONNER, W.H. 1976: Productivity and nutrient export studies in a cyprus swamp and lake system in Louisiana. In Wiley, M. (ed.), *Estuarine processes II*. New York: Academic Press, 255–69.

DAYTON, P.K., ROBILIIARD, G.A. and DE VRIES, A.L. 1969: Anchor ice formation in McMurdo Sound, Antarctica and its biological effects. *Science* 163, 273–5.

DEAN, R.G. 1991: Equilibrium beach profiles: Characteristics and applications. *Journal of Coastal Research* 7, 53–84.

DEAN, R.G. and MAURMEYER, E.M. 1983: Models for beach profile response. In Komar, P.D. (ed.), *Handbook of coastal processes and erosion*. Boca Raton, Florida: CRC Press, 151–66.

DE GROOT, K. 1973: Geochemistry of tidal flat brines at Umm Said, S E Qatar, Persian Gulf. In Purser, B.H. (ed.), *The Persian Gulf*. Berlin: Springer-Verlag, 377–94.

D'ELIA, C.F., BUDDEMEIER, R.W. and SMITH, S.V. 1991: Workshop on coral bleaching, coral reef ecosystems and global changes, report of proceedings. College Park, Maryland: Maryland Sea Grant College Publication, 1–49.

DENNESS, B., CONWAY, B.W., McCANN, D.M. and GRAINGER, P. 1975: Investigation of a coastal landslip at Charmouth, Dorset. *Quarterly Journal of Engineering Geology* 8, 119–40.

DIAMENTE, J.M., PYLE, T.E., CARTER, W.E. and SCHERE, W. 1987: Global change and the measurement of absolute sea level. *Progress in Oceanography* 18, 1–21.

DICKS, B., BAKKE, T. and DION, I.M.T. 1988: Oil exploration and production and oil spills. In Salomons, W., Bayre, B.L., Duursma, E.K. and Forstner, U. (eds), *Pollution of the North Sea*. New York: Springer-Verlag, 524–37.

DIJKEMA, K.S. 1990: Salt and brackish marshes around the Baltic Sea and adjacent parts of the North Sea: Their vegetation and management. *Biological Conservation* 57, 191–210.

DIONNE, J.-C. 1989a: Boulder barricades in the St Lawrence estuary and gulf, Québec, Canada. (abstract) *Geoöko plus* 1, 75.

DIONNE, J.-C. 1989b: An estimate of shore ice action in a Spartina tidal marsh, St Lawrence Estuary, Québec, Canada. *Journal of Coastal Research* 5, 281–93.

DOAKE, C.S.M. and VAUGHAN, D.G. 1991: Break-up of Wordie Ice Shelf, Antarctica. In Kotlyakov, V.M., Ushakov, A. and Glazovsky, A. (eds), *Glaciers–ocean–atmosphere interactions*. IAHS Publication No. 208, 161–5.

DOLAN, R. 1971: Coastal landforms: crescentic and rhythmic. *Geological Society of America, Bulletin* 82, 177–80.

DOLAN, R. and DAVIS, R.E. 1992: An intensity scale for Atlantic coast north east storms. *Journal of Coastal Research* 8, 840–53.

DOLAN, R. and GODFREY, P. 1973: Effects of hurricane Ginger on the barrier islands of North Carolina. *Geological Society of America, Bulletin* 84, 1329–34.

DOLAN, R., VINCENT, L. and HAYDEN, B. 1974: Crescentic coastal landforms. *Zeitschrift für Geomorphologie* 18, 1–12.

DOLAN, R., HAYDEN, B., MAY, S.K. and MAY, P. 1982: Erosion hazards along the Mid-Atlantic coast. In Craig, R.G. and Craft, J.L. (eds), *Applied geomorphology*. London: Allen and Unwin, 165–80.

DONE, T.J. 1992: Phase shifts in coral reef communities and their ecological significance. *Hydrobiologia* 247, 121–32.

DOODY, J.P. 1989: Management for nature conservation. *Proceedings, Royal Society of Edinburgh B* 96, 247–65.

DOODY, J.P. 1992: Salt marsh conservation. In Allen, J.R.L. and Pye, K. (eds), *Saltmarshes: Morphodynamics, conservation and engineering significance*. Cambridge: Cambridge University Press, 80–114.

DOODY, J. 1993: Changing attitudes in coastal conservation. *Enact* 1, 3, 4–6.

DOORNKAMP, J.C., BRUNSDEN, D. and JONES, D.C. (eds), 1980: *Geology, geomorphology and pedology of Bahrain*. Norwich: GeoAbstracts.

DREWRY, D.J. 1991: The response of the Antarctic ice sheet to climatic change. Harris, C.M. and Stonehouse, B. (eds), *Antarctica and global change*. London: Belhaven Press, 90–106.

DUBINSKY, Z. (ed.) 1990: *Coral reefs*. Amsterdam: Elsevier.

DUBOIS, R.N.L. 1992: A re-evaluation of Bruun's rule and supporting evidence. *Journal of Coastal Research* 8, 618–28.

DUNBAR, R.B., LEVENTER, A.R. and STOCKTON, W.L.K. 1989: Biogenic sedimentation in McMurdo Sound, Antarctica. *Marine Geology* 85, 155–79.

DYER, K.R. 1986: *Coastal and estuarine sediment dynamics*. Chichester: Wiley.

EARNEY, F.C.F. 1990: *Marine mineral resources*. London: Routledge.

EBISEMIJU, F.S. 1987: An evaluation of factors controlling present rates of shoreline retrogradation in the western Niger Delta, Nigeria. *Catena* 14, 1–12.

EL-ASHRY, M.T. (ed.) 1977: Air photography and coastal problems. *Benchmark papers in geology*, Vol. 38. Stroudsburg PA: Hutchinson-Ross.

ELLISON, J.G. 1993: Mangrove retreat with rising sea level, Bermuda. *Estuarine, Coastal and Shelf Science* 37, 75–87.

ELVERHØI, A., PFIRMAN, S.L., SOLHEIM, A. and LARSSEN, B.B. 1989: Glaciomarine sedimentation in epicontinental seas exemplified by the northern Barents Sea. *Marine Geology* 85, 225–50.

EMBABI, N.S. 1993: Environmental aspects of geographical distribution of mangrove in the UAE. In Lieth, H. and Al Masoom, A. (eds), *Towards the rational use of high salinity tolerant plants*, Vol. 1. Amsterdam: Kluwer Academic, 45–58.

EMERY, K.O. and KUHN, G.G. 1980: Erosion of rocky shores at La Jolla, California. *Marine Geology* 37, 197–208.

EMERY, K.O. and KUHN, G.G. 1982: Sea cliffs: their processes, profiles and classification. *Geological Society of America, Bulletin* 93, 644–54.

FAIRBANKS, R.G. 1989: A 17000 year glacio-eustatic sea level record: influence of glacial melting rates on the Younger Dryas event and deep-ocean circulation. *Nature* 342, 637–42.

FAIRBRIDGE, R.W. 1983: Isostasy and eustasy. In Smith, D. and Dawson, A.G. (eds), *Shorelines and isostasy*. London: Academic Press, 3–26.

FARMER, B.H. 1983: *An introduction to South Asia*. London: Methuen.

FISHER, O. 1866: On the disintegration of a chalk cliff. *Geological Magazine* 3, 354–6.

FLATHER, R.A. and KHANDKER, H. 1993: The storm surge problem and possible effects of sea level change on coastal flooding in the Bay of Bengal. In Warrick, R.A., Barrow, E.M. and Wigley, T.M.L. (eds), *Climate and sea level change*. Cambridge: Cambridge University Press, 229–45.

FLOOD, P.G. 1986: Sensitivity of coral cays to climatic variations, southern Great Barrier Reef. *Coral Reefs* 5, 13–18.

FÖCKE, J.W. 1978: Limestone cliff morphology on Curaçao (Netherlands Antilles), with special attention to the origin of notches and vermetid/coralline algal surf benches ('cornices', 'trottoirs'). *Zeitschrift für Geomorphologie* 22, 329–49.

FOOKES, P.G. 1978: Middle East — inherent ground problems. *Quarterly Journal of Engineering Geology* 11, 33–49.

FORBES, D.L. and TAYLOR, R.B. 1994: Ice in the shore zone and the geomorphology of cold coasts. *Progress in Physical Geography* 18, 59–90.

FRENCH, J.R 1991: Eustatic and neotectonic controls on saltmarsh sedimentation. In Kraus, N.C., Gingerich, K.J. and Kriebel, D.L. (eds), *Coastal Sediments '91*. New York: American Society of Civil Engineers, 1223–36.

FRENCH, J.R. 1993: Numerical simulation of vertical marsh growth and adjustment to accelerated sea level rise, North Norfolk, UK. *Earth Surface Processes and Landforms* 18, 63–81.

FRENCH, J.R. 1994: Tide-dominated coastal wetlands and accelerated sea-level rise: a NW European perspective. *Journal of Coastal Research, Special Issue*, in press.

FRENCH, J.R. and SPENCER, T. 1993: Dynamics of sedimentation in a tide-dominated backbarrier saltmarsh, Norfolk, UK. *Marine Geology* 110, 315–31.

FRENCH, J.R., SPENCER, T. and REED, D.J. (eds) 1994a: Geomorphic response to sea level rise: existing evidence and future impacts. *Earth Surface Processes and Landforms*, in press.

FRENCH, J.R., SPENCER, T., MURRAY, A.L. and ARNOLD, N.S. 1994b: Geostatistical analysis of sediment deposition in two small tidal wetlands, Norfolk, UK. *Journal of Coastal Research*, in press.

FREY, R.W. and BASAN, P.B. 1985: Coastal salt marshes. In Davis, R.A. (ed.), *Coastal sedimentary environments*. 2nd edition. New York: Springer-Verlag, 225–301.

FRIHY, O.E. 1988: Nile delta shoreline changes: aerial photographic study of a 28 year period. *Journal of Coastal Research* 4, 597–606.

FROOMER, N. 1980: Morphological changes in some Chesapeake Bay tidal marshes resulting from accelerated soil erosion. *Zeitschrift für Geomorphologie Supplementband* 34, 242–54.

GABRIELIDES, G.P., ALZIEU, C., READMAN, J.W., BACLI, E., ABOUL DAHAB, U. and SALIHOGLU, I. 1990: MED POL survey of organotins in the Mediterranean. *Marine Pollution Bulletin* 21, 233–7.

GAGLIANO, S.M., MEYER-ARENDT, K.J. and WICKERM, K.M. 1981: Land loss in the Mississippi river deltaic plain. *Transactions, Gulf Coast Association of Geological Societies* 31, 295–300.

GALT, J.A., LEHR, W.J. and PAYTON, D.L. 1991: Fate and transport of the Exxon Valdez oil spill. *Environmental Science and Technology* 25, 202–9.

GALVIN, C.J. 1968: Breaker type classification on three laboratory benches. *Journal of Geophysical Research* 73, 3651–9.

GALVIN, C.J. 1972: Wave breaking in shallow water. In Meyer, R.E. (ed.), *Waves on beaches*. New York: Academic Press, 413–56.

GARES, P.A. 1992: Topographic changes associated with coastal dune blowouts at Island Beach State Park, New Jersey. *Earth Surface Processes and Landforms* 17, 587–604.

GARRITY, S.D. and LEVINGS, S.C. 1993: Effects of an oil spill on some organisms living on a mangrove (*Rhizophora mangle* L.) roots in low wave energy habitats in Caribbean Panama. *Marine Environmental Research* 35, 257–71.

GEISTER, J. 1977: The influence of wave exposure on the ecological zonation of Caribbean coral reefs. *Proceedings, 3rd International Coral Reef Symposium, Miami* 1, 23–9.

GEORGOS, D. and PERISSORAKIS, C. 1992: Implication of future climatic changes on the inner Thermaikos Gulf. In Jeftic, L., Milliman, J.D. and Sestini, G. (eds), *Climatic change and the Mediterranean*. London: Edward Arnold, 495–534.

GESAMP 1990: *The state of the marine environment*. Oxford: Blackwell Scientific.

GIESEN, W. 1993: Indonesian mangroves: an update on remaining area and main management issues. Paper presented at International Seminar on Coastal Zone Management of Small Island Ecosystems, Ambon, 10pp.

GLAZOVSKY, A.F., MACHERET, Y.Y. and MOSKALEVSKY, M.Y. 1991: Tidewater glaciers of Spitsbergen. In Kotlyakov, V.M., Ushakov, A. and Glazovsky, A. (eds), *Glaciers–ocean–atmosphere interactions*. IAHS Publication No. 208, 229–39.

GLAZOVSKY, A., NASKUND, J.-O. and ZALE, R. 1992: Deglaciation and shoreline displacement on Alexander Land, Franz Josef Land. *Geografisker Annaler* 74A, 283–93.

GLEASON, D.F. and WELLINGTON, G.M. 1993: Ultraviolet radiation and bleaching. *Nature* 365, 836–8.

GLYNN, P.W. 1973: Aspects of the ecology of coral reefs in the western Atlantic region. In Jones, O.A. and Endean, R. (eds), *Biology and geology of coral reefs*, Vol 2. New York: Academic Press, 271–324.

GLYNN, P.W. 1990: Coral mortality and disturbances to coral reefs in the tropical eastern Pacific. In Glynn, P.W. (ed.), *Global ecological consequences of the 1982–3 El Niño Southern Oscillation*. Amsterdam: Elsevier, 55–126.

GLYNN, P.W. 1993: Coral reef bleaching: ecological perspectives. *Coral Reefs* 12, 1–17.

GLYNN, P.W., CORTES, J., GUSANN, H.M. and RICHMOND, R.J. 1988: El Niño (1982–3) associated coral mortality and relationship to sea surface temperature deviations in the tropical eastern Pacific. *Proceedings, 6th International Coral Reefs Symposium, Townsville* 3, 237–43.

GODFREY, P.J. and GODFREY, M.M. 1973: Comparison of ecological and geomorphologic interactions between altered and unaltered barrier island systems in North Carolina. In Coates, D.R. (ed.), *Coastal geomorphology*. Binghampton: SUNY, 239–58.

GOLDBERG, E.D. 1994: *Coastal zone space — Prelude to conflict?* Paris: UNESCO.

GOLDSMITH, F.B. 1975: The sea cliff vegetation of Shetland. *Journal of Biogeography* 2, 297–308.

GOLDSMITH, F.B. 1977: Rocky cliffs. In Barnes, R.S.K. (ed.), *The coastline*. Chichester: Wiley, 237–51.

GOLDSMITH, V. 1985: Coastal dunes. In Davis, R.A. (ed.), *Coastal sedimentary environments*. 2nd edition. New York: Springer-Verlag, 171–236.

GOLDSMITH, V. 1989: Coastal sand dunes as geomorphic systems. *Proceedings, Royal Society of Edinburgh B* 96, 3–15.

GOLDSMITH, V., ROSEN, P. and GERTNER, Y. 1990: Eolian transport measurements, winds and comparisons with theoretical transport in Israeli coastal dunes. In Nordstrom, K.F., Psuty, N.P. and Carter, R.W.G. (eds), *Coastal dunes*. Chichester: Wiley, 79–101.

GOLIK, A. and GERTNER, Y. 1992: Litter on the Israeli coastline. *Marine Environmental Research* 33, 1–15.

GORDON, D.M. 1988: Disturbance to mangroves in tropical-arid Western Australia: hypersalinity and restricted tidal exchange as factors leading to mortality. *Journal of Arid Environments* 15, 117–45.

GOREAU, T.F. 1959: The physiology of skeletal formation in corals. I. A method for measuring the rate of calcium deposition by corals under different conditions. *Biological Bulletin* 116, 59–75.

GOREAU, T.F. and GOREAU, N.L. 1959: The physiology of skeleton formation in corals II. Calcium deposition by hermatypic corals under various conditions in the reef. *Biological Bulletin* 117, 239–50.

GOREAU, T.F., GOREAU, N.L. and GOREAU, T.J. 1979: Corals and coral reefs. *Scientific American*, 238, 111–20.

GOUDIE, A.S. 1985: *Salt weathering*. School of Geography, University of Oxford Research Paper No. 33, 31pp.

GOUDIE, A.S. 1990: *The landforms of England and Wales*. Oxford: Basil Blackwell.

GOUDIE, A.S. (ed.) 1994: *Encyclopaedic dictionary of physical geography*. 2nd edition. Oxford: Basil Blackwell.

GRAINGER, P. and KALAUGHER, P.G. 1987: Intermittent surging movements of a coastal landslide. *Earth Surface Processes and Landforms* 12, 597–603.

GRANT, A. and MIDDLETON, R. 1990: An assessment of metal contamination of sediments in the Humber estuary, UK. *Estuarine, Coastal and Shelf Science* 31, 71–85.

GREEN, C.P. and McGREGOR, D.F.M. 1990: Orfordness: Geomorphological conservation perspectives. *Transactions, Institute of British Geographers* 15, 48–59.

GREGORY, J.W. 1913: *The nature and origin of fiords*. London: John Murray.

GRIEVE, H. 1959: *The great tide*. Chelmsford: Essex County Council.

GRIFFIN, D.A. and LE BLOND, P.H. 1990: Estuarine/ocean exchange controlled by spring-neap tidal mixing. *Estuarine, Coastal and Shelf Science* 30, 275–97.

GRIFFITH, T.W. and ANDERSON, J.B. 1989: Climatic controls of sedimentation in bays and fjords of the northern Antarctic peninsula. *Marine Geology* 85, 181–204.

GRIGG, R.W. 1982: Darwin point: a threshold for atoll formation. *Coral Reefs* 1, 29–34.

GRIGG, R.W. 1992: Coral reef environmental science: truth versus the Cassandra syndrome. *Coral Reefs* 11, 183–6.

GRIGG, R.W. 1994: Science management of the world's fragile coral reefs. *Coral Reefs* 13, 1.

GROVE, J.M. 1988: *The Little Ice Age*. London: Methuen.

GUILCHER, A. 1953: Éssai sur la zonation et la distribution des formes littorales de dissolution du calcaire. *Annales de Geographie* 62, 161–79.

GUILCHER, A. 1980: Observations géomorphologiques sur des littoraux subarctiques de la pointe nord de l'ile Sakhaline (extreme-orient Sovietique). *Revue de Géomorphologie Dynamique* 29, 101–15.

GUILCHER, A. 1981: Cryoplanation littorale et cordons glaciers de basse mer dans la region de Rimousk, cote sud de l'estuaire du St Laurent, Quebec. *Geographie Physique et Quaternaire* 35, 155–69.

GUILCHER, A. 1985: Retreating cliffs in the humid tropics: an example from Paraiba, Northeastern Brazil. *Zeitschrift für Geomorphologie Supplementband* 57, 95–103.

GUILCHER, A., BODERE, J.-C., COUDE, A., HANSOM, J.D., MOIGN, A. and PEULVAST, J.-P. 1986: Le probleme des strandflats en cinq pays de hautes latitudes. Revue de Geologie Dynamique et de Geographie Physique 27, 47–79.

GUILCHER, A. 1988: *Coral reef geomorphology*. Chichester: Wiley.

GUILLÉN, J. and PALANQUES, A. 1993: Longshore bar and trough systems in a microtidal, storm-wave dominated coast — the Ebro delta (Northwestern Mediterranean). *Marine Geology* 115, 239–52.

HALLOCK, P., HINE, A.C., VARGO, G.A., ELROD, J.A. and JAAP, W.C. 1988: Platforms of the Nicaraguan Rise: examples of sensitivity of carbonate sedimentation to excess trophic resources. *Geology* 16, 1104–7.

HANDS, E.B. 1980: Prediction of shore retreat and nearshore profile adjustments to rising sea levels on the Great Lakes. Coastal Engineering Research Centre, *Technical Memorandum* 80–7, 119pp.

HANDS, E.B. 1983: The Great Lakes as a test model for profile responses to sea level changes. In Komar, P.D. (ed.), *Handbook of coastal processes and erosion*. Boca Raton, Florida: CRC Press, 176–89.

HANSEN, J.R. and INGÓLFSSON, A. 1993: Patterns in species composition of rocky shore communities in sub-arctic fjords of eastern Iceland. *Marine Biology* 117, 469–81.

HANSOM, J.D. 1983: Shore platform development in the South Shetland Islands, Antarctica. *Marine Geology* 53, 211–29.

HANSOM, J.D. 1986: Coastal zone management in the United Kingdom. In Ritchie, W., Stone, J.C. and Mather, A.S. (eds), *Essays for Professor R E H Mellor*. Aberdeen: Aberdeen University Press, 323–32.

HANSOM, J.D. and KIRK, R.M. 1989: Ice in the intertidal zone: examples from Antarctica. *Essener Geographische Arbeiten* 18, 211–36.

HARDISTY, J. 1990: *Beaches: form and process*. London: Unwin Hyman.

HARDISTY, J. 1994: Beach and nearshore sediment transport. In Pye, K. (ed.), *Sediment transport and depositional processes*. Oxford: Blackwell Scientific, 219–55.

HARDOY, J.E., MITLIN, D. and SATTERTHWAITE, D. 1992: *Environmental problems in Third World cities*. London: Earthscan.

HARMSWORTH, G.C. and LONG, S.P. 1986: An assessment of saltmarsh erosion in Essex, England with reference to the Dengie Peninsula. *Biological Conservation* 35, 377–87.

HARPER, J.R. 1990: Morphology of the Canadian Beaufort Sea coast. *Marine Geology* 91, 75–91.

HARRISON, S. 1992: A large calving event of Ventisquero San Rafael, southern Chile. *Annals of Glaciology* 38, 208–9.

HAWKINS, J.P. and ROBERTS, C.M. 1993: Effects of recreational SCUBA diving on coral reefs: trampling on reef flat communities. *Journal of Applied Ecology* 30, 25–30.

HAYES, M.O. 1979: Barrier island morphology as a function of tidal and wave regime. In Leatherman, S.P. (ed.), *Barrier islands*. New York: Academic Press, 1–27.

HEIN, F.J. and SYVITSKI, J.P.M. 1989: Sea-floor gouges and pits in deep fjords, Baffin Island. Possible mammalian feeding traces. *Geo-Marine Letters* 9, 91–4.

HEINE, J.N. 1989: Effects of ice scour on the structure of sublittoral marine algal assemblages of St Lawrence and St Matthew Islands, Alaska. *Marine Ecology Progress Series* 52, 253–60.

HENDERSON-SELLERS, A. and ROBINSON, P.J. 1986: *Contemporary climatology*. London: Longman Scientific and Technical.

HEQUETTE, A. and BARNES, P.W. 1990: Coastal retreat and shoreface profile variations in the Canadian Beaufort Sea. *Marine Geology* 91, 113–32.

HERNANDEZ-AVILA, M.L., ROBERTS, H.H. and ROUSE, L.J. 1977: Hurricane-generated waves and coastal boulder rampart formation. *Proceedings, 3rd International Coral Reef Symposium, Miami* 2, 71–8.

HESP, P.A. 1983: Morphodynamics of incipient foredunes in New South Wales, Australia. In Brookfield, M. and Ahlbrant, T. (eds), *Aeolian processes and sediments*. Amsterdam: Elsevier, 325–42.

HESP, P.A. 1988: Surfzone, beach and foredune interactions on the Australian south east coast. *Journal of Coastal Research, Special Issue* 3, 15–25.

HESP, P.A. 1989: A review of biological and geomorphological processes involved in the initiation and development of incipient foredunes. *Proceedings, Royal Society of Edinburgh B* 96, 181–201.

HESP, P.A. 1991: Ecological processes and plant adaptations on coastal dunes. *Journal of Arid Environments* 21, 165–92.

HIGHSMITH, R.C., RIGGS, A.C. and ANTONIO, C.M. 1980: Survival of hurricane-generated coral fragments and a disturbance model of reef calcification/growth rates. *Oecologica* 46, 322–9.

HILTON, M.J. 1989: Management of the New Zealand coastal sand mining industry: Some implications of the Pakiri coastal sand body. *New Zealand Geographer* 45, 14–25.

HINRICHSEN, D. 1990: *Our common seas: Coasts in crisis.* London: Earthscan.

HODGKIN, E.P. 1970: Geomorphology and biological erosion of limestone coasts in Malaysia. *Bulletin, Geological Society of Malaysia* 3, 27–51.

HOEKSTRA, P. 1993: Late Holocene development of a tide induced elongate delta, the Solo delta, East Java. *Sedimentary Geology* 83, 211–33.

HOFFMAN, J.S., KEYES, D. and TITUS, J.G. 1983: *Projecting future sea level rise: Methodology, estimates to the year 2100, and research needs.* United States Environmental Protection Agency Report 230–09–007. Washington DC: Government Printing Office.

HOFFMAN, J.S., WELLS, J.B. AND TITUS, J.G. 1986: Future global warming and sea level rise. In: Sigbjarnason, G. (ed.), *Iceland coastal and river symposium*, Reykavik, Natural Energy Authority, 245–66

HOLLAND, A.F., ZINGMERK, R.B. and DEAN, J.M. 1974: Quantitative evidence concerning the stabilisation of sediment by benthic diatoms. *Marine Biology* 27, 191–6.

HOLLIGAN, P.M. and REINERS, W.A. 1992: Predicting the responses of the coastal zone to global change. In Woodward, F.I. (ed.), *Ecological consequences of global climate change.* Advances in Ecological Research Vol. 22. London: Academic Press, 211–55.

HOPLEY, D. 1982: *Geomorphology of the Great Barrier Reef.* New York: Wiley Interscience.

HOPLEY, D. 1988: Anthropogenic influences on Australia's Great Barrier Reef. *Australian Geographer* 19, 26–45.

HOTTEN, R.D. 1988: Sand mining on Mission Beach, San Diego, California. *Shore and Beach* 56, 18–21.

HOYT, J.H. 1967: Barrier island formation. *Geological Society of America Bulletin* 78, 1125–36.

HUDSON, B.J. 1980: Anthropogenic coasts. *Geography* 65, 194–202.

HUGHES, T.P. 1983: Evolutionary ecology of colonial reef organisms, with particular reference to corals. *Biological Journal, Linnean Society* 20, 39–58.

HUGHES, T.P. 1989: Community structure and diversity of coral reefs: the role of history. *Ecology* 70, 275–9.

HUNG, T.C. and HAN, B.C. 1992: Relationships among the species of Cu, organic compounds and bioaccumulation along the mariculture area of Taiwan. *Science of the Total Environment* 125, 359–72.

HUTCHINSON, J.N. 1983: A pattern in the incidence of major coastal landslides. *Earth Surface Processes and Landforms* 8, 391–8.

HUTCHINSON, S.M. 1993: The magnetic record of particulate pollution in a saltmarsh, Dee Estuary, UK. *The Holocene* 3, 342–50.

HUTCHINSON, J.N., BROMHEAD, E.N. and LUPINI, J.F. 1980: Additional observations on the Folkestone Warren landslides. *Quarterly Journal of Engineering Geology* 13, 1–3.

IBE, A.C. 1988: Nigeria. In Walker, H.J. (ed.), *Artificial structures and shorelines.* Dordrecht: Kluwer Academic, 287–94.

IMBERT, D. and PORTECOP, J. 1986: Étude de la production de litière dans la mangrove de Guadeloupe (Antilles francaises). *Acta Oecologica/Oecologia Plantarum* 7, 379–96.

INMAN, D.L. and NORDSTROM, K.F. 1971: On the tectonic and morphologic classification of coasts. *Journal of Geology* 79, 1–21.

JACOBSON, H.A. 1988: Historical development of the saltmarsh at Wells, Maine. *Earth Surface Processes and Landforms* 13, 475–86.

JAGGER, K.A., PSUTY, N.P. and ALLEN, J.R. 1991: Caleta morphodynamics, Perdido Key, Florida, USA. *Zeitschrift für Geomorphologie Supplementband* 81, 99–113.

JAMES, N.P. and GINSBURG, R.N. 1979: The seaward margin of Belize barrier and atoll reefs. *International Association of Sedimentologists Special Publication 3*. Oxford: Blackwell Scientific.

JAMES, P.A., WHARFE, A.S., PEGG, R.K. and CLARKE, D. 1986: A cation budget analysis of a coastal dune system in north west England. *Catena* 13, 1–10.

JEFTIC, L., MILLIMAN, J.D. and SESTINI, G. (eds) 1992: *Climatic change and the Mediterranean.* London: Edward Arnold.

JELGERSMA, S. and SESTINI, G. 1992: Implications of a future rise in sea level on the coastal lowlands of the Mediterranean. In Jeftic, L., Milliman, J.D. and Sestini, G. (eds), *Climatic change and the Mediterranean.* London: Edward Arnold, 282–303.

JENNER, K.A. and HILL, P.R. 1991: Sediment transport at the Mackenzie Delta-Beaufort Sea interface. In Marsh, P. and Ommanney, C.S.L. (eds), *Mackenzie Delta.* Saskatoon: Environmental Canada, 39–52.

JERWOOD, L.C., ROBINSON, D.A. and WILLIAMS, R.B.G. 1990a: Experimental frost and salt weathering of chalk. I. *Earth Surface Processes and Landforms* 15, 611–24.

JERWOOD, L.C., ROBINSON, D.A. and WILLIAMS, R.B.G. 1990b: Experimental frost and salt weathering of chalk. II. *Earth Surface Processes and Landforms* 15, 699–708.

JIN LIU, YU ZHI-YING and CHEN DE CHANG 1987: The formation and erosion of the old Yellow river submerged delta in northern Jiangsi Province. In Gardiner, V. (ed.), *International Geomorphology 1986*. Chichester: Wiley, 999–1008.

JOHANNES, R.E. *et al.* 1972: The metabolism of some coral reef communities: a team study of nutrient and energy flux at Eniwetok. *Bioscience* 22, 541–3.

JOHN, B.S. and SUGDEN, D.E. 1975: Coastal geomorphology of high latitudes. *Progress in Physical Geography* 7, 53–132.

JOHNSON, D.W. 1919: *Shore processes and shoreline development.* New York: Wiley.

JOKIEL, P.L. and COLES, S.L. 1990: Response of Hawaiian and other Indo-Pacific reef corals to elevated temperature. *Coral Reefs* 8, 155–62.

JONES, M.A.J. and BACON, P.R. 1990: Beach tar contamination in Jamaica. *Marine Pollution Bulletin* 21, 331–4.

JONES, A.M. and BAXTER, J.M. 1985: The use of *Patella vulgata* L. in rocky shore surveillance. In Moore, P.G. and Seed, R. (eds), *The ecology of rocky coasts*. London: Hodder and Stoughton, 265–73.

KAMALUDIN, B.H. 1993: The changing mangrove shorelines in Kuala Kuran, Peninsular Malaysia. *Sedimentary Geology* 83, 187–93.

KANA, W. and AL-SARAWI, M. 1988: Kuwait. In Walker, H.J. (ed.), *Artificial structures and shorelines*. Dordrecht: Kluwer Academic, 264–8.

KANA, T.W., MICHEL, J., HAYES, M.O. and JENSEN, J.R. 1984: The physical impact of sea level rise in the area of Charleston, South Carolina. In Barth, M.C. and Titus, J.G. (eds), *Greenhouse effect and sea level rise*. New York: Van Nostrand Reinhold, 105–50.

KANWISHER, J.W. and WAINWRIGHT, S.A. 1967: Oxygen balance in some reef corals. *Biological Bulletin* 133, 378–90.

KATZ, A. 1989: Coastal resource management in Belize: Potentials and problems. *Ambio* 18, 139–40.

KELLETAT, D. 1980: Formenschatz und Prozessgefuge des 'biokarstes' an der kuste von Nordost-Mallorca (Cala Guya). *Berliner Geographische Studien* 7, 99–113.

KELLETAT, D. 1985: Bio-destructive und bio-konstructive formelemente an den Spanischen Mittelmeerkusten. *Geoökodynamik* 6, 1–20.

KELLETAT, D. 1988: Quantitative investigations on coastal bioerosion in higher latitudes: an example from northern Scotland. *Geoökodynamik* 9, 41–51.

KELLETAT, D. 1989: Biosphere and man as agents in coastal geomorphology and ecology. *Geoökodynamik* 10, 215–52.

KELLETAT, D. 1992: Coastal erosion and protection measures at the German North Sea coast. *Journal of Coastal Research* 8, 699–711.

KEMP, P.H. 1975: Wave asymmetry in the nearshore zone and breaker area. In Hails, J. and Carr, A.P. (eds), *Nearshore sediment dynamics and sedimentation*. Chichester: John Wiley, 47–67.

KIDSON, C and CARR, A.P. 1960: Dune reclamation at Braunton Burrows, Devon. *The Chartered Surveyor* Dec. 1960, 3–8.

KIDSON, C., COLLIN, R.C. and CHISHOLM, N.W.T. 1989: Surveying a major dune system — Braunton Burrows, North west Devon. *Geographical Journal* 155, 94–105.

KINSEY, D.W. 1983: Standards of performance in coral reef primary production and carbon turnover. In Barnes, D.J. (ed.), *Perspectives on coral reefs*. Canberra: Clouston/AIMS, 209–20.

KINSMAN, B. 1965: *Wind waves, their generation and propagation on the ocean surface*. Englewood Cliffs, New Jersey: Prentice-Hall.

KIRK, R.M. 1980: Mixed sand and gravel beaches: morphology, processes and sediments. *Progress in Physical Geography* 4, 189–210.

KIRKBY, M.J. 1984: Modelling cliff development in South Wales: Savigear reviewed. *Zeitschrift für Geomorphologie* 28, 405–26.

KLEIN, G. DE V. 1985: Intertidal flats and intertidal sand bodies. In Davis, R.A. (ed.), *Coastal sedimentary environments*. New York: Springer-Verlag, 187–224.

KNOWLTON, N., LAN, J.C. and KELLER, B.D. 1990: Case study of natural population collapse: Post-hurricane predation on Jamaican staghorn corals. *Smithsonian Contributions to the Marine Sciences* 31, 1–25.

KNOX, G.A. 1983: The living resources of the Southern Ocean: A scientific overview. In Vicuña, F.O. (ed.), *Antarctic resources policy*. Cambridge: Cambridge University Press, 21–60.

KOCH, M.S., MENDELSOHN, I.A. and McKEE, K.L. 1990: Mechanism for the hydrogen sulphide-induced growth limitation in wetland macrophytes. *Limnology and Oceanography* 35, 399–408.

KOMAR, P.D. 1976: *Beach processes and sedimentation*. Englewood Cliffs, New Jersey: Prentice-Hall.

KOMAR, P.D. 1983a: Beach processes and erosion: An introduction. In Komar, P.D. (ed.), *CRC Handbook of coastal processes and erosion*. Boca Raton, Florida: CRC Press, 1–20.

KOMAR, P.D. 1983b: The erosion of Siletz Spit, Oregon. In Komar, P.D. (ed.), *CRC Handbook of coastal processes and erosion*. Boca Raton, Florida: CRC Press, 65–76.

KOMAR, D. and SHIH, S.-M. 1993: Cliff erosion along the Oregon coast: A tectonic-sea level imprint plus local controls by beach processes. *Journal of Coastal Research* 9, 747–65.

KRAFT, J.C., ALLEN, E.A., BELKNAP, D.F., JOHN, C.J. and MAURMEYER, E.M. 1979: Processes and morphologic evolution of an estuarine and coastal barrier system. In Leatherman, S.P. (ed.), *Barrier islands*. New York: Academic Press, 149–84.

KRIEBEL, D.L. and DEAN, R.G. 1985: Numerical simulation of time-dependent beach and dune erosion. *Coastal Engineering* 9, 221–45.

KÜHLMANN, D.H.H. 1988: The sensitivity of coral reefs to environmental pollution. *Ambio* 17, 13–21.

KUHN, G.G. and SHEPARD, F.P. 1984: *Sea cliffs, beaches and coastal valleys of San Diego County*. Berkeley: University of California Press.

LANG, J. 1973 Interspecific aggression by scleractinian corals. 2. why the race is not only to the swift. *Bulletin of Marine Science* 23, 260–79.

LAVRENTIDES, G. 1993: Dry coastal ecosystems of Greece. In Maarel, E. van der (ed.), *Dry coastal ecosystems*. Amsterdam: Elsevier, 429–42.

LEATHERMAN, S.P. 1989: Impacts of accelerated sea level rise on beaches and coastal wetlands. In White, J.C. (ed.), *Global climatic change linkages*. Amsterdam: Elsevier, 43–57.

LEATHERMAN, S.P. 1993: Coastal change. In Gurney, R.J., Foster, J.L. and Parkinson, C.C. (eds), *Atlas of satellite observations related to global change*. Cambridge: Cambridge University Press, 327–40.

LE CAMPION-ALSUMARD, T. 1975: Étude experimentale de la colonisation d'éclats de calcite par les cyanophycées endolithes marines. *Cahiers Biologie Marine* 16, 177–85

LE CAMPION-ALSUMARD, T. 1979: Le biokarst marin: role des organismes perforants. Actes du Symposium International sur l'érosion karstique, UIS, Aix en provence, 133–40.

LE CAMPION-ALSUMARD, T., ROMANO, J.C., PEYROT CLAUSADE, M., LE CAMPION, J. and PAUL, R. 1993: Influence of some coral reef communities on the calcium carbonate budget of Tiahura Reef (Moorea, French Polynesia). *Marine Biology* 115, 685–93.

LEWIS, J.R. 1964: *The ecology of rocky shores*. London: English University Press.

LEWIS, J.R. 1977: Rocky foreshores. In Barnes, R.S.K. (ed.), *The coastline*. London: Wiley, 147–58.

LEY, R.G. 1977: The influence of lithology on marine karren. *Abhandlungen zur karst und Hohlenkunde Reiche A, Heft* 15, 81–100.

LEY, R.G. 1979: The development of marine karren along the Bristol Channel coastline. *Zeitschrift für Geomorphologie Supplementband* 32, 75–89.

LINDÉN, O. 1990: Human impact on tropical coastal zones. *Nature and Resources* 26, 3–11.

LINS, H.F. 1985: Storm-generated variations in nearshore beach topography. *Marine Geology* 62, 13–29.

LITTLE, C. and METTAM, C. 1994: Rocky shore zonation in the Rance tidal power basin. *Biological Journal of the Linnean Society* 51, 169–82.

LMER CO-ORDINATING COMMITTEE 1992: Understanding changes in coastal environments: The LMER program. *EOS, Transactions American Geophysical Union* 73, 481–5.

LORING, D.H. and ASMUND, G. 1989: Heavy metal contamination of a Greenland fjord system by mine wastes. *Environmental Geology and Water Science* 14, 61–71.

LOVRIC, A.Z. and USLU, T. 1993: Dry coastal ecosystems of Turkey. In Maarel, E. van der (ed.), *Dry coastal ecosystems*. Amsterdam: Elsevier, 443–62.

LOYA, Y. 1976: Recolonisation of Red Sea corals affected by natural catastrophes and man made perturbations. *Ecology* 57, 278–89.

LUCAS, P.H.L. 1972: *Protected landscapes*. London: Chapman and Hall.

LUKAS, K.J. 1979: The effects of marine microphytes on carbonate substrata. *Scanning Electron Microscopy 1979* II, 447–55.

LUSCZYNSKI, N.J. and SWARZENSKI, W.V. 1966: Salt-water encroachment in southern Nassau and southern Queens counties, Long Island, New York. *USGS Water Supply Paper*, 1–76.

McCANN, A.M. 1994: The Roman port of Cosa. *Scientific American*, June 1994, 92–9.

McCANN, S.B. and DALE, J.E. 1986: Sea-ice break up and tidal flat processes, Frobisher Bay, Baffin Island. *Physical Geography* 7, 168–80.

McCAVE, I.N. 1970: Deposition of fine-grained suspended sediment from tidal currents. *Journal of Geophysical Research* 75, 4151–8.

McDOWELL, D.M. and O'CONNOR, B.A. 1977: *Hydraulic behaviour of estuaries*. London: Macmillan.

McGOWN, A., ROBERTS, A.G. and WOODROW, L.K.R. 1988: Geotechnical and planning aspects of coastal landslides in the UK. In Bonnard, C. (ed.), *Landslides*, Vol. 2. Rotterdam: Balkema, 1201–6.

McGREEVY, J.P. 1985: A preliminary scanning electron microscope study of honeycomb weathering of sandstone in a coastal environment. *Earth Surface Processes and Landforms* 10, 509–18.

McINTIRE, W.G. and WALKER, H.J. 1964: Tropical cyclones and coastal morphology in Mauritius. *Annals, American Association of Geographers* 54, 582–96.

McKAY, P.J. and TERICH, J.A. 1992: Gravel barrier morphology: Olympic National Park, Washington State, USA. *Journal of Coastal Research* 8, 813–29.

McLACHLAN, A. 1988: Dynamics of an exposed beach/dune coast, Algoa Bay, South East Africa. *Journal of Coastal Research, Special Issue* 3, 91–5.

McLACHLAN, A. 1990: The exchange of materials between beach and dune systems. In Nordstrom, K.F., Psuty, N.P. and Carter, R.W.G. (eds), *Coastal dunes*. Chichester: Wiley, 201–16.

McLACHLAN, A. 1991: Ecology of coastal dune fauna. *Journal of Arid Environments* 21, 229–44.

McROY, C.P. and HELFFERICH, C. (eds) 1977: *Seagrass ecosystems*. New York: Marcel Dekker.

MAHTAB, F.U. 1992: The delta regions and global warming: impact and response strategies for Bangladesh. In Schmandt, J. and Clarkson, J. (eds), *The regions and global warming*. New York: Oxford University Press, 28–43.

MALLOCH, A.J.C. 1993: Dry coastal ecosystems of Britain: Cliffs. In Maarel, E. van der (ed.), *Dry coastal ecosystems*. Amsterdam: Elsevier, 229–44.

MANN, K.E. 1982: *Ecology of coastal waters: A systems approach*. Oxford: Blackwell Scientific.

MARIÑO, M.J. 1992: Implications of climatic change on the Ebro delta. In Jeftic, L., Milliman, J.D. and Sestini, G. (eds), *Climatic change and the Mediterranean*. London: Edward Arnold, 304–27.

MARSH, W.M. 1983: *Landscape planning*. Reading, Mass.: Addison-Wesley.

MARTIN, M.H. 1990: A history of *Spartina* on the Avon coast. *Proceedings, Bristol Naturalists' Society* 50, 47–56.

MARTIN, M.H. and BECKETT, C.L. 1990: Heavy metal pollution in the Severn Estuary. *Proceedings, Bristol Naturalists' Society* 50, 105–12.

MARTINI, J.P. (ed.) 1986: *Canadian inland seas*. Amsterdam: Elsevier.

MASSELINK, G. and SHORT, A.D. 1993: The effect of tide range on beach morphodynamics and morphology: A conceptual beach model. *Journal of Coastal Research* 9, 785–806.

MASSON, P.J. 1990: The interactions of the physical and biotic components on a reclaimed coastal foredune. *Journal of Arid Environments* 19, 251–9.

MATHEWSON, C.C. and COLE, W.L 1982: Geomorphic processes and land use planning, South Texas barrier islands. In Craig, R.G. and Craft, J.L. (eds), *Applied geomorphology*. London: Allen and Unwin, 131–47.

MATSUKURA, Y. and MATSUOKA, N. 1991: Rates of tafoni weathering on uplifted shore platforms in Nojima-zaki, Boso Peninsula, Japan. *Earth Surface Processes and Landforms* 16, 51–6.

MAY, V.J. 1977: Earth cliffs. In Barnes, R.S.K. (ed.), *The coastline*. London: John Wiley, 215–35.

MEGANCK, R.A. 1991: Coastal parks as development catalysts: A Caribbean example. *Ocean and Shoreline Management* 15, 25–36.

MEIER, M.F. 1984: Contributions of small glaciers to global sea level. *Science* 226, 1418–21.

MESSIER, D., INGRAM, R.G. and ROY, D. 1986: Physical and biological modifications in response to La Grande hydroelectric complex. In Martini, J.P. (ed.), *Canadian inland seas*. Amsterdam: Elsevier, 403–24.

MIDDLEMISS, F.A. 1983: Instability of chalk cliffs between the South Foreland and Kingsdown, Kent, in relation to geological structure. *Proceedings, Geologists' Association* 94, 115–22.

MILLER, L., CHENEY, R.F. and DOUGLAS, B.L. 1988: GEOSAT altimeter observations of Kelvin waves and the 1986–7 El Niño. *Science* 239, 52–4.

MILLIMAN, J.D. 1973: Caribbean coral reefs. In Jones, O.A. and Endean, R. (eds), *Biology and geology of coral reefs*, Vol. 1. New York: Academic Press, 1–50.

MILLIMAN, J.D., JEFTIC, L. and SESTINI, G. 1992: The Mediterranean Sea and climatic change — An overview. In Jeftic, L., Milliman, J.D. and Sestini, G. (eds), *Climatic change and the Mediterranean*. London: Edward Arnold, 1–14.

MOLLAH, M.A. 1993: Geotechnical conditions of the deltaic alluvial plains of Bangladesh and associated problems. *Engineering Geology* 36, 125–40.

MOLNIA, B.F. 1985: Processes on a glacier-dominated coast, Alaska. *Zeitschrift für Geomorphologie Supplementband* 57, 141–53.

MOLVAER, J. and SKEI, J. 1988: Fjords. In Salomons, W., Bayne, B.L., Duursma, E.K. and Forstner, V. (eds), *Pollution of the North Sea*. Berlin: Springer-Verlag, 474–88.

MOREIRA, M.E.S.A. 1992: Recent salt marsh changes and sedimentation rate in the Sado estuary, Portugal. *Journal of Coastal Research* 8, 631–40.

MORTON, R.A. 1979: Temporal and spatial variations in shoreline changes and their implications, examples from the Texas Gulf coast. *Journal of Sedimentary Petrology* 49, 1101–12.

MORTON, R.A. 1988: Interactions of storms, seawalls and beaches of the Texas coast. In Kraus, N.C. and Pilkey, O.H. (eds), The effects of seawalls on the beach. *Journal of Coastal Research, Special Issue* 4, 113–34.

MOTTERSHEAD, D.N. 1989: Rates and patterns of bedrock denudation by coastal salt spray weathering: A seven year record. *Earth Surface Processes and Landforms* 14, 383–98.

MUENCH, R.D., PEASE, C.H. and SALO, S.A. 1991: Oceanographic and meteorological effects on autumn sea ice distribution in the western Arctic. *Annals of Glaciology* 15, 171–7.

MUNK, W.H. and SARGENT, M.S. 1954: Adjustment of Bikini Atoll to ocean waves. *US Geological Survey Professional Paper* 260C, 275–80.

MURTY, T.S., FLATHER, R.A. and HENRY, R.F. 1986: The storm surge problem in the Bay of Bengal. *Progress in Oceanography* 16, 195–233.

MUSTOE, G.E. 1982: The origin of honeycomb weathering. *Geological Society of America, Bulletin* 93, 108–15.

NAKAMURA, M. 1985: Assessment of land-based sources of pollution in the East Asian Seas: preliminary assessment results and the prospects for full assessment exercise. In Kato, I., Kumamoto, N., Matthews, W.H. and Suhaimi, A. (eds), *Environmental pro-*

tection and coastal zone management in Asia and the Pacific. Tokyo: University of Tokyo Press, 183–201.

NAKASHIMA, L.D. and MOSSA, J. 1991: Responses of natural and seawall-backed beaches to recent hurricanes on the Bayou Laforche headland, Louisiana. *Zeitschrift für Geomorphologie* 35, 239–56.

NELSON, J.G. 1993: Conservation and use of Mai Po marshes, Hong Kong. *Natural Areas Journal* 13, 215–19.

NEUMANN, A.C. and MACINTYRE, I. 1985: Reef response to sea level rise: Keep-up, catch-up or give-up. *Proceedings, 5th International Coral Reef Congress, Tahiti* 3, 105–10.

NEWBERRY, J. and SIVA SUBRAMANIAM, A. 1978: Middle East — sewerage projects for coastal towns of the Libyan Arab Republic. *Quarterly Journal of Engineering Geology* 11, 101–12.

NEWMAN, W.S. and FAIRBRIDGE, R.W. 1986: The management of sea-level rise. *Nature* 320, 319–21.

NGOILE, M.A.K. and SHUNULA, J.P. 1992: Status and exploitation of the mangrove and associated fishery resources in Zanzibar. *Hydrobiologia* 247, 229–34.

NICHOLS, R.L. 1961: Characteristics of beaches formed in polar climates. *American Journal of Science* 259, 694–708.

NICHOLS, M.M. and BIGGS, R.B. 1985: Estuaries. In Davis, R.A. (ed.), *Coastal sedimentary environments*. New York: Springer-Verlag, 77–186.

NIOLNE, B., DIOUF, A., WILLER, B., NDOUR, F., CISS, M. and BA, D. 1992: Monitoring heavy metal pollution of the mangrove oysters of Senegal. *Revue Internationale d'Oceanographie Medicale* 105–106, 71–8.

NIXON, S.W. 1980: Between coastal marshes and coastal waters — a review of twenty years of speculation and research into the role of salt marshes in estuarine productivity and water chemistry. In Hamilton, R. and McDonald, K.B. (eds), *Estuarine and wetland processes*. New York: Plenum, 437–525.

NORDSTROM, K.F. and McCLUSKEY, J.M. 1985: The effects of houses and sand fences on the eolian sediment budget at Fire Island, New York. *Journal of Coastal Research* 1, 39–46.

NORDSTROM, K.F., McCLUSKEY, J.M. and ROSEN, P.S. 1986: Aeolian processes and dune characteristics of a developed shoreline: Westhampton beach, New York. In Nickling, W.G. (ed.), *Aeolian geomorphology*. Boston: Allen and Unwin, 131–47.

NORTON, T.A. 1985: The zonation of seaweeds on rocky shores. In Moore, P.G. and Seed, R. (eds), *The ecology of rocky shores*. London: Hodder and Stoughton, 7–21.

NORTON, D. and WELLER, G. 1984: The Beaufort Sea: Background, history and perspective. In Barnes, P.W., Schell, D.M. and Reimnitz, E. (eds), *The Alaskan Beaufort Sea*. Orlando: Academic Press, 3–19.

NUNN, P.D. 1994: *Oceanic islands*. Oxford: Basil Blackwell.

NUMMEDAL, D., HINE, A.C. and BOOTHROYD, J.C. 1987: Holocene evolution of the south central coast of Iceland. In Fitzgerald, D.M. and Rosen, P.S. (eds), *Glaciated coasts*. San Diego, Academic Press, 115–50.

ODUM, E.P. 1971: *Fundamentals of ecology*. 3rd edition. Philadelphia: W B Saunders.

ODUM, W.E. and HEALD, E.J. 1972: Trophic analysis of an estuarine mangrove community. *Bulletin of Marine Science* 22, 671–738.

ODUM, H.T. and ODUM, E.P. 1955: Trophic structure and productivity of a windward reef coral community on Eniwetok Atoll. *Ecological Monographs* 25, 291–320.

OENEMA, O. and DeLAUNE, R.D. 1988: Accretion rates in saltmarshes in the eastern Scheldt, southwest Netherlands. *Estuarine, Coastal and Shelf Science* 26, 379–94.

OERLEMANS, J. 1993: Possible changes in the mass balance of the Greenland and Antarctic ice sheets and their effects on sea level. In Warrick, R.A., Barrow, E.M. and Wigley, T.M.L. (eds), *Climate and sea level change*. Cambridge: Cambridge University Press, 144–62.

ORFORD, J.D. and CARTER, R.W.G. 1982: Crestal overtop and washover sedimentation on a fringing sandy gravel barrier coast, Carnsore Point, south east Ireland. *Journal of Sedimentary Petrology* 52, 265–78.

ORHAN, E.A. 1989: Engineering geological considerations in a salt dome region surrounded by sebkha sediments, Saudi Arabia. *Engineering Geology* 26, 215–32.

OTERI, A.U. 1983: Delineation of saline intrusion in the Dungeness shingle aquifer using surface geophysics. *Quarterly Journal of Engineering Geology* 16, 43–5.

OTT, M.L., HARASMA, M.S., BROEKMAN, R.A. and ROZEMA, J. 1993: Relation between heavy metal concentrations in salt marsh plants and soil. *Environmental Pollution* 82, 13–22.

OTVOS, E.G. 1979: Barrier island erosion and history of migration, North Central Gulf Coast. In Leatherman, S.P. (ed.), *Barrier islands*. New York: Academic Press, 291–319.

OUESLATI, A. 1992: Salt marshes in the Gulf of Gabes (south eastern Tunisia): their morphology and recent dynamics. *Journal of Coastal Research* 8, 727–33.

OWENS, J.S. and CASE, G.O. 1908: *Coast erosion and foreshore protection*. London: St Bride's Press.

OWENS, E.H., BOEHM, D., HARPER, J.R. and ROBSON, W. 1987: Fate and persistence of crude oil stranded on a sheltered beach. *Arctic* 40, 109–23.

OYEGUN, C.U. 1991: Spatial and seasonal aspects of shoreline changes at Forcados Beach, Nigeria. *Earth Surface Processes and Landforms* 16, 293–305.

PAPATHANASSIOU, E. and ZENELOS, A. 1993: A case of recovery in benthic communities following a reduction in chemical pollution in a Mediterranean ecosystem. *Marine Environmental Research* 36, 131–52.

PARNELL, K.E. 1988: Physical process studies in the Great Barrier Reef Marine Park. *Progress in Physical Geography* 12, 209–36.

PATERSON, D.M. and UNDERWOOD, G.J.C. 1990: The mudflat ecosystem and epipelic diatoms. *Proceedings, Bristol Naturalists' Society* 50, 74–82.

PEARSON, T.H. 1988: Energy flow through fjord systems. In Jansson, B.-O. (ed.), *Coastal-offshore ecosystem interactions*. Berlin: Springer-Verlag, 188–207.

PELTO, M.S. and WARREN, C.R. 1991: Relationship between tidewater glacier calving velocity and water depth at the calving front. *Annals of Glaciology* 15, 115–18.

PENLAND, S., BOYD, R. and SUTER, J.R. 1988: The transgressive depositional systems of the Mississippi delta plain: A model for barrier shoreline and shelf sand development. *Journal of Sedimentary Petrology* 58, 932–49.

PENTREATH, R.J. 1987: The interaction with suspended settled sedimentary materials of long-lived radionuclides discharged into UK coastal waters. *Continental Shelf Research* 7, 1457–69.

PERILLO, G.M.E., DRAPEAU, G., PICCOLO, M.C. and CHAOUQ, N. 1993: Tidal circulation pattern on a tidal flat, Minas Basin, Canada. *Marine Geology* 112, 219–36.

PETHICK, J.S. 1981: Long term accretion rates on tidal salt marshes. *Journal of Sedimentary Petrology* 51, 571–7.

PETHICK, J.S. 1984: *An introduction to coastal geomorphology*. London: Edward Arnold.

PETHICK, J.S. 1992: Salt marsh geomorphology. In Allen, J.R.L. and Pye, K. (eds), *Salt marshes*. Cambridge: Cambridge University Press, 41–62.

PETHICK, J.S. 1993: Shoreline adjustments and coastal management: physical and biological processes under accelerated sea level rise. *Geographical Journal* 159, 162–8.

PFIRMAN, S.L. and SOLHEIM, A. 1989: Subglacial meltwater discharge in the open marine tidewater glacial environment: observations from Nordaustlandet, Svalbard Archipelago. *Marine Geology* 86, 265–81.

PHILANDER, S.G. 1990: *El Niño, La Nina and the Southern Oscillation*. San Diego: Academic Press.

PHILLIPS, J.D. 1986: Coastal submergence and marsh fringe erosion. *Journal of Coastal Research* 2, 427–36.

PILKEY, O.H. 1991: Coastal erosion. *Episodes* 14, 46–57.

PILKEY, O.H., YOUNG, R.S., RIGGS, S.R., SMITH, S., WU, H. and PILKEY, W.D. 1993: The concept of a shoreface profile of equilibrium: a critical review. *Journal of Coastal Research* 9, 255–278.

PINDER, D.A. and WITHERICK, M.E., 1990: Port industrialisation, urbanizations and wetland loss. In Williams, M. (ed.), *Wetlands: A threatened landscape*. (Institute of British Geographers Special Publication 25). Oxford: Blackwell, 234–66.

PIRAZZOLI, P.A. 1993: Global sea-level changes and their measurement. *Global and Planetary Change* 8, 135–48.

PISAREVSKAYA, L.G. and POPOV, I.K. 1991: Free-drifting icebergs and thermohaline structure. In Kotlyakov, V.M., Ushakov, A. and Glazovsky, A. (eds), *Glaciers–ocean–atmosphere interactions*. IAHS Publication No. 208, 447–54.

PITTS, J. 1983: Geomorphological observations as aids to the design of coastal protection works on a part of the Dee estuary. *Quarterly Journal of Engineering Geology* 16, 291–300.

PITTS, J. 1986: The form and stability of a double undercliff: an example from south west England. *Engineering Geology* 22, 209–16.

PLANT, N. and GRIGGS, G.V.S. 1990: Coastal landslides caused by the October 17, 1989 earthquake, Santa Cruz, California. *Geology* 43, 75–84.

PLUIS, J.L.A. and DE WINDER, B. 1990: Natural stabilisation. *Catena Supplement* 18, 209–23.

POSTMA, H. 1967: Sediment transport and sedimentation in the estuarine environment. In Lauff, G.H. (ed.), *Estuaries*. American Association for the Advancement of Science Publication 83, 158–79.

POTTER, P.E. 1984: South American beach sand and plate tectonics. *Nature* 311, 645–8.

POWELL, R.D. and ELVERHØI, A. (eds) 1989: Modern glacimarine environments: Glacial and marine controls of modern lithofacies and biofacies. *Marine Geology, Special Issue* 85, 101–20.

PRINGLE, A.W. 1985: Holderness coast erosion and the significance of ords. *Earth Surface Processes and Landforms* 10, 107–24.

PSUTY, N.P. and MOREIRA, M.E.S.A. 1992: Characteristics and longevity of beach nourishment at Praia de Rocha, Portugal. *Journal of Coastal Research* 8, 660–76.

PURDY, E. 1974: Reef configurations: cause and effect. In Laporte, L.F. (ed.), *Reefs in time and space.* Tulsa, Oklahoma: Society of Economic Paleontologists and Mineralogists Special Publication 18, 9–76.

PURSER, B.H. and COREAU, J.-P. 1973: Aragonitic supratidal encrustations on the Trucial Coast, Persian Gulf. In Purser, B.H. (ed.), *The Persian Gulf.* Berlin: Springer-Verlag, 343–76.

PURSER, B.H. and EVANS, G. 1973: Regional sedimentation along the Trucial Coast, S. Persian Gulf. In Purser, B.H. (ed.), *The Persian Gulf.* Berlin: Springer-Verlag, 211–32.

PYE, K. 1980: Beach salcrete and aeolian sand transport: evidence from North Queensland. *Journal of Sedimentary Petrology* 50, 257–61.

PYE, K. 1983: Formation and history of Queensland coastal dunes. *Zeitschrift für Geomorphologie Supplementband* 45, 175–204.

PYE, K. 1990: Physical and human influences on coastal dune development between the Ribble and Mersey estuaries, N W England. In Nordstrom, K.F., Psuty, N.P. and Carter, R.W.G. (eds), *Coastal dunes.* Chichester: Wiley, 339–59.

RANWELL, D.S. and BOAR, R. 1986: *Coast dune management guide.* Huntingdon: Institute of Terrestrial Ecology.

RASOWO, J. 1992: Mariculture development in Kenya: alternatives to siting ponds in the mangrove ecosystem. *Hydrobiologia* 247, 209–14.

RECLUS, E. 1873: *The oceans, atmosphere and life.* New York: Harper and Brothers.

REED, D.J. 1988: Sediment dynamics and deposition on a retreating coastal salt marsh. *Estuarine, Coastal and Shelf Science* 26, 67–79.

REED, D.J. 1990: The impact of sea level rise on coastal salt marshes. *Progress in Physical Geography* 14, 465–81.

REED, D.J. 1991: Ponds and bays: natural processes of coastal marsh erosion in the Mississippi deltaic plain, Louisiana, USA. *Zeitschrift für Geomorphologie Supplementband* 81, 41–51.

REIMOLD, R.J. 1977: Mangals and salt marshes of eastern United States. In Chapman, V.J. (ed.), *Wet coastal ecosystems.* Amsterdam: Elsevier, 157–66.

RÉTIÈRE, C. 1994: Tidal power and the aquatic environment of La Rance. *Biological Journal of the Linnean Society* 51, 25–36.

RICHARDS, K.S. and LORRIMAN, N.R. 1987: Basal erosion and mass movement. In Anderson, M.G. and Richards, K.S. (eds), *Slope stability.* Chichester: Wiley, 331–57.

RICHMOND, R.H. 1990: The effects of ENSO on the dispersal of corals and other marine organisms. In Glynn, W. (ed.), *Global ecological consequences of the 1982–3 El Niño Southern Oscillation.* Amsterdam: Elsevier, 127–40.

RITCHIE, W. and PENLAND, S. 1990: Aeolian sand bodies of the south Louisiana coast. In Nordstrom, K.F., Psuty, N.P. and Carter, R.W.G. (eds), *Coastal dunes*. Chichester: John Wiley, 105–27.

ROBERTS, H.H. 1974: Variability of reefs with regard to changes in wave power around an island. *Proceedings, 2nd International Coral Reef Symposium, Brisbane* 2, 497–512.

ROBERTS, H.H., MURRAY, S.P. and SUHAYDA, J.M. 1977: *Physical processes on a fore reef shelf environment*. Proceedings, 3rd International Coral Reef Symposium, Miami 2, 507–16.

ROBERTS, H.H., ROUSE, L.J. Jr, WALKER, N.D. and HUDSON, J.H. 1982: Cold-water stress in Florida Bay and Northern Bahamas: a product of cold-air outbreaks. *Journal of Sedimentary Petrology* 52, 145–55.

ROBERTS, H.H., LUGO-FERNANDEZ, A., CARTER, B. and SIMMS, M. 1988: Across-reef flux and shallow subsurface hydrology in modern reefs. *Proceedings, Sixth International Coral Reef Symposium, Townsville* 2, 509–15.

ROBERTS, H.H., WILSON, P.A. and LUGO-FERNANDEZ, A. 1992: Biologic and geologic resonses to physical processes: examples from modern reef systems of the Caribbean-Atlantic region. *Continental Shelf Research* 12, 809–34.

ROBINSON, L.A. 1977: Marine erosive processes at the cliff foot. *Marine Geology* 23, 257–71.

ROBINSON, D.A. and JERWOOD, L.C. 1987: Subaerial weathering of chalk shore platforms during harsh winters in south east England. *Marine Geology* 77, 1–14.

RODRIGUEZ, A. 1981: Marine and coastal environmental stress in the wider Caribbean region. *Ambio* 10, 283–94.

ROMAN, C.T., ABLE, K.W., LAZZARI, M.A. and HECK, K.L. 1990: Primary productivity of angiosperm and macroalgal dominated habitats in a New England salt marsh: A computer analysis. *Estuarine, Coastal and Shelf Science* 30, 35–45.

ROSE, H.E. 1990: Birds of the Avon shore. *Proceedings, Bristol Naturalists' Society* 50, 95–104.

ROSEN, P. 1979: Aeolian dynamics of a barrier island system. In Leatherman, S.P. (ed.), *Barrier islands*. New York: Academic Press, 81–98.

ROSEN, B.R. 1981: The tropical high diversity enigma — the coral's eye view. In Forey, P.L. (ed.), *The evolving biosphere*. London: BMNH/Cambridge University Press, 103–29.

ROUGERIE, F. and WAUTHY, B. 1993: The endo-upwelling concept: from geothermal convection to reef construction. *Coral Reefs* 12, 19–30.

ROYAL COMMISSION ON COAST EROSION 1907: *First report into coast erosion and the reclamation of tidal lands in the UK*. London: H.M.S.O.

RUCH, P., MIRMAND, M., JOUANNEAU, J.-M. and LATOUCHE, C. 1993: Sediment budget and transfer of suspended sediment from the Gironde Estuary to Cap Ferret Canyon. *Marine Geology* 111, 109–19.

RUDDLE, K. 1989: Solving the common-property dilemma: village fisheries rights in Japanese coastal waters. In Berkes, F. (ed.), *Common property resources*. London: Belhaven Press, 168–84.

RUDKIN, H. 1992: Stonebarrow Hill. In Allison, R.J. (ed.), *The coastal landforms of West Dorset. Geologists' Association Guide* 47, 50–61.

RUSSELL, R.J. 1963: Recent recession of tropical cliffy coasts. *Science* 139, 9–15
RUTIN, J. 1992: Geomorphic activity of rabbits on a coastal sand dune, de Blink dunes, the Netherlands. *Earth Surface Processes and Landforms*, 17, 85–94.
RUZ, M.H. and ALLARD, M. 1994: Foredune development along a subarctic emerging coastline, eastern Hudson Bay, Canada. *Marine Geology* 117, 57–74.
SAENGER, P. and SIDDIQI, N.A. 1993: Land from the sea: the mangrove afforestation program of Bangladesh. *Ocean and Coastal Management* 20, 23–39.
SAHAGIAN, D.L., SCHWARTZ, F.W. and JACOBS, D.K. 1994: Direct anthropogenic contributions to sea level rise in the 20th century. *Nature* 367, 54–7.
SANCETTA, C. 1989: Sediment transport by fecal pellets in British Columbian fjords. *Marine Geology* 89, 331–46.
SANDERS, J.E. and KUMAR, I.N. 1975. Evidence of a shoreface retreat and in-place 'drowning' during Holocene submergence of barriers, shelf off Fire Island, New York. *Geological Society of America, Bulletin*, 86, 65–76.
SANGUIN, A.-I. 1989 Aux marges de l'Oekoumene Canadien: le Parc National Pacific Rim (Colombie Britannique). *Hommes et Terres du Nord* 3, 135–41.
SANJAUME, G. and PARDO, J. 1991: Dune regeneration on a previously destroyed dunefield, Devesa del Saler, Valencia, Spain. *Zeitschrift für Geomorphologie Supplementband* 81, 125–34.
SARGENT, M.C. and AUSTIN, T.S. 1954: Biologic economy of coral reefs. *US Geological Survey, Professional Paper* 260E, 293–300.
SARRE, R.D. 1989a: Aeolian sand drift from the intertidal zone on a temperate beach: Potential and actual rates. *Earth Surface Processes and Landforms* 14, 247–58.
SARRE, R.D. 1989b: The morphological significance of vegetation and relief on coastal foredune processes. *Zeitschrift für Geomorphologie Supplementband* 73, 17–31.
SASEKUMAR, A., CON, V.C., LEH, M.U. and D'CRUZ, R. 1992: Mangroves as a habitat for fish and prawns. *Hydrobiologia* 247, 195–207.
SAVIGEAR, R.A.G. 1952: Some observations on slope development in South Wales. *Transactions, Institute of British Geographers* 18, 31–52.
SCHNEIDER, J. and TORUNSKI, H. 1983: Biokarst on limestone coasts; morphogenesis and sediment production. *Marine Ecology* 4, 45–63.
SCOFFIN, T.P. 1987: *An introduction to carbonate sediments and rocks.* Glasgow: Blackie.
SCOFFIN, T.P., STEARN, C.W., BOUCHER, D., FRYDAL, P.M, HAWKINS, C.M., HUNTER, I.G. and MacGEACHY, J.K. 1980: Calcium carbonate budget of a fringing reef on the west coast of Barbados. II Erosion, sediments and internal structure. *Bulletin of Marine Science* 30, 475–508.
SCOR WORKING GROUP '89 1991: The response of beaches to sea level changes: A review of predictive models. *Journal of Coastal Research* 7, 895–921.
SCOTT, P.J.B., RISK, M.J. and CARRIQUIRY, J.D. 1988: El Niño, bioerosion and the survival of east Pacific reefs. *Proceedings, 6th International Coral Reef Symposium, Townsville*, 2, 517–20.
SEMENIUK, V. 1981: Long term erosion of the tidal flats, King Sound, N Western Australia. *Marine Geology* 43, 21–48.
SEN GUPTA, R., ALI, M., BHUIYAN, A.L., SIVASINGAM, P.M., SUBASINGHE, S. and

TISMIZI, N. 1990: *State of the marine environment in the South Asian Seas region.* UNEP Regional Seas Reports and Studies No. 123.

SESTINI, G. 1992: Implications of climatic changes for the Po delta and Venice lagoon. In Jeftic, L., Milliman, J.F. and Sestini, G. (eds), *Climatic change and the Mediterranean.* London: Edward Arnold, 428–94.

SHENNAN, I. 1993: Geographical information systems and future sea level rise. In Warrick, R.A., Barrow, E.M. and Wigley, T.M.L. (eds), *Climate and sea level change.* Cambridge: Cambridge University Press, 215–28.

SHEPARD, F.P. 1963: *Submarine geology.* New York: Harper and Row.

SHINN, E.A. 1973: Carbonate coastal accretion in an area of longshore transport, N E Qatar, Persian Gulf. In Purser, B.H. (ed.), *The Persian Gulf.* Berlin: Springer-Verlag, 179–92.

SHORT, A.D. 1988: Wave, beach, foredune and mobile dune interactions in south east Australia. In Psuty, N.P. (ed.), Dune–beach interaction: Proceedings of a special session. *Journal of Coastal Research, Special Issue* 3, 5–9.

SHORT, A.D. and HESP, P.A. 1982: Wave, beach and dune interactions in south eastern Australia. *Marine Geology* 48, 259–84.

SILVESTER, R. 1974: *Coastal engineering.* Amsterdam: Elsevier.

SIMS, P. and TERNAN, L. 1988: Coastal erosion, protection and planning in relation to public policies — a case study from Downderry, South east Cornwall. In Hooke, J.M. (ed.), *Geomorphology in environmental planning.* Chichester: Wiley 231–44.

SMITH, S.H. 1988: Cruise ships: a serious threat to coral reefs and associated organisms. *Ocean and Shoreline Management* 11, 231–48.

SMITH, R.I.L. 1993: Dry coastal ecosystems of Antarctica. In Maarel, E. van der (ed.), *Dry coastal ecosystems.* Amsterdam: Elsevier, 51–72.

SMITH, S.V. 1973: Carbon dioxide dynamics: a record of organic carbon production, respiration, and calcification in the Eniwetok reef flat community. *Limnology and Oceanography* 18, 106–20.

SMITH, S.V. 1978: Coral-reef area and the contribution of reefs to processes and resources of the World's oceans. *Nature* 273, 225–6.

SMITH, S.V. and KINSEY, D.W. 1978: Calcification and organic carbon metabolism as indicated by carbon dioxide. In Stoddart, D.R. and Johannes, R.E. (eds), *Coral reefs: research methods.* Paris: UNESCO, 469–84.

SNEAD, R.E. 1980: *World atlas of geomorphic features.* New York: Van Nostrand Reinhold.

SNEAD, R.E. 1985: Bangladesh. In Bird, E.C.F. and Schwartz, M.L. (eds), *The world's coastline.* New York: Van Nostrand Reinhold, 761–5.

SNEDAKER, S.C. 1993: Impact on mangroves. In Maul, G.A. (ed.), *Climatic change in the inter-Americas sea.* London: Edward Arnold, 282–305.

SNODGRASS, D., GROVES, G.W., HASSELMANN, K.F., MILLER, G.R., MUNK, W.H. and POWERS, W.H. 1966: Propagation of ocean swell across the Pacific. *Philosophical Transactions, Royal Society of London* 259A, 431–97.

SOROKIN, Y.I. 1990: Aspects of trophic relations, productivity and energy balance in coral reef ecosystems. In Dubinsky, Z. (ed.), *Coral reefs.* Amsterdam: Elsevier, 401–10.

SOUCHEZ, R.A. and LORRAIN, R.D. 1991: *Ice composition and glacier dynamics*. Berlin: Springer-Verlag.

SOURNIA, A. 1976: Primary production of sands in the lagoon of an atoll and the role of formanifera symbionts. *Marine Biology* 37, 29–32.

SOUTHWARD, A.J. and SOUTHWARD, E.C. 1978: Recolonisation of rocky shores in Cornwall after use of toxic dispersants to clean up the Torrey Canyon oil spill. *Journal of the Fisheries Research Board of Canada* 35, 682–706.

SPENCER, T. 1988: Limestone coastal morphology. *Progress in Physical Geography* 12, 66–101.

SPENCER, T. 1994: Tropical coral islands — an uncertain future? In Roberts, N. (ed.), *The changing global environment*. Oxford: Blackwell, 190–209.

SPLETTSTOESSER, J.F. 1985: Note on rock striations caused by Penguin feet, Falkland Islands. *Arctic and Alpine Research* 17, 107–11.

STAMP, D.L. 1939: Recent coastal changes in south eastern England V: Some economic aspects of coastal loss and gain. *Geographical Journal* 93, 496–503.

STANLEY, D.J. and CHEN, Z. 1993: Yangtze delta, eastern China I Geomorphology and subsidence of Holocene depocenter. *Marine Geology* 112, 1–11.

STANLEY, D.J. and WARNE, A.G 1993: Nile delta: Recent geological evolution and human impact. *Science* 260, 628–34.

STEARN, C.W., SCOFFIN, T.P. and MARTINDALE, W. 1977: Calcium carbonate budget of a fringing reef on the west coast of Barbados. I. — zonation and productivity. *Bulletin of Marine Science* 27, 479–510.

STEHLI, F.G. and WELLS, J.W. 1971: Diversity and age patterns in hermatypic corals. *Systematic Zoology* 20, 115–26.

STEVENSON, J.C., WARD, L.G. and KEARNEY, M.S. 1986: Vertical accretion in marshes with varying rates of sea level rise. In Wolfe, D.A. (ed.), *Estuarine variability*. Orlando: Academic Press, 241–60.

STEVENSON, J.C., WARD, L.G. and KEARNEY, M.S. 1988: Sediment transport and trapping in marsh systems: implications of tidal flux studies. *Marine Geology* 80, 37–59.

STOCKER, T.F. 1994: Climate change; the variable ocean. *Nature* 367, 221–3.

STODDART, D.R. 1969: Ecology and morphology of Recent Coral Reefs. *Biological Reviews* 44, 433–98.

STODDART, D.R. 1985: Symposium No. 3 — Hurricane effects on coral reefs: conclusion. *Proceedings, 5th International Coral Reef Congress, Tahiti*, 3, 349–50.

STODDART, D.R. 1990: Coral reefs and islands and predicted sea level rise. *Progress in Physical Geography* 14, 521–36.

STRAIN, P.M. 1986: The persistence and mobility of a light crude oil in a sandy beach. *Marine Environmental Research* 19, 49–76.

STUMPF, R.P. 1983: The process of sedimentation on the surface of a salt marsh. *Estuarine, Coastal and Shelf Science* 17, 495–508.

SUGDEN, D. 1982: *Arctic and Antarctic: A modern geographic synthesis*. Oxford: Blackwell.

SUGDEN, D. 1991: The stepped response of ice sheets to climatic change. In Harris, C.M. and Stonehouse, B. (eds), *Antarctica and global climatic change*. London: Belhaven Press, 107–14.

SUMMERFIELD, M.A. 1991: *Global geomorphology*. London: Longman.

SUNAMURA, T. 1982: A predictive model for wave-induced cliff erosion, with application to Pacific coasts of Japan. *Journal of Geology* 90, 167–78.

SUNAMURA, T. 1983: Processes of sea cliff and platform erosion. In Komar, P.D. (ed.), *Handbook of coastal processes and erosion*. Boca Raton, Florida: CRC Press, 233–65.

SUNAMURA, T. 1988: Projection of future coastal cliff recession under sea level rise induced by the greenhouse effect: Nii-Jima Island, Japan. *Transactions, Japanese Geomorphological Union* 9, 17–33.

SUNAMURA, T. 1992: *Geomorphology of rocky coasts*. Chichester: Wiley.

SUTER, J.R., MOSSA, J. and PENLAND, S. 1989: Preliminary assessments of the occurrence and effects of utilisation of sand and gravel aggregate resources of the Louisiana inner shelf. *Marine Geology* 90, 31–7.

SWAN, S.B. ST.C. 1971: Coastal geomorphology in a humid tropical low energy environment: the islands of Singapore. *Journal of Tropical Geography* 33, 43–61.

SYVITSKI, J.P.M. 1989: On the deposition of sediment within glacier-influenced fjords: oceanographic controls. *Marine Geology* 85, 301–29.

SYVITSKI, J.P.M., BURRELL, D.C. and SKEI, J.M. 1987: *Fjords: Processes and products.* New York: Springer-Verlag.

TALAVERA, H.C. 1985: Impacts of population pressure on coastal zones and settlement patterns in the Philippines. In Kato, I., Kumamoto, N., Matthews, W.H. and Suhaimi, A. (eds), *Environmental protection and coastal zone management in Asia and the Pacific.* Tokyo: University of Tokyo Press, 175–82.

TAYLOR, R.B. and McCANN, S.B. 1983: Coastal depositional landforms in Northern Canada. In Smith, D. and Dawson, A.G. (eds), *Shorelines and isostasy*. London: Academic Press, 53–75.

THOLLOT, P. 1992: Importance of mangroves for the reef fish fauna from New Caledonia. *Cybium* 16, 331–4.

THOM, B.G. 1967: Mangrove ecology and deltaic geomorphology: Tabasco, Mexico. *Journal of Ecology* 55, 301–43.

THOMAS, M.L.H. 1994: Littoral communities and zonation on rocky shores in the Bay of Fundy, Canada: an area of high tidal range. *Biological Journal of the Linnean Society* 51, 149–56.

THOMAS, R.H. 1986: Future sea level rise and its early detection by satellite remote sensing. In Titus, J.G. (ed.) *Effects of changing stratospheric ozone and global climate*, Vol. 4: Sea level rise, UNEP/EPA (USA), 19–36.

TITUS, J.C. 1990: Greenhouse effect, sea level rise and barrier islands: case study of Long Beach Island, New Jersey. *Coastal Management* 18, 65–90.

TITUS, J.G. 1991: Greenhouse effect and coastal wetland policy: How Americans could abandon an area the size of Massachusetts at minimum cost. *Environmental Management* 15, 39–58.

TITUS, J., PARK, R.A., LEATHERMAN, S.P., WEGGEL, J.R., GREENE, M.S., MANSEL, P.W., BROWN, S., GAUNT, C., TREHAN, M. and YOHE, G. 1991: Greenhouse effect and sea level rise: the cost of holding back the sea. *Coastal Management* 19, 171–204.

TJIA, H.D. 1985: Notching by abrasion on a limestone coast. *Zeitschrift für Geomorphologie* 29, 367–72.

TOLBA, M.K. and EL-KHOLY, O.A. (eds) 1992: *The world environment, 1972–1992*. London: Chapman and Hall.

TOMLINSON, M.J. 1978: Middle East — highway and airfield pavements. *Quarterly Journal of Engineering Geology* 11, 65–73.

TOOLEY, M.J. 1990: The chronology of coastal dune development in the UK. *Catena Supplement* 18, 81–8.

TRENBERTH, K.E. and SHEA, D.J. 1987: On the evolution of the Southern Oscillation. *Monthly Weather Review* 115, 3078–96.

TRENHAILE, A.S. 1983: The development of shore platforms in high latitudes. In Smith, D. and Dawson, A.G. (eds), *Shorelines and isostasy*. London: Academic Press, 77–93.

TRENHAILE, A.S. 1987: *The geomorphology of rock coasts*. Oxford: Oxford University Press.

TRICART, J. 1972: *The landforms of the humid tropics*. London: Longman.

TRUDGILL, S.T. 1976: The marine erosion of limestone on Aldabra Atoll, Indian Ocean. *Zeitschrift für Geomorphologie Supplementband* 26, 164–200.

TRUDGILL, S.T. 1985: *Limestone geomorphology*. London; Longman.

TRUDGILL, S.T. 1987: Bioerosion of intertidal limestone, County Clare, Eire. 3: Zonation, process and form. *Marine Geology* 74, 111–21.

TURNER, K. and DAGLEY, J. 1993: What price sea walls? *Enact* 1,3, 8–9.

TURNER, R.E. and RABALAIS, N.N. 1994: Coastal eutrophication near the Mississippi River delta. *Nature* 368, 619–21.

TWILLEY, R.R. 1988: Coupling of mangroves to the productivity of estuarine and coastal waters. In Jansson, B.-O. (ed.), *Coastal-offshore ecosystem interactions*. Berlin: Springer-Verlag, 155–87.

TWILLEY, R.R., CHEN, R.H. and HARGIS, T. 1992: Carbon sinks in mangroves and their implications to carbon budget of tropical coastal ecosystems. *Water, Air and Soil Pollution* 64, 265–88.

UK DEPARTMENT OF THE ENVIRONMENT 1993: *Managing the coast*. London: HMSO.

UNEP 1993: *Environmental Data Report* 1993–4. Oxford: Blackwell.

VAN DER MAAREL, E. (ed.) 1993: *Dry coastal ecosystems. Polar regions and Europe*. Amsterdam: Elsevier.

VAN DER MEULEN, F. 1990: European dunes: consequences of climatic change and sea level rise. *Catena Supplement* 18, 209–23.

VAN DIJK, H.W.J. 1989: Ecological impact of drinking-water production in Dutch coastal dunes. In van der Meulen, E., Jungerius, P.D. and Visser, J. (eds), *Perspectives in coastal dune management*. The Hague: S.P.B. Academic, 163–82.

VAN DYKE, J.M. 1991: Protected marine areas and low-lying atolls. *Ocean and Shoreline Management* 16, 87–160.

VAN EERDT, M.M. 1987: The influence of basic soil and vegetation parameters on salt marsh cliff strength. In Gardiner, V. (ed.), *International Geomorphology 1986* I. Chichester: Wiley, 1073–86.

VAN KATWIJK, M.M., MEIER, N.F., VAN LOON, R., VAN HOVE, E.M., GIESEN, W.B.J.T., VAN DER VELDE, G. and DEN HARTOG, C. 1993: Subaki river sediment load and coral stress: correlation between sediments and condition of the Malindi–Watamu reefs in Kenya (Indian Ocean). *Marine Biology* 117, 675–83.

VILES, H.A. 1987: Blue-green algae and terrestrial limestone weathering on Aldabra Atoll: A light and scanning electron microscope study. *Earth Surface Processes and Landforms* 12, 319–30.

VILES, H.A. (ed.) 1988: *Biogeomorphology*. Oxford: Basil Blackwell.

WADHAMS, P. 1991: Atmosphere–ice–ocean interactions in the Antarctic sea ice. In Harris, C.M. and Stonehouse, B. (eds), *Antarctica and global climatic change*. London: Belhaven Press, 65–81.

WALKER, H.J. (ed.) 1988a: *Artificial structures and shorelines*. Dordrecht: Kluwer Academic.

WALKER, H.J. 1988b: Alaska. In Walker, H.J. (ed.), *Artificial structures and shorelines*. Dordrecht: Kluwer Academic, 489–98.

WALKER, H.J. 1991: Bluff erosion at Barrow and Wainwright, Arctic Alaska. *Zeitschrift für Geomorphologie Supplementband* 81, 53–61.

WALKER, H.J. and MOSSA, J. 1986: Human modification of the shoreline of Japan. *Physical Geography* 7, 116–39.

WALKER, N.D., ROBERTS, H.H., ROUSE, L.J. Jr and HUH, O.K. 1982: Thermal history of reef-associated environments during a record cold-air outbreak event. *Coral Reefs* 1, 83–8.

WALKER, H.J., COLEMAN, J.M., ROBERTS, H.H. and TYE, R. 1987: Wetland loss in Louisiana. *Geografiska Annaler* 69A, 189–200.

WALSH, J.E. 1977: Exploitation of mangal. In Chapman, V.J. (ed.), *Wet coastal ecosystems*. Amsterdam: Elsevier, 347–62.

WARD, L.G., MOSLOW, T.F. and FINKELSTEIN, K. 1987: Geomorphology of a tectonically active, glaciated coast, south-central Alaska. In Fitzgerald, D.M. and Rosen, P.S. *Glaciated coasts*. San Diego: Academic Press, 1–31.

WARREN, J.K. 1989: *Evaporite sedimentology*. New Jersey: Prentice Hall.

WARREN, C.R. 1993: Rapid recent fluctuations of the calving San Rafael glacier, Chilean Patagonia: Climatic or non-climatic? *Geografiska Annaler* 75A, 111–15.

WARRICK, R.A. and FARMER, G. 1990: The greenhouse effect, climatic change and rising sea level — implications for development. *Transactions, Institute of British Geographers* 15, 5–20.

WARRICK, R.A. and OERLEMANS, J. 1990: Sea level rise. In Houghton, J.T., Jenkins, G.J. and Ephraums, J.J. (eds), *Climatic change: The IPCC scientific assessment*. Cambridge: Cambridge University Press, 257–81.

WARRICK, R.A., BARROW, E.M. and WIGLEY, T.M.L. (eds), 1993: *Climate and sea level change*. Cambridge: Cambridge University Press.

WASHBURN, A.L. 1980: *Geocryology*. London: Edward Arnold.

WELLINGTON, G.M. 1982: Depth zonation of corals in the Gulf of Panama: control and facilitation by resident reef fishes. *Ecological Monographs* 52, 223–41.

WELLS, J.W. 1957: Corals. *Geological Society of America, Memoirs* 67, 1087–104.

WELLS, S.M. (ed.) 1988: *Coral reefs of the world*. Gland: IUCN, 3 volumes.

WELLS, J.T., ADAMS, C.E., PARK, Y.-A. and FRANKENBERG, E.W. 1990: Morphology, sedimentology and tidal channel processes on a high-tide-range mudflat, west coast of South Korea. *Marine Geology* 95, 111–30.

WEST, R.C. 1977: Tidal salt-marsh and mangal formations of Middle and South America. In Chapman, V.J. (ed.), *Wet coastal ecosystems*. Amsterdam: Elsevier, 193–213.

WHEELER, W.H. 1902: *The sea-coast*. London: Longmans, Green and Co.

WICKREMATHE, H.J.M. 1985: Environmental problems of the coastal zone in Sri Lanka. *Economic Review* 10, 8–10.

WIEGEL, R.L. 1964: *Oceanographic engineering*. Englewood Cliffs, New Jersey: Prentice Hall.

WIGLEY, T.M.L. 1992: Future climate of the Mediterranean basin with particular emphasis on changes in precipitation. In Jeftic, L., Milliman, J.D. and Sestini, G. (eds), *Climatic change and the Mediterranean*. London: Edward Arnold, 15–44.

WIGLEY, T.M.L. and RAPER, S.C.B. 1992: Implications for climate and sea level of revised IPCC emissions scenarios. *Nature* 357, 293–300.

WIGLEY, T.M.L. and RAPER, S.C.B. 1993: Future changes in global mean temperature and sea level. In Warrick, R.A., Barrow, E.M. and Wigley, T.M.L. (eds), *Climate and sea level change*. Cambridge: Cambridge University Press, 111–33.

WILLETS, B.B. 1989: Physics of sand movement in vegetated dune systems. *Proceedings, Royal Society of Edinburgh B* 96, 37–49.

WILLIAMS, R.B.G. 1988: The biogeomorphology of periglacial environments. In Viles, H.A. (ed.), *Biogeomorphology*. Oxford: Basil Blackwell, 222–53.

WILLIAMS, E.H. Jr and BUNKLEY-WILLIAMS, L. 1990: The world-wide coral reef bleaching cycle and related sources of coral mortality. *Atoll Research Bulletin* 335.

WILLIAMS, A.T. and DAVIES, P. 1980: Man as a geological agent: the sea cliffs of Llantwit Major, Wales, UK. *Zeitschrift für Geomorphologie Supplementband* 34, 129–41.

WILLIAMS, M.A.J., DUNKERLEY, D.L., DE DECKKER, P., KERSHAW, A.P. and STOKES, T. 1991: *Quaternary environments*. London: Edward Arnold.

WILLIAMS, S.J., PENLAND, S. and ROBERTS, H.H. In press. Processes affecting coastal wetland loss in the Louisiana delta plain. *Coastal Zone*, 1993 conference.

WITMAN, J.D. 1992: Physical disturbance and community structure of exposed and protected reefs: a case study from St John, U S Virgin Islands. *American Zoologist* 32, 641–54.

WONG, P.P. 1990: The geomorphological basis of beach resort sites — some Malaysian examples. *Ocean and Shoreline Management* 13, 127–47.

WONG, P.P. 1992: The newly reclaimed land. In Gupta, P. and Pitts, J. (eds), *Physical adjustments in a changing landscape*. Kent Ridge, Singapore: Singapore University Press, 243–58.

WOO, M.-K., LEWCOWICZ, A.G. and ROUSE, W.R. 1992: Response of the Canadian permafrost environment to climatic change. *Physical Geography* 13, 287–317.

WOOD, M.E., KELLEY, J.T. and BELKNAP, D.F. 1989: Patterns of sediment accumulation in the tidal marshes of Maine. *Estuaries* 12, 237–46.

WOOD, J.M., HORWITZ, P. and COX, H. 1992: Levels of heavy metals in dated sediments from Lindisfarne Bay, Derwent Estuary, Tasmania. *Science of the Total Environment* 125, 253–62.

WOODLEY, J.D. *et al.* 1981: Hurricane Allen's impact on Jamaican coral reefs. *Science* 214, 749–55.

WOODROFFE, C.D. 1990: The impact of sea level rise on mangrove shorelines. *Progress in Physical Geography* 14, 483–520.

WOODROFFE, C.D. 1993: Late Quaternary evolution of coastal and riverine plains of South-East Asia and Northern Australia — an overview. *Sedimentary Geology* 83, 163–75.

WOODWORTH, P.L. 1990: Search for acceleration in records of European mean sea-level. *International Journal of Climatology* 10, 129–43.

WORLD RESOURCES INSTITUTE 1992: *World Resources 1992–3*. New York: Oxford University Press.

WRIGHT, L.D. 1985: River deltas. In Davis, R.A. (ed.), *Coastal sedimentary environments*. New York: Springer-Verlag, 1–76.

WRIGHT, L.D. and SHORT, A.D. 1984: Morphodynamic variability of surf zones and beaches: A synthesis. *Marine Geology* 56, 93–118.

YAMAGUCHI, M. 1975: Sea level fluctuations and mass mortalities of reef animals in Guam, Mariana Islands. *Micronesica* 11, 227–43.

YESNER, D.R. 1988: Island biogeography and prehistoric human adaptation on the southern coast of Maine (USA). In Bailey, G. and Partington, J. (eds), *The archaeology of prehistoric coastlines*. Cambridge: Cambridge University Press, 53–63.

YONGE, C.M. 1930: Studies on the physiology of corals. I. Feeding mechanisms and food. *Scientific Reports, Great Barrier Reef Expedition*. London: BMNH 1, 13–57.

YOUNG, I.I. 1991: Critical ecosystems and nature conservation in Antarctica. In Harris, C.M. and Stonehouse, B. (eds), *Antarctica and global climatic change*. London: Belhaven Press.

YOUNG, H.R. and NELSON, C.S. 1988: Endolithic biodegradation of cool-water skeletal carbonates on Scott shelf, North western Vancouver Island, Canada. *Sedimentary Geology, Special Issue* 60, 251–67.

YOUSSEF, E.S.A.A. 1991: A note on the geomorphology of the coastal plain between Al Hudaydah and Al Salif Peninsula, Red Sea Coast, Yemen Arab Republic. *Geographical Journal* 157, 71–3.

ZEDLER, J.B. 1988: Restoring diversity in saltmarshes: Can we do it? In Wilson, E.O. (ed.), *Biodiversity*. Washington: National Academy Press, 317–25.

ZIMMERMAN, L. and KELLETAT, D. 1984: Die schleichende Ölpest in Mittelmeer — ihre Auswirkungen auf ökologie und küstenformung als Beispiel für die Beeinflussung natürlicher Prozesse durch anthropogenes Fehlverhalten. *Geoökodynamik* 5, 77–98.

ZMARZLY, D.L., STEBBINS, T.D., PASKO, D., DUGGAN, R.M. and BARWICK, K.L. 1994: Spatial patterns and temporal succession in soft bottom macroinvertebrate assemblages surrounding an ocean outfall on the southern San Diego Shelf: relation to anthropogenic and natural events. *Marine Biology* 118, 293–307.

INDEX

ablation 55–6, *Fig. 2.17*
abrasion 120, 138
Acanthaster 48, 240, 250–3
  see also Crown-of-Thorns starfish
accretionary
  deficit 177, *Fig. 5.12*
  rates 181, 236–7
  surplus 177, *Fig. 5.12*
accumulation 56, *Fig. 2.17*
Acropora 224, 225, 238
aeolian chute 95
aerobic respiration 274
Agent Orange 204
aggregates,
  dredging of 14
  mining for 76
air photography 15, 110
Alaska 26
algae,
  calcareous 136, 143, 214
  brown 144
  fucoid 144, 274
  green bubble 249–50
algal,
  crusts 90
  mats 168, 176, 197
  ridges 224
allochthonous sediment 170
Ammophila 71, 72, 88
amphidronic point 43–4, *Fig. 2.14*
anaerobic respiration 274
anchor ice 270
anoxic sediments 47
Antarctica 13, 26, 55–6, 254, 256–8, 261, 262, 264–5, *Fig. 7.3*, 268–9, 276–7, 282, 285–8
aquaculture 11, 157, 187, 291
Arabian Gulf 174–6
Arctic 11, 254, 256–8
armoured beach surfaces 74
asexual budding 216
asthenosphere 21
Aswan High Dam 75, 302
Atlantic Ocean, 235
  North 26, 54
  reefs 224–5, 237–8
Australia 22, 52, 60, 64, 75, 83, 169, 170, 172, 174, 187, 192, 233, 236
autochthonous sediment 170

backwash 32, 34, *Fig. 2.10*, 76
bacteria 47, 65, 67, 168, 214
Bahrain 50
Bangkok 57
Bangladesh 169, 293–9, 309
Barbados 244–6
barrier,
  beaches 50, 60
  islands 60, 64, 79, 85, 90, 92, 93–102, 201
  cyclical changes to *Fig. 3.13*
  erosion 95
  natural 91
  management 91, 96–8
  morphodynamics 97
  morphologies 93–6, *Plate 3.4*
  restoration of 83
  wetlands 159
  spits 159
basal erosion 117, 149
Bay,
  of Bengal 1, 169, 293–9
  of Fundy 43
beach cusps 34, 39
beach,
  dissipative 36, 37, 65, 73, 76, 90, *Table 2.4*, *Plate 2.3*
  face retreat 84
  fill, 97
  nourishment 97
  reflective 34,36, *Table 2.4*, 37, 65, 73
beach — *contd*
  ridges 64
  rock 76
  sand, mineralogy of 62
  water table 76, 85
beaches,
  cold coast 263
  distribution of 60, *Fig. 3.1*
  ecology 64–7
  erosion 77, 97
  gravel 62
  interactions with dunes 73–5
  maintenance 100
  morphodynamics 32–41, 65, 68
  new 83–4
  nourishment of 77, 81–3, 86, 101
  pollution 86–7
  raised 261
  tar on 87
  tourism on 302
  types 63
  zeta form 63
benthic,
  organisms 44, 47–8, 64, 274
  microflora 65
bioconstructions 119
biodegradation, of oil 285
biodiversity,
  of corals 207
  loss 305
bioerosion,
  see biological erosion
biogenic pelletization 158
biogeomorphology 3, 7, 9, *Fig. 1.4*, 10, 17, 18, 48, 54, 90, 108
biological,
  erosion 119–20, 138, 140, 143, 148, 217, 220, 222, 240, 265–7
  weathering 136, 138

bioturbation 192, 274
birds, 64, 72, 121, 190, 197, 198, 204, 205
and oil spills 284–5
Black Ven 125
Blackwater Estuary 194–6, 197
blasting, of cliffs 152–3, *Fig. 4.15*
bleaching, of corals 235, 238–40, 241
blowouts 68, 69, 72, *Table 3.1*, 90, 107
blue-green algae, 138, 142, 143, 222, 265
see also cyanobacteria
Blue Plan 308–9
blue slipper 124
boat channel 236
boring organisms 136, 139
boulder barricades 261
Braunton Burrows, Devon 15, 69, 90
breakwaters 75, 76, 78, 79–80, 86, 104
breaking waves 30–32
Brunsden, Denys 125–7
Bruun rule 84–5, 181

calcification 214
calcium carbonate,
content of beach sand 62
content of dunes 71
production of reefs 217
Canada 22, 43, 161, 254, 256, 259, 261, 263–4, 275, 276, 277
carbon, global storage of 172
Cardiff Bay barrage 184
Caribbean 14, 224, 225–30, 233, 240
'catch-up' reefs 222–3
cays 224
cesspools 244
char lands 298
chemical defoliants 204
Chesapeake Bay 161, 162, 186–7
Chesil Beach 62
Chile 257
China Sea 51
clay 158
cliff, 110
cold coast 263, 265
drainage 147–8
erosion 146
failure 122–31
instability 118
cliff — *contd*
morphologies 113–17
recession rates 117, 118, 124, 131, 153–5
retreat 126, 261, *Fig. 4.9*
stabilization 147–8
coastal,
aquifer, 50
classification and plate tectonics 20, *Table 2.1*
erosion 1, 2, 15
flooding 15
lowland management 310–12
plain 3, *Fig. 1.1*, 11, 13, *Table 2.1*, *Table 2.2*
protection 157, 190
retreat 264
Coastal ecosystems,
diversity of 47
productivity of 44, 47
Coastal Embankment Project 299
coastal zone 3, *Fig.1.1*, 11, 20
ownership of 13
population of 289–90
management, 2, 14–15, 289, 291, 293
see also shoreline management
management act, USA 14–5
cockle dredging 198
cohesion 158
cold coasts, 254–88
human impacts on 275–7
collision coasts 20, *Table 2.2*
community metabolism,
of reefs 210–1
competition 72, 140
condominium 155
condoms 86
continental shelf 3, 20, 43, 62, *Table 2.1*
eutrophication of 291
human uses of 13, *Table 1.3*
copper pollution 192
coral reefs, 206–53
degradation of, 241–7
pollution of 291
coral skeletons 211–2
corals,
aggression, 230–2
autotrophic corals 214
branching 217, 230, 233
carnivorous 213–4
defensive mechanisms of 231–2
depth limitation of 218–20
diversity of 218
corals — *contd*
growth rates of 216–20, 236
hermatypic corals 207, *Fig. 6.1*
light 217–20
massive 213, 216–7, 230, 233, 244
metabolism 238
mortality of 238, 240, 251
scleractinian corals 207
sexual reproduction in 216
subaerial exposure 220
temperature 236–40
use for building 244
Coriolis force 43, 52, 161
corniche 136
costs,
of sea level rise *Table 3.5*
of coastal defence *Table 3.6*
co-tidal lines 44, *Fig. 2.14*
crabs 173, 205
Crown-of-Thorns starfish 250–3
crust formation 176
cryoplanation 262
currents,
alongshore 30
rip *Table 2.4*, 65
cuspate forelands 63
cyanobacteria 47, 65, 264, 268
see also blue-green algae
cyclones 1, 125
see also hurricanes

dams 75, 186
damselfish 232–3, *Fig. 6.9*
deforestation 55, 57, 243, 291, 297
deltas, 160, 162–3, 170, 294, *Fig. 5.3*
estuaries, *and* 163
fjord-head 274
deposit feeders 48
deposition *Table 2.1*
detrital food chains 213
detritus 47
diamonds 76
diatoms,
in sea ice 265, 268
on beaches 64, 65, 67, *Table 2.4*
on mudflats 165–6
on salt marshes 168
diesel 286
see also oil
digital terrain models 15
dissipative beach 36, 37, 65, 73, 76, 90, *Table 2.4*, *Plate 2.3*

Distant Early Warning (DEW), line 276
disturbance, 144
  controlled 106
  regime 90
Doldrums 26
dredging 75, 79, 104
drift alignment of beaches 63
drumstick, shaped barriers 94
dunes,
  distribution of 60, *Fig. 3.1*
  Dutch dunes 106–9
  ecology of 71–3
  erosion of 69, 90, *Table 3.1*
  infiltration works 107
  interactions with beaches 73–5
  management of 87–90, *Table 3.2*
  migration of 93
  morphodynamics 69–71
  origins 67–8
  over-management 106
  parabolic 68
  planting, of 90
  stability of 69, 93
  transgressive 69, 73
  types 68–9
Dungenness 50, 62, 63, 67, 83
dykes 204

earthquakes 125, 259, 277, 285, 293
eat outs 169
ebb tide, 161
  deltas 93–4
ecotourism,
  see green tourism
El Nino-Southern Oscillation (ENSO) events 51–3, 55, 131, 149, 220, 238–9, 241, 291
emergence *Fig. 2.2*
emergent coasts 19, 22
encrusting organisms 119, 136, 139
endo-upwelling 213
energy flow, in reefs *Fig. 6.3*
engineered retreat 98–102
England, 13, 15, 22, 50, 62, 69, 79–80, 81, 90, 117, 122–3, 146–7, 157, 161, 163, 188
  Dorset 17, 85, 125, 147
  eastern 128, *Fig. 4.9*, 166, 181, 191
  Essex, marshes *Fig. 5.6*, 181, 194, *Fig. 5.10*, *Fig. 5.13*, 197

England — *contd*
  southern 83, 122, 147
epontic 265
equilibrium beach profile 84–5
erosion, *Table 2.1*
  coastal 76, 85
  marine erosion 126
  of armour blocks 17
  of beaches 307–8
  of cliffs 117
  of mudflats 165
  of the Nile delta 302
  of permafrost bluffs 278
  of salt marshes 181
  of sabkhas 193
  of salt marshes 181
  of sandy coasts 75, 103
  of wetlands 184
estuaries, 159–62
  fully mixed 159–61
  stratified 159–61
estuarine sedimentation 162
European Union 86
eustasy 21, 54
eutrophication 214, 291, 303–5, 308
evaporite minerals 174
Exclusive Economic Zones 13, *Fig. 1.5*
exposure rating 64–5
Exxon Valdez 283–6

faecal pellets 47, 48, 158, 275
fences 88, *Table 3.2*
fertilisers 87, *Table 3.1*
filter feeders 65
fish, 14, 64, 265
  farming 187
  herbivorous 232
  and mangroves 173–4
fjords, 254, 259–61, 270–5
  circulation of 271, *Fig. 7.6*
  ecology of 274–5
  geochemistry of 274
  hydrology of 271
  pollution of 276
  stagnant 273
  well-mixed 273
flagellates 65
flocculation 158
Flood Action Plan 299
flood,
  deltas 93–4
  tide 161
flooding 51, 104, 291, 306
flows 125
flushing rates 270
fluvial inputs 49

food webs,
  beach 65, 67
  dune 73
  rocky littoral 143
forebulge 21
foredunes 68, 69, 73, 95
fore-reef 225–8, 230
France 149, 161, 192
frazil ice 256
freeze-thaw activity 263
French, Jon 177, 194
frost 138
fungi 168

gabions *Table 3.2*
Galvin's breaker coefficient 32, *Fig. 2.9*
gamma ratio 30–1
gannets,
  necks used as shoes 144
  beaks used as pegs 144
Geographical Information Systems (GIS) 16–7
geoid 54
Ghyben-Herzberg principle 50
glacial,
  coasts 267–70
  deposits 60, 113, 259
  -interglacial cycles 114–6
  margin sediments 64
glaciers 55
glacio-isostasy 21, 75
Global Sea Level Observing System (GLOSS) 55
global warming, 54–6
  cold coasts, *and* 280–2
  corals, *and* 239–40
  rocky coasts, *and* 149–50
golf courses 93, 146
Grand Cayman Island *Plate 3.3*, 225–30, *Plate 6.5*
gravel, 62, 128
  beaches 67
  forelands 67
grazing 139, 157, 169, 191
Great Barrier Reef 206, 224, 228, 233, 236, 243, 246–7, 251–2
Greece 87, 121
Greenland 55–6, 261, 268, *Table 7.2*, 279
green tourism 308
gross primary productivity, of reefs 207, 213
groundwater, 50, 58, 73, 131
  Bangladesh, in 298
  cliff failure, and 146, 149
  extraction 57, 202

groundwater — *contd*
dune 87, 93, 107, *Fig. 3.16*
pumping of 57
in sabkhas 176
saline 305
groynes, 76, 80–81, *Fig. 3.9a*, 86, *Table 3.2*, 97, 105
angled 81, *Fig. 3.9b*
Gulf of Mexico 75, 233

halophytes 165
hard engineering 76–7
harvesting 144
Hawaii 247–50
hay making 191
hazards 15, 153, 155, 277, 282, 291, 306, 309
heavy metal pollution 191–2, 279
Heritage Coast 151
hexachlorobenzene (HCB) 279–80
Holocene 68, 103, 172 179, 193, 196, 203, 223, 224
honeycombs 118
humid tropics 113
Hurricane,
Agnes 162
Allen 241
Andrew 64, 233–5
Betsy 233
Donna 233
Gilbert 235
Ginger 92
Hugo 235
hurricanes 77, 91, 95, 96, *Fig. 3.13*, 233
Hydro-electric power 276

ice, 256–8
cliff 261
foot 262, 263
scour 267
sheet 21, 55, 282
shelves 56–7, 257, 258, 268
streams 56
iceberg calving 257–8, 282
Iceland 60, 277, *Fig 7.7*
imposex, in whelks 305
Indian Ocean 224, 233
Indonesia 181, 187, 189
infralittoral fringe 142
inlets 79, 91, 93–5, 97
Inman, D.L. 20, *Table 2.2*, 24
in situ raising of barrier islands 98–102
intermediate,
beach states 74
disturbance hypothesis 233
interstitial,
flora 47, 64, 72
food chain 67
intertidal zone, 11, 140
human use of *Table 1.3*
invertebrate macrofauna 64, 65
isostatic 23, 259
isostasy 54
Israel 69–70, 86–7
Italy 75, 186
IUCN 15, *Table 1.4*

Jamaica 87
Japan 75, 78, 81, 146, 150
jetties 75, 78–9, *Fig. 3.7*, 81
jointing 118
jokulhlaups 277
Jump-Off Joe 153–5

'k' selected species 275, *Table 7.4*
karst 223
kelp 48, 76, 143, 267
Kemp's phase difference 32
keystone,
predator 270
species 143
krill 270
Kuroshio current 51
Kuwait 76, 84

lagoons 64, 93, 103–5, 265, 303, 308
Lagos harbour 103–4, *Fig. 3.15*
land claim, 11, 197
see also reclamation
landslides 125, 146
land use changes 57
lapies, 136
see also marine karren
levées, 101, *Table 3.5*, 201, 202
see also sea wall
lichens 67, 121–2, 139, 142, 265
limpet 143, 144
lithosphere 21
litter dynamics 173
littoral zone 140, *Fig. 4.13*
Llantwit Major 151–3
log spiral beaches 63
Long Island, New York 50
Long Beach Island, New Jersey 100–1
longshore,
bars 37–8, *Plate 2.4*
currents 39, 65
drift 40, 79, 81, 85, 103, 151
sediment transport 75, 76, 95

Mackenzie delta 263–4
macroalgae 241
macrobenthos 65
macroscopic food chain 65
macrotidal 44, 133
Malaysia 63, 186
mangrove swamps, 104, 169–74
environmental settings 169–72, *Fig. 5.4*
mangroves,
adaptations 172–3
afforestation 298–9
deforestation 184
human impacts on 186, *Table 5.1*, *Table 5.2*, 203–5, 291
losses of 181–4, 297
reefs, and 241
sea level rise, and 196–7
marginal sea coasts 20
mariculture 187
marinas 184
marine karren 136
mass movements 113, 117, *Fig. 4.4*, 124–5
mean sea level 54
Mediterranean 44, 119, 136, 300–9, 310
meiofauna 67, 87
Mekong delta 203–5
Mersey Estuary 161, *Fig. 5.3*
mesotidal 44, *Fig. 3.12*, 133
micro-atolls 220
microbial loop 65
microbiological contamination 87
microtidal 44, *Fig. 3.12*, 133, 300
midlittoral zone *Fig. 4.13*, 142, 149
Millepora 224, 225
minerogenic marshes 167
Mississippi delta 64, 157, 160, 198–201
models,
of wetland growth 177–9, *Fig. 5.8*, 194
molluscs,
bivalve 142
browsing 120

monsoon 51, 103, 293
moon,
  tide-producing effect 41
morphodynamics,
  of beaches 32–41
  of dunes 71
  of mass movements *Fig. 4.7*
  of mudflats 165
motu 236
mud, 157–8
  balls 159
mudflats, 165–6
  salt marshes, *and* 166, 167
mudflows *Fig. 4.4*, 125
mudslides 126
multi year ice 256

Namibia 76
nature conservation 107
nearshore zone 84
Netherlands, 13, 22, 71, 181, 184
  see also Dutch
net primary productivity, (*see also* gross primary productivity),
  of beaches 67
  of lagoon 265
  of mangroves 173
  of salt marshes 168
  of sea ice 270
New Zealand 19, 117, 254
Nigeria 19, 117, 254
Niger delta 103, 157, 169, 184
nilas 256
Nile delta 75, 302, 308
nitrogen 67, 213, 249
no protection strategy 98–102
Nordstrom, Karl F 20, *Table 2.2*, 24
North Sea 23, 63
Norway 259, *Table 7.2*, 279–80
notch 119–120, 136
nutrient cycling 15, 139, 143
nutrients,
  in mangroves 173
  in reefs 206, 211–13, 214, 216
  in salt marshes 168–9
  soil 71, 107
  sources and sinks 48
  wetland 156

ocean,
  thermal expansion of 55

offshore bar 104
oil, 202
  extraction 56, 157, 184
  pollution 87, 148–9, 192, 242, 302
  resources 103, 276
  spill 283–5
oligotrophic waters 207
opportunistic patterning of vegetation 172
ords 128
Orfordness 62, 63
organic flocculation 158
organogenic marshes 166
outwelling hypothesis 168
overwash 91, 94, 100
oyster beds 157

Pacific Ocean, 52, *Fig. 2.16*, 233, 240
  Central 207, 228
  East Central *Plate 2.2*
  North Central 247
  South 22
pancake ice 256
Papua New Guinea 52, 54
partially mixed estuaries 169–61
pebbles 60
pelagic 44
penguins 267, 270, 288
periodic settling model 157
permafrost 258, 262–7
Pethick, John 166, 194–5
petroleum, 286
  see also oil
phase shifts 241
Philippines 157, 291
phosphorous 67, 71, 213, 249
photosynthesis of reef communities 207–8, 211
phytoplankton 47, 64, 207
piers 75
planulae 216, 225, 250
plateformes a vasques 136
plunging waves *Fig. 2.9*, 34, *Fig. 2.10*, 103
pollution,
  in Antarctica 286–8
  in Bangladesh 297
  in the Mediterranean 302–5, 308–9
  of fjords 274, 278–80, 282
  of rocky shores 143, 148–9
polychaete worms 65, 87, 274
polychlorinated biphenyls (PCBs) 192, 303
polyps *Fig. 6.2*, 213, 215–6

polyp bail-out 216
population 11, 275, 289–90, 301
Porolithon 224
ports 75, 104, 157, *Fig. 5.11*
Portugal 83, 146, 166
potatoes,
  fermented 86
primary producers 65
protozoans 67

quarrying 144
Quaternary 21, 203

'r' selected species 274, *Table 7.4*
rabbits 71, *Table 3.1*
railways 145
ramps 133
Rance estuary 149
reclamation, 11, 92, 104, 157, 184–5, 190, 305
  see also land claim
recreation 107, 151
reef,
  accretion 217–23
  communities, 225–36
    respiration of 209
  photosynthesis of 207–8, 211
  flat 224
reefs,
  current-dominated *Table 6.3*
  'give-up' 222–3
  hard and soft 240–1
  Indo-Pacific 224–5, 250
  'keep-up' 222–3
  management 244–7
  wave-dominated *Table 6.3*
regression 103
relative tidal range 36, 37, *Fig. 2.11*
reservoirs 57
resilience 7, *Table 1.2*, 205, 250, 282, 288
resistance, 7
  see also sensitivity
rice cultivation 204, 296
rivers 49
roads 145
rock,
  hardness *Fig. 4.3*
  strength 118, *Fig. 4.5*
rockfalls 118, *Fig. 4.4*, 125, 145–6, 151–3
rockslides 131
rocky coast evolution 113, *Fig. 4.3*

rollover 159
rotational slides 125, *Fig. 4.4*, 126
rotational slumps 128
runzel marks 165
Russia 276

sabkhas 174–6, *Fig. 5.5*, 181, 193
salcrete,
  see salt crusts 74–5
saline intrusion 50, 202, 296, 298, 299, 305
salinity 187
salt,
  crusts 74–5
  evaporator 187, *Table 5.1*
  flats 174
  marshes, 166–9
    cold coasts 263
    ecology of 167–9
    erosion and accretion 186–7
    growth rates 177
    human impacts on 188–9
  production 157, 187–8
  spray 49, 73, 121, 131, 264
  tolerance 71, 121
  water,
    in estuaries 159–61
  weathering 49, 118, 138
  wedge 159–61
sand,
  cays 224, 236
  dunes,
    erosion of *Plate 3.1*
  fences 97–8, *Table 3.3*
  mining 75–6, *Table 3.1*
  pumping 79
  transport, 73
    aeolian 73, 97
    rate 40
    vegetation, *and* 70–1
satellite imagery 15, 110
Savigear, R.A.G. 116
scale *Table 1.1*, *Table 2.1*
scientific research 276, 286–8
schlerochronology 216
Scolt Head Island 166, 168, 177, *Fig. 5.6*, 181, *Fig. 5.9*
Scotland 62, 63, 161
seagrasses 213, 241, 243
sea ice 256–8, 262, 267, 268, 277, 282
sea level,
  Bangladesh, *and* 297–8
  barrier islands, *and* 98–102
sea level — *contd*
  beaches, *and* 84–6
  change, 19, 114
  cliffs, *and* 131
  cold coasts, *and* 280–2
  deltas, *and* 163
  dunes, *and* 93, 106
  fall, 68
  history 22–4
  rise 2, 57–8, *Fig. 2.18*, 64, 67–8, 75, 76, 149
  surface temperatures, 51–2
  the Mediterranean, *and* 306–7
  wetlands, *and* 176–81, 193–7
sea urchins 48, 240
sea wall, 76, 77–8, 81, 86, 87, 117, 186, 197, 278
  and barrier islands 98–102
sea water,
  geochemistry of 48–9, *Table 2.5*
seaweeds 148
sediment,
  availability 163
  budget *Plate 1.1*, 6, *Fig. 1.3*
  cell 4
  circulations 15
  discharge 186
  erodibility 165
  loss 81
  movements 4, 10, 51, 85, 97
  production 148
  sinks 242
  sources 19, 64, 75
  stability 165
  supply 170, 259, 302
  transport,
    and currents 39
    and tide 36–7
    and waves 38–9
    longshore 41
    and jetties 78–9
sedimentation,
  effects on reefs 242–3
  rates in fjords 270
  tidewater glacier 268
sensitivity 7, *Table 1.2*, 9, 205
set back,
  scheme 197
  zones 100, 105
sewage 86, 191, 243, 249–50, 280, 302–3, *Fig. 8.5*
Seychelles 19, 114
shingle, *Fig. 4.8*
  see also gravel
shoreline,
  management 7
  retreat 84
shore,
  platforms, 110–2, 133–44
    ecology of 139–44
    erosion of 133–6
    and ice 262–3
  zone, 4, *Fig. 1.2*, 6, 11, 20, *Table 2.3*
    ownership of 13–4
shrimp ponds 157, 187
sill 271–3
Singapore 117
soft engineering 77, 88
soils,
  geotechnical problems of 297
solution,
  of limestone 119–120
South Africa 65–6, *Plate 3.2*, 72–3, *Plate 3.4*, 145
South America 52, 62
Southern Oscillation 52, *Fig. 2.16*
Spain 80
Spartina 9, 166, 168, 179, 190–1, 198, 263
species diversity 72
Spinifex 71, 72
spits 63
sponges 139, 220, 240
spur and groove 225
Spurn Head 81
Stella's sea cow 48
steric effect 55
Stonebarrow Hill 125–8
storm surges 51, 294, 296, 298
storms 51, 91, 128, 162, 236, 241
strandflat 259, 261
sub-horizontal platforms 133
submarine canyons and waves 30
submergence *Fig. 2.2*, 163
submergent coasts 19, 22
subsidence 57, 104, 202, 305–6
subtidal zone, 140
  human uses of *Table 1.3*
sun bathing,
  by polar bears 262
sun, tide-producing effect 41
Sundarbans 294, 296, 298
supralittoral zone 140–6
supratidal zone 140
surf,
  scaling factor 34, 37, *Fig. 2.11*

surf — *contd*
zone 65
skipping 95
surging,
glaciers 268
waves 32, *Fig. 2.9*, 34, *Fig. 2.10*, *Plate 2.2*
swash, 32, 34, *Fig. 2.10*
alignment of beaches 63, 81
swell dominated coasts 63
tafoni 118
tambak 187
tectonic activity,
and cliffs 113, 131
and coastal geomorphology 259
tetrapods 146
thaw slumping 280
thermokarst 261, 263
tidal,
bulge 41, *Fig. 2.13*, 43
channel 165, 167, 169, 176
currents,
in estuaries 161
gauge records 54–5, 57
power schemes 149
range 43, 44, 60, 73
resonance 43
tides, 41–4, 161
red 297
tidewater glacier 257, 258, 268, *Fig. 7.5*
topples 118, *Fig. 4.4*, 125
Torrey Canyon 149
tourism, 275
and Antarctica 276–7
in the Mediterranean 301
and reefs 243–4
trailing edge coasts 20
transgression, marine 103
tributylin (TBT) 305
trophic levels 209
trottoir 136
Trudgill, Steven 142
tundra vegetation 264–5
turbidity maximum 161

ultraviolet (UV) radiation 239, 240
undercliff 124, 126
UNEP 15
urbanization 96, 249, 275
USA, 14, 62, 161, 290–1, 312

USA — *contd*
Alaska 60, 254, 256, 257, 258–9, 261, *Table 7.2*, 275, 276, 277–8, 284–6
Atlantic (East) Coast 39, 51, 60, 64, 84, 95, 96, 97, 159
California 60, 78–9, 146, 198
Florida 22, 83
Georgia 168, 191–2
Louisiana 74, 76, 96, *Fig. 3.13*, 163, 189, 198–203
Maine 100
New Jersey 69, 83, 100–101
North Carolina 38, 83, 90
Oregon 128–30, *Fig. 4.10*, *Plate 4.4*, 153–5
Pacific (West) Coast 39, 60
South Carolina 50
Texas 77, 95–6, 100
Washington State 62

vegetation,
dunes, *and* 70–71
marsh accretion, *and* 168
planting 97
removal in wetlands 189–90
rockfalls, *and* 145–6
slope stability, *and* 133, 147–8
succession 172
Venice 57, 305–6, 307, 308
vermetids 136, 139
Very Long Baseline Interferometry (VLBI) 55
Vietnam 203–5

Wales 62, *Fig. 3.5*, 77
Walker circulation 52
war 242
washover 93–4
water,
column microbes 65
supply,
from dunes 106–8
wave,
celerity 26, 28, 29, 30
crests 29, 30
energy 28, 29, 30, 38, 68, *Fig. 3.3*
height 26, 28, *Fig. 2.5*, 29, 30, 31

wave — *contd*
length 26, *Fig. 2.5*, 29
period 26, 29, 32
power 28, 39–40, 228, *Fig. 6.7a*
rays 30, 31
refraction 30, *Fig. 2.8*
shoaling transformations *Fig. 2.6*
spectrum 26
steepness 29
wave-dominated barrier 93–4, *Fig. 3.12*
waves, 24–41
breaking 30–32
edge *Table 2.4*
gravity waves 26
infragravity waves *Fig 2.3*, 38–9
spilling waves 32, *Fig. 2.9*, 34
swell waves 26
weapons testing 242
weathering 113, 117–8, 138
wedge failures 118
weirs 202
wetlands,
construction, *and* 187
environmental settings 159–64
margins 181
open coast 163–4
reclamation 305
types 165–76
whelks,
penis growth in female 305
surfing, 65
wildlife 305
Woodroffe, Colin 169–72
world wave environments *Fig. 2.4*

Zanzibar 190
zonation,
ecological on beaches 65, *Fig. 3.2*
vegetation on cliffs 121, *Fig. 4.6*
vegetation on marshes 168
on limestone shore *Fig. 4.12*
rocky shore 140–3
of seaweeds 143
zooplankton 64, 213, 214
zooxanthellae, 209–11, 215, 216, 220, 238